Railway Ecology

Luís Borda-de-Água · Rafael Barrientos
Pedro Beja · Henrique M. Pereira
Editors

Railway Ecology

Editors

Luís Borda-de-Água
CIBIO/InBIO, Centro de Investigação em Biodiversidade e
    Recursos Genéticos
Universidade do Porto
Vairão
Portugal

and

CEABN/InBIO, Centro de Ecologia Aplicada "Professor Baeta
    Neves", Instituto Superior de Agronomia
Universidade de Lisboa
Lisboa
Portugal

Rafael Barrientos
CIBIO/InBIO, Centro de Investigação em Biodiversidade e
    Recursos Genéticos
Universidade do Porto
Vairão
Portugal

and

CEABN/InBIO, Centro de Ecologia Aplicada "Professor Baeta
    Neves", Instituto Superior de Agronomia
Universidade de Lisboa
Lisboa
Portugal

Pedro Beja
CIBIO/InBIO, Centro de Investigação em Biodiversidade e
    Recursos Genéticos
Universidade do Porto
Vairão
Portugal

and

CEABN/InBIO, Centro de Ecologia Aplicada "Professor Baeta
    Neves", Instituto Superior de Agronomia
Universidade de Lisboa
Lisboa
Portugal

Henrique M. Pereira
German Centre for Integrative Biodiversity Research (iDiv)
    Halle-Jena-Leipzig
Leipzig
Germany

and

CIBIO/InBIO, Centro de Investigação em Biodiversidade e
    Recursos Genéticos
Universidade do Porto
Vairão
Portugal

and

CEABN/InBIO, Centro de Ecologia Aplicada "Professor Baeta
    Neves", Instituto Superior de Agronomia
Universidade de Lisboa
Lisboa
Portugal

ISBN 978-3-319-57495-0   ISBN 978-3-319-57496-7   (eBook)
DOI 10.1007/978-3-319-57496-7

Library of Congress Control Number: 2017938143

Printed on acid-free paper

This Springer imprint is published by Springer Nature
The registered company is Springer International Publishing AG
The registered company address is: Gewerbestrasse 11, 6330 Cham, Switzerland

Foreword

Eureka! The first book on railway ecology has arrived. This pioneering edited volume, written by an international cast, brings together and synthesizes today's "state-of-the-science and application." Building on the diverse European railway network, the approach is global, with work also from North America, South America, and Asia. The ecological effects of railways are central, while the mitigation of effects appears throughout the book.

Rich scientific highlights are compelling: wildlife mortality patterns; effects on biodiversity; the barrier effect; lots on ungulates, bears, bats, songbirds, and waterbirds. But there is also information on plants, elephants, amphibians, kites, gazelles, and many more, e.g., detecting mortality "hotspots," and rail-side (verge) habitats, as well as case studies worldwide.

Mitigations to minimize impacts emerge as a motif: wildlife crossing approaches; noise reduction; reducing outward noise propagation; alerting wildlife >2 seconds before an expected animal-train collision; reducing avian collisions and electrocution by the elevated net of wires; restricting access; and decreasing vibrations are also discussed in detail.

Many nations are actively adding track, while high-speed trains are spreading... for good reason. Rail transport is more efficient than road traffic at moving people and goods, and there is more fuel efficiency; less greenhouse gas emission; less unhealthful air pollution; less traffic congestion; and less land consumed. Most importantly, it is also safer.

Furthermore, people personally depend on, and are affected by, trains. Trains take commuters to work daily. Every day we depend on long freight trains successfully carrying coal and oil. Many of us ride trains to other cities. Small towns often wither without trains stopping at their stations. Industry and jobs depend on trains bringing heavy resources, such as grain, coal, and minerals. We also buy heavy products, including autos, brought by train from factories. Train noise and their vibrations degrade our neighborhoods near the tracks. Infrequent chemical spills pollute water bodies and neighborhoods. Children wave at the engineer, and gleefully count the wagons/cars of long freight trains.

Railway ecology really differs from road ecology. Thus much wildlife moves along rail corridors. At any location, noise from the trains is usually infrequent, loud, and at high-and-low frequencies. Wildlife collisions are invisible to passengers. The rail corridor is typically narrow. Its linear and gentle-curve route repeatedly slices through landscape patterns. Persistent herbicides are widely and intensively used. Considerable vibrations alter soil structure and fauna. Little-used rail sidings and short stretches connected to local industries support successional vegetation and animals. Trains go faster, but stopping requires a long time and distance.

Yet road ecology concepts developed in the past three decades apply to many aspects of railway systems. The first major road-ecology conference took place in Orlando, Florida, in 1995, the same year that my book, *Land Mosaics*, appeared, with 12 pages on road ecology, but only about seven sentences and seven references to railway ecology. Two decades later, in *Urban Ecology: Science of Cities*, I wrote four pages with 46 references to railway ecology. Meanwhile, in 1998 we wrote a 24-page *Annual Review of Ecology and Systematics* article on road ecology, followed in 2003 by the first relatively comprehensive book (481 pages and 1,078 references), Road Ecology: Science and Solutions, written by 14 co-authors. Twelve years later (2015), R. van der Ree, co-editors, and international leaders in the field produced an updated synthesis, *Handbook of Road Ecology*, with numerous mitigation and planning solutions.

Road ecology became mainstream for both scientists and government in only three decades. The International Conference of Ecology and Transportation (ICOET) and InfraEco Network Europe (IENE) provide a home base for both road ecology and railway ecology. And today railway ecology is on a roll. Many early papers appeared in the 1990s, but the most recent (in the last 5-10 years) show an impressive increase in quality research articles published. This book, *Railway Ecology*, is poised to spark the next growth phase.

Also, the book highlights and suggests many of the ecological research frontiers awaiting us. Think of the role of rail corridors in urban areas, and in intensive agricultural land: distinctive and predominant railway species; heterogeneous railside habitats; vegetation and fauna; railyards; freight trains and rails linked to local industries; pollutant distributions from diesel, steam, and electric engines; train noise and wildlife; interactions between the rail corridor and the sequence of adjoining habitats/land uses; species movement and dispersal along railways; the barrier effect and genetic variation on opposite sides of railways; diverse effects on air, soil, and water; railway impacts on populations; and, finally, regional railway networks rather than rail locations or segments.

This book will catalyze new ecological research, new mitigations, and better planning, construction, and maintenance. Better government policies and company practices will evolve.

Railway Ecology is a treasure chest, bulging with insights, many of them previously unseen. Relish the pages ahead, jump into the field, and increase the body of research. Become part of the solution: put the mitigations to work for ecological railways and for our land.

February 2017   Richard T.T. Forman
                                                                           Harvard University

Preface

Rationale and Purpose

The transportation of people and goods is a critical part of the economy. However, a judicious choice of the means of transportation can ameliorate the impacts of economic activities on the environment. In this respect, railways can play a major role, as they provide far more cost-effective and energetically efficient ways of transporting passengers and freight than motor vehicles and airplanes. Hence, railways can contribute to the global efforts for curbing the emission of greenhouse gases and thus help achieve the goals set by the Convention on Climate Change, and the recent Paris Agreement to reduce the forecasted rise in global temperatures in the twenty-first century. Calls have thus been made to consider railways as the backbone of sustainable transportation and to increase their share in relation to more polluting modes of transport.[1] However, as with many other economic endeavors, railways have environmental impacts ranging from several forms of pollution to

---

[1]Sadler, K., 2015. Rail to play key role in strategy for tackling climate change. http://www.europeanrailwayreview.com/24816/rail-industry-news/rail-to-play-key-role-in-strategy-for-tackling-climate-change/.

wildlife mortality. Therefore, a careful assessment of railways' impacts on nature and of mitigation measures is in order.

Trains are not cars, and railway tracks are not roads. These assertions are obvious, but they are often ignored when assessing the potential and actual biodiversity impacts of railways, which are normally equated to those of roads. This is regrettable because roads and railways differ in many respects, which can strongly influence their impacts. For instance, traffic on railways tends to be less intense than on roads, but trains often travel at higher speeds than cars. Roads and railways also have different physical structures, especially in the case of electrified railways, where overhead lines along the rail tracks can represent an additional source of impacts. All these differences are likely to affect wildlife responses to roads and railways, and hence their impacts on, for example, animals' behavior, mortality, and landscape connectivity.

A quick survey of the literature shows that studies on the biodiversity impacts of railways have greatly lagged behind those of roads. This is probably because the road network is much larger than the railway network, but probably also because the impacts of roads are more easily observed by ecologists and the general public. For instance, the safety and economic costs associated with the risk of collision with a wild or domesticated animal is more easily perceived in a road than in a railway, although the latter are also very significant: an accident with an animal is likely to go unnoticed by passengers in a train, with the remarkable exception of the train driver, but the same will not happen so easily in a car. Even the legal frameworks under which roads and railways operate can be different, conditioning the way the environment in the vicinity can be protected and which mitigation measures can be applied.

All these differences add to a number of specificities that clearly set railways apart from roads, and thus a paradigm shift is needed whereby the impacts of railways on biodiversity are considered on their own. This book aims at contributing to this shift, filling a gap in the literature about the impacts of transportation on biodiversity. We have brought together 44 researchers from 12 countries, from North and South America to Asia and Europe. We aimed at combining—in a single volume—the most relevant information that has been produced on the interactions between railways and wildlife, and to illustrate such interactions with a set of carefully chosen case studies from around the world. We have tried to produce a comprehensive volume that should be of interest to researchers and practitioners alike, including the staff of railway and consultancy companies that deal with the environmental challenges of railway planning, construction, and operation every day.

Although this book addresses several environmental problems raised by railways, we would like its main message to be one of hope. Indeed, while we expect societies to keep looking for more efficient means of transportation, we also expect that efficiency to become ever more synonymous with environmental efficiency, as societies seek to mitigate impacts or even improve biodiversity. We hope that our book can help achieve this goal, contributing to the mitigation of the negative impacts of railways on biodiversity, thereby improving the sustainability and environmental benefits of this mode of transportation.

Organization

We have divided the book into two parts. The first aims at reviewing the main ideas and methods related to the identification, monitoring, and mitigation of railway impacts on biodiversity, with emphasis on wildlife mortality, barrier effects, biological invasions, and the effects of other railway environmental impacts such as noise and chemical pollution. We begin by setting the scene (Chap. 1–Railway Ecology), framing railways in their economic context, and providing several examples of how railways can impact biodiversity. In Chap. 2 (Wildlife Mortality on Railways), Sara Santos and her colleagues provide a comprehensive review of mortality rates on railways, paying particular attention to the sources of bias when estimating mortality rates. In Chap. 3 (Methods for Monitoring and Mitigating Wildlife Mortality on Railways), Filipe Carvalho and his colleagues have further developed the theme of wildlife mortality, focusing on monitoring and mitigation. They discuss the application to railways of mitigation procedures routinely used in road ecology, and provide an overview of wildlife crossing structures and their role in reducing mortality and barrier effects. In Chap. 4 (Railways as Barriers for Wildlife: Current Knowledge and Future Steps), Rafael Barrientos and Luís Borda-de-Água examine the barrier effects of railways, reviewing the evidence and providing an overview on procedures used to quantify barrier effects, with emphasis on genetic methods and individual-based computer simulations. They then discuss mitigation measures and their effectiveness, providing guidelines for monitoring and mitigation. In Chap. 5 (Aliens on the Move: Transportation Networks and Non-native Species), Fernando Ascensão and César Capinha examine the role of railways in the spread of invasive species, showing that land transportation systems have greatly contributed to species introductions with high economic and environmental costs. They discuss measures to decrease the risk of alien species introductions, paying special attention to verges and associated vegetation corridors. In Chap. 6 (Railway Disturbances: Types, Effects on Wildlife and Mitigation Measures), Silva Lucas and her colleagues look at biodiversity impacts caused by railways due to noise, air, soil and water pollution, as well as soil erosion. They conclude that impacts are species-specific, depending largely on species traits, and that impacts can be minimized through improvements in the railway structure and the implementation of mitigation measures.

The second part of the book provides a set of case studies from around the world that illustrate the impacts of railways on wildlife and ways to reduce those impacts. Reflecting the strong interest in the topic of wildlife mortality, four chapters focus on the patterns of mortality resulting from collisions with trains and railway structures, each of which suggests several mitigation measures to reduce such mortality. In Chap. 7 (Bird Collisions in a Railway Crossing a Wetland of International Importance − Sado Estuary, Portugal), Carlos Godinho and colleagues examine the risk of aquatic bird mortality due to railway bridge crossing wetland habitats, by combining data from surveys on carcasses and the observation of bird movements. They have found that this bridge had a low risk for aquatic birds, as only a few dead birds were found. In addition, less than 1% of 27,000 bird movements observed over 400 hours crossed the area of collision risk. In Chap. 8 (Cross-scale Changes in Bird Behavior around a High-speed Railway: from Landscape Occupation to Infrastructure Use and Collision Risk), Malo and colleagues examine bird mortality on a high-speed railway line crossing a rural landscape in central Spain. They found that the species commonly associated with rural and open spaces tended to avoid the railway, while those that are already associated with man-made structures were attracted by the railway. The latter species were those most exposed to train collisions; indeed, using video cameras in the trains' cockpits to analyze birds' responses to incoming trains, they observed that birds become habituated to the presence of trains, a behavior that leads to increased mortality.

Collision with large mammals is also a matter of concern in railway ecology, and this is dealt with in the two following chapters. In Chap. 9 (Relative Risk and Variables Associated with Bear and Ungulate Mortalities Along a Railroad in the Canadian Rocky Mountains), Dorsey and colleagues report the findings from a 21-year data set of train crashes with elk (*Cervus elaphus*), deer (*Odocoileus* spp.), American black bears (*Ursus americanus*) and grizzly bears (*U. arctos*) in Banff and Yoho National Parks, Canada. They found that mortality hotspots were affected by species abundances, train speed and the characteristics of the infrastructure. In Chap. 10 (Railways and Wildlife: a Case Study of Train-Elephant Collisions in Northern West Bengal, India), Roy and Sukumar report on the problems of elephant collision with trains in India, describing the spatial and temporal variations in mortality and relating these to elephant behavior. They found that mortality occur mainly at night and in well-defined hotspots, that males are the most susceptible to train collisions, and that mortality seems to be associated with the periods of elephants raiding of crops.

The next four chapters deal with problems related to habitat loss due to the railway infrastructure and its impacts, and the loss of connectivity due to barrier effects. In Chap. 11 (Assessing Bird Exclusion Effects in a Wetland Crossed by a Railway), Carlos Godinho and colleagues examined the extent to which aquatic birds were excluded from wetland habitats close to a railway bridge. They found that bird densities were similar in areas both close to and far from the bridge, and thus there were no noticeable exclusion effects on the wetland bird community. In Chap. 12 (Evaluating the Impacts of a New Railway on Shorebirds: a Case Study in Central Portugal), Tiago Múrias and colleagues addressed a similar problem, using comprehensive monitoring of data on bird numbers, behavior and breeding during the pre-construction, construction, and post-construction phases of a new railway line. Using a Before-After-Control-Impact (BACI) design, they found that the abundance of breeding and wintering shorebirds was reduced in saltpans close to the railway in the post-construction phase. In Chap. 14 (Fragmentation of Ungulate Habitat and Great Migrations by Railways in Mongolia), Ito and colleagues evaluate the impact of the Ulaanbaatar-Beijing Railway on the great migrations of ungulates in the largest grassland in the world – Mongolia's Gobi-steppe ecosystem. They found that fencing along the railway represents a source of mortality and a barrier to ungulate movement that prevent the long-distance movements required to find food during the harsh winters of Mongolia.

The final four chapters describe novel approaches to reduce and mitigate the biodiversity impacts of railways and also some positive impacts of railways on biodiversity. In Chap. 15 (Railway Ecology − Experiences and Examples in the Czech Republic), Keken and Kušta provide an overview of the impacts and management policies of railways in the Czech Republic. They describe the current Czech railway network and the plans for future expansion, discussing several management options for preventing accidents with animals and mitigating other environmental impacts. In Chap. 16 (Ecological Roles of Railway Verges in Anthropogenic Landscapes: a Synthesis of Five Case Studies in Northern France), Vandevelde and Penone focus on railway verges and their potential positive impacts on the environment. They show that verges may provide habitats for grass plants, bats, and orthopteran, as well as functional connectivity to plants, thereby counteracting some of the negative effects of large-scale urbanization. In Chap. 17 (Wildlife Deterrent Methods for Railways – an Experimental Study), Seiler and Olsson discuss management options for reducing ungulate collisions with trains in Sweden. They review trends in ungulate collisions with trains since 1970 and their economic costs, and then introduce and discuss a new crosswalk design with a deterrent system, where animals are encouraged to leave the railway shortly before trains arrive. Finally, in Chap. 18 (Commerce and Conservation in the Crown of the Continent), Waller describes how a partnership between a railway company, the government and public stakeholders allowed the recovery of the grizzly bear (*Ursus arctos*) in Glacier National Park in the U.S. They describe the mortality problem that resulted from grizzly bears feeding on grain spilled from derailed cars and becoming habituated to the presence of trains, and how this was dealt with through the partnership of a range of stakeholders.

In the final Chap. 19, we wrap up the key messages of the book. We briefly consider the future of railway ecology and give several recommendations on how to mitigate their impacts on the environment.

We hope this book inspires scientists and practitioners to develop approaches to make railways increasingly biodiversity-friendly. But we would also like to dedicate this book to those working for railway companies that have done their best to ameliorate the impacts of railways on the environment and made railways one of the most sustainable modes of transportation.

Lisboa, Portugal   Luís Borda-de-Água
Lisboa, Portugal   Rafael Barrientos
Lisboa, Portugal   Pedro Beja
Halle, Germany   Henrique M. Pereira

Acknowledgements

We would like to thank Infraestruturas de Portugal, and in particular João Morais Sarmento, Ana Cristina Martins, Cândida Osório de Castro, Graça Garcia and Luísa Vales de Almeida without whom this book would not have been possible. We would also like to thank Sasha Vasconcelos for her willingness to help with our constant questions about English. Finally, we would like to thank Margaret Deignan at Springer for her invaluable help in putting this book together.

Contents

Editors and Contributors

About the Editors

Luís Borda-de-Água is a researcher in ecology and conservation biology at CIBIO-InBIO, University of Porto (Portugal) collaborating with the Infraestructuras de Portugal Biodiversity Chair. He took his Ph.D. in ecology at Imperial College, London, UK, having studied before electrical engineering and physics. His background had led him to work mainly on theoretical and computational aspects of ecology. Presently he divides his research activity between studies on global biodiversity patterns and the impacts of linear infrastructures on wildlife.

Rafael Barrientos is a conservation biologist who received his Ph.D. in 2009. He has a back-ground on the impacts of linear infrastructures on wildlife, including road-kills, collision with power lines or the genetics of fragmented populations. He has participated in several management projects, for instance from the Spanish National Parks Autonomous Agency, OAPN. Also, he has participated in purely research projects in Spanish institutions like the Arid Zones Experimental Station, the National Museum of Natural History or the Castilla-La Mancha University. Since 2015 he has a postdoctoral position at the Infraestructuras de Portugal Biodiversity Chair at CIBIO-InBIO, Universidade do Porto (Portugal).

Pedro Beja is a senior researcher in ecology and conservation biology at CIBIO-InBIO, University of Porto (Portugal), where he holds since 2012 the EDP Chair in Biodiversity and leads the ApplEcol—Applied Population and Community Ecology research group. He has over 25 years of professional experience, with work carried out in the public administration, private environmental consultancy, academia, and research institutions. His main research interests are in the area of biodiversity conservation in human-dominated landscapes, including agricultural, forest and freshwater systems. He has published over one hundred scientific papers on a variety of topics related to the conservation and management of biodiversity in Europe, Africa and South America.

Henrique M. Pereira is a leading expert on global biodiversity change. He has worked both as a researcher and as a practitioner, having served as the Director of Peneda-Gerês National Park and as the coordinator of the Portugal Millennium Ecosystem Assessment. Since 2013, he is the Professor of Biodiversity Conservation at iDiv—German Center for Integrative Biodiversity Research of the Martin Luther University Halle-Wittenberg and holds since 2015 the Infraestruturas de Portugal Biodiversity Chair at CIBIO-InBIO, Universidade do Porto (Portugal). He is the Chair of the Biodiversity Observation Network of the Group on Earth Observations and co-chair of the Expert Group on Scenarios and Models from the Intergovernmental Platform on Biodiversity and Ecosystem Services. He has published over one hundred scientific papers and reports on biodiversity issues.

Contributors

Fernando Ascensão CIBIO/InBIO, Centro de Investigação em Biodiversidade e Recursos Genéticos, Universidade do Porto, Vairão, Portugal; CEABN/InBIO, Centro de Ecologia Aplicada "Professor Baeta Neves", Instituto Superior de Agronomia, Universidade de Lisboa, Lisboa, Portugal

Rafael Barrientos CIBIO/InBIO, Centro de Investigação em Biodiversidade e Recursos Genéticos, Universidade do Porto, Vairão, Portugal; CEABN/InBIO, Centro de Ecologia Aplicada "Professor Baeta Neves", Instituto Superior de Agronomia, Universidade de Lisboa, Lisboa, Portugal

Pedro Beja CIBIO/InBIO, Centro de Investigação em Biodiversidade e Recursos Genéticos, Universidade do Porto, Vairão, Portugal; CEABN/InBIO, Centro de Ecologia Aplicada "Professor Baeta Neves", Instituto Superior de Agronomia, Universidade de Lisboa, Lisboa, Portugal

Luís Borda-de-Água CIBIO/InBIO, Centro de Investigação em Biodiversidade e Recursos Genéticos, Universidade do Porto, Vairão, Portugal; CEABN/InBIO, Centro de Ecologia Aplicada "Professor Baeta Neves", Instituto Superior de Agronomia, Universidade de Lisboa, Lisboa, Portugal

Pedro Brum Faculdade de Ciências Socias e Humanas da Universidade Nova de Lisboa, Lisboa, Portugal

César Capinha Global Health and Tropical Medicine Centre (GHTM), Instituto de Higiene e Medicina Tropical (IHMT), Universidade Nova de Lisboa, Lisboa, Portugal

Filipe Carvalho CIBIO/InBIO, Centro de Investigação em Biodiversidade e Recursos Genéticos, Universidade de Évora, Évora, Portugal; Department of Zoology and Entomology, School of Biological and Environmental Sciences, University of Fort Hare, Alice, South Africa

Ramon Gomes de Carvalho Centro Brasileiro de Estudos em Ecologia de Estradas, Universidade Federal de Lavras, Lavras, Brazil

Cândida Osório de Castro Direção de Engenharia e Ambiente, Infraestruturas de Portugal, Almada, Portugal

Luísa Catarino LabOr—Laboratory of Ornithology, ICAAM—Instituto de Ciências Agrárias e Ambientais Mediterrânicas, University of Évora, Évora, Portugal

Céline Clauzel LADYSS, UMR 7533 CNRS, Sorbonne Paris Cité, University Paris-Diderot, Paris, France; TheMA, CNRS, University Bourgogne Franche-Comté, Besançon Cedex, France

Anthony Clevenger Western Transportation Institute/College of Engineering, Montana State University, Bozeman, MT, USA

Benjamin P. Dorsey Parks Canada, Revelstoke, BC, Canada

Eladio L. García de la Morena SECIM, Servicios Especializados de Consultoría e Investigación Medioambiental, Madrid, Spain

Carlos Godinho LabOr—Laboratory of Ornithology, ICAAM—Instituto de Ciências Agrárias e Ambientais Mediterrânicas, Universidade de Évora, Évora, Portugal

David Gonçalves CIBIO/InBIO, Centro de Investigação em Biodiversidade e Recursos Genéticos, Universidade do Porto, Vairão, Portugal; Departamento de Biologia, Faculdade de Ciências, Universidade do Porto, Vairão, Portugal

Clara Grilo Centro Brasileiro de Estudos em Ecologia de Estradas, Universidade Federal de Lavras, Lavras, Brazil; Setor Ecologia, Departamento Biologia, Universidade Federal de Lavras, Lavras, Brazil

Jesús Herranz Terrestrial Ecology Group, Departamento de Ecología, Universidad Autónoma de Madrid, Madrid, Spain

Israel Hervás Terrestrial Ecology Group, Departamento de Ecología, Universidad Autónoma de Madrid, Madrid, Spain

Takehiko Y. Ito Arid Land Research Center, Tottori University, Tottori, Japan

Z. Keken Department of Applied Ecology, Faculty of Environmental Sciences, Czech University of Life Sciences Prague, Suchdol, Czech Republic

T. Kušta Department of Game Management and Wildlife Biology, Faculty of Forestry and Wood Sciences, Czech University of Life Sciences Prague, Suchdol, Czech Republic

Badamjav Lhagvasuren Institute of General and Experimental Biology, Mongolian Academy of Sciences, Ulaanbaatar, Mongolia

Ricardo Jorge Lopes CIBIO/InBIO, Centro de Investigação em Biodiversidade e Recursos Genéticos, Universidade do Porto, Vairão, Portugal

Rui Lourenço LabOr—Laboratory of Ornithology, ICAAM—Instituto de Ciências Agrárias e Ambientais Mediterrânicas, Universidade de Évora, Évora, Portugal

Priscila Silva Lucas Centro Brasileiro de Estudos em Ecologia de Estradas, Universidade Federal de Lavras, Lavras, Brazil

Juan E. Malo Terrestrial Ecology Group, Departamento de Ecología, Universidad Autónoma de Madrid, Madrid, Spain

João T. Marques Unidade de Biologia da Conservação, Departamento de Biologia, Universidade de Évora, Évora, Portugal; CIBIO/InBIO, Centro de Investigação em Biodiversidade e Recursos Genéticos, Universidade de Évora, Évora, Portugal

Cristina Mata Terrestrial Ecology Group, Departamento de Ecología, Universidad Autónoma de Madrid, Madrid, Spain

António Mira Unidade de Biologia da Conservação, Departamento de Biologia, Universidade de Évora, Évora, Portugal; CIBIO/InBIO, Centro de Investigação em Biodiversidade e Recursos Genéticos, Universidade de Évora, Évora, Portugal

Tiago Múrias CIBIO/InBIO, Centro de Investigação em Biodiversidade e Recursos Genéticos, Universidade do Porto, Vairão, Portugal

Mattias Olsson EnviroPlanning AB, Gothenburg, Sweden

C. Penone Institute of Plant Sciences, University of Bern, Bern, Switzerland

Henrique M. Pereira CIBIO/InBIO, Centro de Investigação em Biodiversidade e Recursos Genéticos, Universidade do Porto, Vairão, Portugal; CEABN/InBIO, Centro de Ecologia Aplicada "Professor Baeta Neves", Instituto Superior de Agronomia, Universidade de Lisboa, Lisboa, Portugal; German Centre for Integrative Biodiversity Research (iDiv) Halle-Jena-Leipzig, Leipzig, Germany

Lisa J. Rew Land Resources and Environmental Sciences, Montana State University, Bozeman, MT, USA

Mukti Roy Asian Nature Conservation Foundation, c/o Centre for Ecological Sciences, Indian Institute of Science, Bangalore, Karnataka, India

Pedro Salgueiro Unidade de Biologia da Conservação, Departamento de Biologia, Universidade de Évora, Évora, Portugal; CIBIO/InBIO, Centro de Investigação em Biodiversidade e Recursos Genéticos, Universidade de Évora, Évora, Portugal

Sara M. Santos CIBIO/InBIO, Centro de Investigação em Biodiversidade e Recursos Genéticos, Universidade de Évora, Évora, Portugal

Andreas Seiler Department of Ecology, Swedish University of Agricultural Sciences, SLU, Riddarhyttan, Sweden

Masato Shinoda Graduate School of Environmental Studies, Nagoya University, Nagoya, Japan

Raman Sukumar Centre for Ecological Sciences, Indian Institute of Science, Bangalore, Karnataka, India

Atsushi Tsunekawa Arid Land Research Center, Tottori University, Tottori, Japan

J.-C. Vandevelde UMR 7204 (MNHN-CNRS-UPMC), Centre d'Ecologie et de Sciences de la Conservation (CESCO), Paris, France

John S. Waller Glacier National Park, West Glacier, MT, USA

List of Figures

List of Tables

Part I
Review

Chapter 1
Railway Ecology

**Luís Borda-de-Água, Rafael Barrientos, Pedro Beja
and Henrique M. Pereira**

Abstract Railways play a major role in the global transportation system. Furthermore, railways are presently being promoted by several governments thanks to their economic and environmental advantages relative to other means of transportation. Although railways have clear advantages, they are not free of environmental problems. The objective of this book is to review, assess, and provide solutions to the impacts of railways on wildlife. We have divided the impacts of railways on biodiversity into four main topics: mortality, barrier effects, species invasions, and environmental disturbances, with the latter ranging from noise to chemical pollution. Railways share several characteristics with roads and with power lines when the trains are electric. Therefore, much can be learned from studies on the impacts of roads and power lines, taking into account, however, that in railways, the two are often combined. Besides the similarities with roads and power lines, railways have specific characteristics. For instance, railways have lower traffic intensity but trains usually have much higher speeds than road vehicles, and the electric structures in railways are typically lower than in most power lines. Thus, railways pose specific challenges and require specific mitigation measures, justifying calling the study of its impacts on biodiversity "railway ecology."

Keywords Barrier effects · Disturbances · Invasions · Linear infrastructures · Mortality · Railway ecology

L. Borda-de-Água (✉) · R. Barrientos · P. Beja · H.M. Pereira
CIBIO/InBIO, Centro de Investigação em Biodiversidade e Recursos Genéticos,
Universidade do Porto, Campus Agrário de Vairão, Rua Padre Armando Quintas,
4485-661 Vairão, Portugal
e-mail: lbagua@gmail.com

L. Borda-de-Água · R. Barrientos · P. Beja · H.M. Pereira
CEABN/InBIO, Centro de Ecologia Aplicada "Professor Baeta Neves", Instituto Superior de
Agronomia, Universidade de Lisboa, Tapada da Ajuda, 1349-017 Lisboa, Portugal

H.M. Pereira
German Centre for Integrative Biodiversity Research (iDiv) Halle-Jena-Leipzig, Deutscher
Platz 5e, 04103 Leipzig, Germany

Introduction

The most basic definition of a railway is "a prepared track which so guides the vehicles running on it that they cannot leave the track" (Lewis 2001). According to this definition, railways were already used by the Greeks and Romans. However, the concept of railway as we know it today—that is, rails made of iron, with trains composed of several wagons pulled by one or more locomotives running on specified timetables, forming nationwide or international networks—is an invention of the early nineteenth century (Lewis 2001).

Since its inception, the technology of rails and trains has evolved considerably. Two main forces have driven this evolution: the reduction of costs and the increase in safety (Shabana et al. 2008; Flammini 2012; Profillidis 2014). The reduction of costs has been achieved by increasing energy efficiency and, simultaneously, increasing the speed and size of the trains. According to the International Energy Agency and the International Union of Railways, from 1975 until 2012, the energy used by passenger/km decreased by 62%, while the energy to transport cargo by tonne-km decreased by 46% (Railway Handbook 2015). This increase in energy efficiency was accompanied by a 60% reduction in CO_2 emissions for passenger and 41% for freight transportation (Railway Handbook 2015). Three countries— China, India, and Russia, with extensive railway networks, and where associated environmental impacts are likely to be important—exemplify these achievements. China boasts 60% (27,000 km) of the total high-speed lines in the world and has the lowest rate of energy consumption per passenger-km (67 kJ/passenger-km). India has the lowest CO_2 emissions per passenger-km (10 g CO_2/passenger-km) and the lowest rate of energy consumption per tonne of goods transported (102 kJ/tonne-km). Russia has the lowest CO_2 emissions per tonne of goods transported (9 g CO_2/tonne-km) (Railway Handbook 2015).

Increases in railway safety have also been substantial, not only for passengers, railway workers and freight, but also for the human populations living in the vicinity of the railways (Shabana et al. 2008; Flammini 2012; Profillidis 2014). The safety achieved in the railway sector is particularly impressive when compared to road safety. According to a report by the European Railway Agency (2013), the fatality risk in the period 2008–2010 measured as the number of fatalities per billion passenger-km is 0.156 to railway passengers and 4.450 for car occupants. In fact, following the same report, transport safety is only surpassed by the airline industry with 0.101 fatalities per billion passenger-km.

Besides economic and safety advantages, there is general agreement that railways have several environmental advantages relative to roads. We highlight two of them: Firstly, railways are less pollutant than roads because the metal-to-metal contact characteristic of railways considerably reduces rolling resistance; thus, a diesel-powered train is more energy efficient than the equivalent number of road vehicles. In addition, an electric-powered train is not a source of direct emissions of greenhouse gases and other air pollutants, and even indirect emissions can be negligible when the electricity is cleanly produced (Chandra and Agarwal 2007;

Profillidis 2014). Secondly, railways require less land occupancy than other means of transportation, and land use is perhaps the main driver of biodiversity loss globally (Pereira et al. 2012). For example, Profillidis (2014) notices that the "high-speed Paris–Lyon line (a distance of 427 km) occupies as much space as the Paris airport at Roissy." This is not a mere detail; it is already an important issue in highly human-populated areas, and it is becoming important globally as the human population grows and pressure on available arable land increases (Profillidis 2014).

More recently, the protection of the habitats crossed by railways and their wildlife has become a main factor to be taken into consideration when designing new railways or maintaining existing ones (Clauzel et al. 2013; Profillidis 2014), associated with an increased societal awareness of the importance of biodiversity (Pereira et al. 2012). However, compared to other transportation systems, such as roads, less is known about the impact of railways on wildlife, as well as its specificities. Whereas there is a large body of research on road ecology, much less exists on railway ecology (Popp and Boyle 2017). Therefore, as the global railway network increases, and more countries promote railways over road or air transportation of people and goods, we feel that a review of the state-of-knowledge in railway ecology is needed. Railway ecology is an emerging field, but with scarce (and scattered) information about its effects on biodiversity (e.g., Dorsey et al. 2015; Popp and Boyle 2017). This present book deals with the impacts of railways on biodiversity along four main topics: wildlife mortality; habitat loss and exclusion of species from their habitats; barrier effects; and exotic species invasions and impacts of other environmental disturbances caused by railways.

Although railway ecology shares several characteristics with road ecology, it also has some specificities. For instance, traffic on railways tends to be lower than on roads, but the speed of the vehicles can be much greater. Railway ecology can also benefit from studies on the impacts of power lines, these are present in railways when trains are powered by electricity, but their height is usually lower than that of other power lines. Therefore, although railways share some characteristics with other linear infrastructures, they also have some particularities that warrant independent consideration. In this book we highlight the characteristics that railways share with other linear infrastructures, identify which measures can be applied to railways, and show what makes the impacts of railways unique, as well as the required mitigation measures.

The impacts of railways on wildlife have received less attention than those of roads probably because one of its major impacts, vehicle-animal collisions, is not visible to the general public (Wells et al. 1999; Cserkész and Farkas 2015). Ordinarily, only the train crews are aware of the animal mortality caused by collisions, as railway right-of-ways have typically restricted access (Wells et al. 1999). In some cases, researchers have studied the impacts caused by railways combined with those of roads (e.g., Vos et al. 2001; Ray et al. 2002; Proctor et al. 2005; Arens et al. 2007; Li et al. 2010), and some studies have found similar impacts of both networks, namely, wildlife collisions (e.g., Cserkész and Farkas 2015). However, often only the road impacts are highlighted, probably because the road network tends to be more spatially developed than the railway network.

Frequently, roads and railways are co-aligned along the same corridor (e.g., Proctor et al. 2005; Li et al. 2010). In fact, the co-occurrence of roads and railways is an important aspect to have in consideration, as wildlife response to one infrastructure can condition the response to the other. For example, in a study in the USA, Waller and Servheen (2005), found that radio-collared grizzly bears (*Ursus arctos*) crossed a highway-railway corridor at night, presumably to avoid high diurnal highway traffic, but when railway traffic was heavier, there was a behavior that led to higher mortality rates on the railway than on the road. In another study in Canada, Clevenger and Waltho (2005) found that black bears (*U. americanus*) and cougars (*Puma concolor*) tended to cross a highway along wildlife passes far from railway tracks, while wolves (*Canis lupus*) preferred to use crossing structures close to the railway.

It is important to acknowledge the different ways that roads and railways impact wildlife. We consider the following five to be the most relevant: First (1) Traffic flow is much lower on railways, with several case studies showing that the number of trains moving through railways per unit of time are about 0.2–1.6% of the number of cars moving through nearby roads (Gerlach and Musolf 2000; Waller and Servheen 2005; Xia et al. 2007; Kušta et al. 2015); (2) railway traffic flow is characterized by long traffic-free intervals—in some cases there is no nighttime railway traffic (e.g., Rodríguez et al. 1997; Pérez-Espona et al. 2008)—although noise and vibrations produced by trains are higher than those produced by cars (Dorsey et al. 2015; Kociolek et al. 2011); (3) railways have lower wildlife mortality, possibly because of lower traffic flow (Cserkész and Farkas 2015; Kušta et al. 2015), but this assertion should be taken carefully, because figures for roads may be inflated as they are far more widespread than railways (Pérez-Espona et al. 2008; Yang et al. 2011) (4) railway corridors are narrower than those of roads, which imply lower loss of habitats when a new line is built (e.g., Gerlach and Musolf 2000; Tremblay and St. Clair 2009); finally, although some maintenance practices that use pollutants (potential disturbers) are shared by railways and roads, like de-icing or the application of herbicides on verges, the impact of vehicles, the most important source of chemical pollutants (Forman et al. 2003), is lower in railways because many trains have electric engines.

Especially interesting for our purposes are the studies comparing railway impacts with those of roads, as they highlight their similarities and differences. These differences and similarities, however, tend to be species-specific. For instance, Gerlach and Musolf (2000) found that in Germany and Switzerland, a 25-year-old highway contributed to genetic substructuring in bank voles (*Clethrionomys glareolus*), while a 40-year-old railway and a 25-year-old country road did not. In a similar vein, railways seemed to have no strong effects on the red deer (*Cervus elaphus*) population's genetic differentiation in the UK, as differentiation was the same with or without railways, while roads were identified as gene-flow barriers to this species (Pérez-Espona et al. 2008). Railways had a low impact is the latter example probably because they were not parallel to orographic barriers, they were relatively sparse, and they had a low traffic flow (Pérez-Espona et al. 2008). In tune with the previous work, Yang et al. (2011) found that roads contributed to the genetic isolation of Chinese populations of Przewalski's gazelles (*Procapra przewalskii*), but railways had no influence on genetic differentiation, probably because of their low traffic flow and the presence of wildlife passes. A similar result was found for roe deer (*Capreolus capreolus*) in Switzerland, although in this case, the authors suggested that the differences could be due to highways being fenced and railways not, as traffic flow was similar (Hepenstrick et al. 2012). In Canada, Tremblay and St. Clair (2009) showed that railways were more permeable to forest song bird movements than were roads, likely due to their narrower width and lower traffic. Indeed, the authors found that the gap size in the vegetation was the most important factor constraining forest bird movement, especially when the gap was larger than 30 m. As a final example, in their study of New England cottontails (*Sylvilagus transitionalis*) in the USA, Fenderson et al. (2014) concluded that major highways limited dispersal, whereas railways and power lines corridors acted as dispersal facilitators.

Railways are more environmentally friendly than road vehicle transportation, but this does not mean that their negative impacts should be ignored. Therefore, while acknowledging that there is a wide range of situations where priority should be given to the development of railways, or to the maintenance of existing ones, it is also crucial to take into account the impacts on the habitats transversed by these infrastructures, and on the wildlife populations occurring therein. However, we believe that these impacts can be considerably reduced once they are identified, and once the decision-makers are willing to pursue the required mitigation measures. In the next chapters, we will first review the impacts of railways on biodiversity (mortality, exclusion and barrier effects, introduction and dispersal of exotic species and pollution) and then present several case studies with a view to identifying problems and proposing strategies to mitigate railways negative effects.

References

Arens, P., van der Sluis, T., van't Westende, W. P. C., Vosman, B., Vos, C. C., & Smulders, M. J. M. (2007). Genetic population differentiation and connectivity among fragmented Moor frog (*Rana arvalis*) populations in The Netherlands. *Landscape Ecology, 22,* 1489–1500.

Chandra, S., & Agarwal, M. M. (2007). *Railway engineering.* New Delhi: Oxford University Press.

Clauzel, C., Girardet, X., & Foltête, J.-C. (2013). Impact assessment of a high-speed railway line on species distribution: Application to the European tree frog (*Hyla arborea*) in Franche-Comté. *Journal of Environmental Management, 127,* 125–134.

Clevenger, A. P., & Waltho, N. (2005). Performance indices to identify attributes of highway crossing structures facilitating movement of large mammals. *Biological Conservation, 121,* 453–464.

Cserkész, T., & Farkas, J. (2015). Annual trends in the number of wildlife-vehicle collisions on the main linear transport corridors (highway and railway) of Hungary. *North-Western Journal of Zoology, 11,* 41–50.

Dorsey, B., Olsson, M., & Rew, L. J. (2015). Ecological effects of railways on wildlife. In R. van der Ree, D. J. Smith, & C. Grilo (Eds.), *Handbook of Road ecology* (pp. 219–227). West Sussex: Wiley.

European Railway Agency. (2013). *Intermediate report on the development of railway safety in the European Union.* http://www.era.europa.eu/document-register/documents/spr%202013%20final%20for%20web.pdf

Fenderson, L. E., Kovach, A. I., Litvaitis, J. A., O'Brien, K. M., Boland, K. M., & Jakuba, W. J. (2014). A multiscale analysis of gene flow for the New England cottontail, an imperiled habitat specialist in a fragmented landscape. *Ecology and Evolution, 4,* 1853–1875.

Flammini, F. (Ed.). (2012). *Railway safety, reliability, and security: Technologies and systems engineering: Technologies and systems engineering.* IGI Global.

Forman, R. T. T., Sperling, D., Bissonette, J. A., Clevenger, A. P., Cutshall, C. D., Dale, V. H., et al. (2003). *Road ecology, science and solutions.* Washington, DC: Island Press.

Gerlach, G., & Musolf, K. (2000). Fragmentation of landscape as a cause for genetic subdivision in bank voles. *Conservation Biology, 14,* 1066–1074.

Hepenstrick, D., Thiel, D., Holderegger, R., & Gugerli, F. (2012). Genetic discontinuities in roe deer (*Capreolus capreolus*) coincide with fenced transportation infrastructure. *Basic and Applied Ecology, 13,* 631–638.

Kociolek, A. V., Clevenger, A. P., St. Clair, C. C., & Proppe, D. S. (2011). Effects of road networks on bird populations. *Conservation Biology, 25,* 241–249.

Kušta, T., Keken, Z., Ježek, M., & Kůta, Z. (2015). Effectiveness and costs of odor repellents in wildlife-vehicle collisions: A case study in Central Bohemia, Czech Republic. *Transportation Research Part D, 38,* 1–5.

Lewis, M. J. T. (2001). Railways in the Greek and Roman world. In *Early railways. A selection of papers from the first international early railways conference* (pp. 8–19).

Li, Z., Ge, Ch., Li, J., Li, Y., Xu, A., Zhou, K., et al. (2010). Ground-dwelling birds near the Qinghai-Tibet highway and railway. *Transportation Research Part D, 15,* 525–528.

Pereira, H. M., Navarro, L. M., & Martins, I. S. (2012). Global biodiversity change: The bad, the good, and the unknown. *Annual Review of Environment and Resources, 37,* 25–50.

Pérez-Espona, S., Pérez-Barbería, F. J., Mcleod, J. E., Jiggins, C. D., Gordon, I. J., & Pemberton, J. M. (2008). Landscape features affect gene flow of Scottish Highland red deer (*Cervus elaphus*). *Molecular Ecology, 17,* 981–996.

Popp, J. N., & Boyle, S. P. (2017). Railway ecology: Underrepresented in science? *Basic and Applied Ecology, 19,* 84–93.

Proctor, M. F., McLellan, B. N., Strobeck, C., & Barclay, R. M. R. (2005). Genetic analysis reveals demographic fragmentation of grizzly bears yielding vulnerably small populations. *Proceedings of the Royal Society of London. Series B, 272,* 2409–2416.

Profillidis, V. A. (2014). *Railway Management and Engineering.* Burlington, VT: Ashgate Publishing Ltd.

Railway Handbook. (2015). *Energy consumption and CO_2 emissions.* http://www.uic.org/IMG/pdf/iea-uic_2015-2.pdf

Ray, N., Lehmann, A., & Joly, P. (2002). Modeling spatial distribution of amphibian populations: A GIS approach based on habitat matrix permeability. *Biodiversity and Conservation, 11,* 2143–2165.

Rodríguez, A., Crema, G., & Delibes, M. (1997). Factors affecting crossing of red foxes and wildcats through non-wildlife passages across a high-speed railway. *Ecography, 20,* 287–294.

Shabana, A. A., Zaazaa, K. E., & Sugiyama, H. (2008). *Railroad vehicle dynamics: A computational approach.* Boca Raton, FL: CRC Press.

Tremblay, M. A., & St. Clair, C. C. (2009). Factors affecting the permeability of transportation and riparian corridors to the movements of songbirds in an urban landscape. *Journal of Applied Ecology, 46,* 1314–1322.

Vos, C. C., Antonisse-De Jong, A. G., Geodharts, P. W., & Smulders, M. J. M. (2001). Genetic similarity as a measure for connectivity between fragmented populations of the moor frog (*Rana arvalis*). *Heredity, 86,* 598–608.

Waller, J. S., & Serveen, C. (2005). Effects of transportation infrastructure on grizzly bears in northwestern Montana. *Journal of Wildlife Management, 69*, 985–1000.

Wells, P., Woods, J. G., Bridgewater, G., & Morrison, H. (1999). Wildlife mortalities on railways; Monitoring methods and mitigation strategies. In G. Evink, P. Garrett, & D. Zeigler (Eds.), *Proceedings of the third international conference on wildlife ecology and transportation* (pp. 237–246). Tallahassee, FL: Florida Department of Transportation.

Xia, L., Yang, Q., Li, Z., Wu, Y., & Feng, Z. (2007). The effect of the Qinghai-Tibet railway on the migration of Tibetan antelope *Pantholops hodgsonii* in Hoh-xil National Nature Reserve, China. *Oryx, 41*, 352–357.

Yang, J., Jiang, Z., Zeng, Y., Turghan, M., Fang, H., & Li, C. (2011). Effect of anthropogenic landscape features on population genetic differentiation of Przewalski's Gazelle: Main role of human settlement. *PLoS ONE, 6*, e20144.

Chapter 2
Current Knowledge on Wildlife Mortality in Railways

Sara M. Santos, Filipe Carvalho and António Mira

Abstract Wildlife mortality on roads has received considerable attention in the past years, allowing the collection of abundant data for a wide range of taxonomic groups. On the contrary, studies of wildlife mortality on railway tracks are scarce and have focused primarily on a few large mammals, such as moose and bears. Nevertheless, many species are found as victims of collisions with trains, although certain taxonomic groups, such as amphibians and reptiles, and/or small bodied species are reported infrequently and their mortality is probably underestimated. However, no assessment of population impacts is known for railways.

Keywords Wildlife mortality · Railway collisions · Collision risk

Introduction

One of the most obvious impacts of railways on wildlife is direct mortality from collisions with trains (Davenport and Davenport 2006; Dorsey et al. 2015; Forman et al. 2003; van der Grift 1999). In addition to collisions, mortality can also occur due to electrocution, wire strikes and rail entrapment (Dorsey et al. 2015; SCV 1996). In fact, some species of small body size can become trapped between the rails and die from dehydration or hunger (Budzic and Budzic 2014; Kornilev et al.

S.M. Santos (✉) · F. Carvalho · A. Mira
CIBIO/InBIO, Centro de Investigação em Biodiversidade e Recursos Genéticos, Universidade de Évora, Pólo de Évora, Casa do Cordovil 2° Andar, 7000-890 Évora, Portugal
e-mail: smsantos@uevora.pt

A. Mira
Unidade de Biologia da Conservação, Departamento de Biologia, Universidade de Évora, Mitra, 7002-554 Évora, Portugal

F. Carvalho
Department of Zoology and Entomology, School of Biological and Environmental Sciences, University of Fort Hare, Private Bag X1314, Alice 5700, South Africa

© The Author(s) 2017
L. Borda-de-Água et al. (eds.), *Railway Ecology*,
DOI 10.1007/978-3-319-57496-7_2

11

2006). Nevertheless, collisions are the most common cause of railway mortality (Dorsey et al. 2015).

Wildlife mortality on roads has received considerable attention thanks to abundant data for a wide range of taxonomic groups and different continents. By contrast, studies on wildlife mortality on railway tracks are scarce, have focused primarily on large mammals, and are concentrated on a few countries, such as Canada and Norway (Dorsey et al. 2015; Gundersen and Andreassen. 1998; Seiler and Helldin 2006; van der Grift 1999). The mammal species receiving the most attention are frequently the larger ones, such as moose, bears or elephants as they cause more damage to trains, disrupt the normal operation of the train network, or hold higher conservation and economic status (Huijser et al. 2012; Rausch 1958; van der Grift 1999). Here we review and discuss the mortality impacts on different taxa.

Mammal Mortality

Mortality of mammals due to train collisions can be of considerable importance. In multispecies surveys (Heske 2015; SCV 1996; van der Grift 1999) of all vertebrate recorded, the approximate proportion of mammals found dead on rail tracks ranges from 26% (Netherlands), to 36% (USA) and 38% (Spain).

Train mortality can have large impacts on mammal populations, particularly for species that are already endangered, species with large home ranges and low density populations, and species with low reproductive rate (van der Grift and Kuijsters 1998; van der Grift 1999). The highest mortality numbers are usually found at sections where rail lines intersect important mammal habitats or migration routes (Child 1983; Gundersen and Andreassen 1998; van der Grift 1999).

Many species are victims of collisions. The body size of the mammal species that are killed varies greatly, ranging from small insectivores (as the hedgehog *Erinaceus europaeus*) and small carnivores (such as the Virginia opossum *Didelphis virginiana*), to large carnivores (such as the grizzly and brown bear *Ursus arctos* and the American lynx *Lynx canadensis*), to ungulates (such as moose *Alces alces* and deer *Cervus elaphus*), and even elephants (*Elephas maximus*) (Child 1983; Gibeau and Herrero 1998; Gibeau and Heuer 1996; Gundersen et al. 1998; Gundersen and Andreassen 1998; Heske 2015; Krofel et al. 2012; Rausch 1958; Singh et al. 2001; van der Grift and Kuijsters 1998). Still, most existing studies focus on large ungulates and carnivores, and most lack quantitative data (van der Grift 1999).

The High Frequency of Collision with Ungulates

Train accidents with ungulates are common worldwide. From Canada and Alaska to Norway and Sweden, moose commonly intercept and travel along the rails, being frequently killed (Child 1983; Eriksson 2014; Gundersen and Andreassen 1998; Huijser et al. 2012; Modafferi and Becker 1997). In British Columbia, Canada, the annual loss of moose to train collisions in the winters between 1969 and 1982 was estimated to range from several hundred to more than 1000 animals (Child 1983). In Norway, the number of moose killed in train collisions has increased from ca. 50 moose per year in the 1950s to a yearly average of 676 during 1990–1996. This has resulted in economic costs due to material damages and loss of income due to the lower number of hunting licenses issued (Gundersen and Andreassen 1998). Jaren et al. (1991) estimated these costs at around $2900 per train-killed moose in Norway. On average, 87 moose are killed along the Rørosbanen railway (Norway) each year, which sums to a socio-economic loss of ca. $250,000 (Jaren et al. 1991).

Although in Alaska moose are considered common species within their range, train kills accounted for 24% of known deaths, being the second cause of death in a studied population (Modafferi and Becker 1997). During the winter of 1989–1990 there was a 35% reduction of the moose population in lower Susitna Valley, Alaska, due to a combined effect of non-natural mortality and poor winter survival. Although train collisions alone were not the only reason for that reduction, they were the main cause of non-natural mortality (61.5%; 351 individuals). There was a reduction of about 70% of moose numbers along this railway between 1984 and 1991, presumably due to high mortality on both the railway and the highway (Modafferi and Becker 1997).

Some behaviours can contribute to this high mortality rate on railways. Moose annually migrate from traditional summer areas to lower elevation winter areas with snowfall determining the onset of these seasonal movements (Child 1983). Wherever railways intercept and/or are parallel to areas occupied by moose, large concentrations of animals may use the right-of-way of rails. Moreover, in winters with above average snowfall, moose use the railways as travelling routes more often, being more vulnerable to collisions (Child 1983; Gundersen et al. 1998; Gundersen and Andreassen 1998; Rausch 1958). About 79% of moose train accidents in Norway occurred from December to March, possibly due to their migratory behaviour (Gundersen and Andreassen 1998).

As observed by Child (1983), moose can return rapidly to train tracks even after a collision experience. A typical fatal event occurs when the animal attempts to leave the tracks, but rapidly returns to the rail, probably because it is free of snow, and then try to escape the approaching train by running on the rail. Sometimes moose show an aggressive behaviour, standing in front of the train and attacking the locomotive, with fatal consequences (Child 1983).

The factors influencing moose collision risk were studied by Gundersen and Andreassen (1998) who found that trains running at night, in the morning or in the evening had a higher risk of moose collision than daytime trains. The probability of collision was also higher during nights of full moon. This could be partially explained by the higher moose activity in those periods (Gundersen and Andreassen 1998). In a parallel study, it was found that the collisions increased in winter with increasing snow depth and ambient temperatures below 0 °C, and were located in the outlets of side valleys (Gundersen et al. 1998). The increase in the food availability close to the railways due to logging activities also increased the number of moose collisions (Gundersen et al. 1998).

Other ungulates are also killed frequently by trains. The elk (*C. canadensis*) and the bighorn sheep (*Ovis canadensis*) were the most frequent victims in Jasper National Park, Canada, while deer (*Odocoileus virginianus*) was the most common casualty in Mount Robson Provincial Park, Canada (Huijser et al. 2012). In Norway, although moose is the most reported victim of railway traffic, the roe deer (*Capreolus capreolus*) is also a common one with 12.4% of all mammals recorded from 1993 to 1996, while reindeer (*Rangifer tarandus*) and muskox (*Ovibus moshatus*) are killed more infrequently, with 2.8% and 0.17% respectively (Gundersen et al. 1998). The roe deer was particularly affected in the Czech Republic with 0.8 kills/km in 2009 (Kušta et al. 2014). In the Iberian Peninsula, although some deer and roe deer are reported to be killed by trains, the wild boar (*Sus scrofa*) is the most common ungulate victim (SCV 1996).

In addition, besides possible population decreases, railway kills may cause shifts in the age structure of populations. Huggard (1993) showed that American elks in Banff National Park, Canada, whose ranges overlap roads or railways are less likely to reach old age than animals away from these infrastructures. Additionally, Huggard (1993) also found differences in body condition of American elk predated by wolves and animals killed on the road or railway: elks killed by wolves were in significantly poorer condition than those killed on road or railways, suggesting a non-natural selection by trains and cars (Huggard 1993).

Train Accidents with Bears

Bears are often the most frequently reported carnivore killed by trains in Central Europe and North America. In Montana, USA, 29 grizzly bears were killed on 109 km section of a railway track between 1980 and 2002 (Waller and Servheen 2005), while nine brown bears were killed on the Ljubljana–Trieste railway, Slovenia, between 1992 and 1999 (Kaczensky et al. 2003). Train accidents with bears were also reported in the Abruzzo mountains, Italy (Boscagli 1987), and in Croatia, where 70% of all traffic killed bears occurred along the Zagreb–Rijeka railway (Huber et al. 1998). In Mount Robson Provincial Park, Canada, the black bear (*U. americanus*) was the carnivore with the highest mortality caused by trains, with a higher number of collisions in railways than on the highways (Huijser et al. 2012). In British Columbia, Canada, 13 black bears were killed on a 15 km railway section between 1994 and 1996 (Munro 1997; Wells et al. 1999).

Why are so many bears killed by trains? Historically, grizzly bears have been attracted to railways by grain leaked from trains along the tracks or that accumulated at sites of repeated derailments. Such concentration of food led grizzly bears being struck and killed by trains at these sites (Waller and Servheen 2005). However, grain spills have been reduced through the years and recent research suggests that most bears struck by trains are young individuals, which are unlikely to have acquired the behaviour of feeding on spilled grains (Kaczensky et al. 2003; Krofel et al. 2012). Some collisions happened because bears had used the railway as a movement route (Kaczensky et al. 2003), or were attracted by food resources in railway verges (Gibeau and Herrero 1998; Waller and Servheen 2005). When both railways and highways are present, however, bears seem to cross railways mostly at night, when highway traffic volume decreases, but when railway traffic volume increases (Waller and Servheen 2005). Apparently, bears have learned to avoid the periods of higher risk in highways but in the process have become more exposed to railway collisions, as suggested by different mortality rates between the two infrastructures (Waller and Servheen 2005).

Even though the railways may pose significant mortality risks for bears, not all populations are threatened by train accidents. This is the case of Slovenian bear population that appears to be increasing in spite of 6.6% of annual bear mortality being caused by railway collisions (Krofel et al. 2012).

The Case of Asian Elephants

In India the railway is quite extensive and train collisions with elephants have been identified as a conservation concern (Deka and Sarma 2012; Singh et al. 2001). From 1987 to 2001, 18 elephants were killed in train accidents in a section of 23 km of rail in the Rajaji National Park, India (Singh et al. 2001). These numbers accounted for 45% of total elephant mortality in the studied area for the same period (Singh et al. 2001). In addition, the Indian Forest Department records show that railway trains were responsible for killing at least 35 elephants in the Assam region between 1990 and 2006 (Deka and Sarma 2012).

Maximum elephant mortality occurred during the summer months of high temperatures and low rainfall. The high temperatures and water scarcity appeared to be the deciding factors forcing elephants to cross the rail tracks during the late dry season when water sources on the southern side of rail had dried up (Singh et al. 2001).

What About the Collisions with Other Mammal Species?

For most European countries most mammals killed in train collisions are of small size and efforts have been made to document all species accidents (SCV 1996; van der Grift 1999). In Netherlands, 38% of collisions comprises lagomorphs (*Lepus europaeus* and *Oryctolagus cuniculus*), and 30% carnivores (including the badger *Meles meles,* red fox *Vulpes vulpes* and stoat *Mustela erminea*). Small percentages were registered for ungulates (roe deer and wild boar: 9%), small insectivores (hedgehog and mole *Talpa europaea*: 6%) and rodents (red squirrel *Sciurus vulgaris,* muskrat *Ondatra zibethicus* and brown rat *Rattus norvegicus*: 4%) (van der Grift 1999). In the Czech Republic, 73.5% of the recorded collisions include roe deer, 20.4% European hare (*Lepus europaeus*), 4.1% wild boar and 2% red fox (Kušta et al. 2011).

In Spain, 18.4% of mammal casualties from trains were carnivores (excluding domestic species), mainly fox and stone marten (*Martes foina*). Ungulates (mostly wild boar) and lagomorphs were also frequently killed by trains, 14.4% each (SCV 1996). Bats may be also killed by trains, especially on railway tunnels and sections close to old buildings, but no data are available to document this for railways (SCV 1996). Information for roads indicates that bats are very difficult to detect (Santos et al. 2011). Thus, bat mortality on railways can be high but have been ignored.

In general, most reported mammal victims are common species, which suggests that the effects on population levels should be small for species with large and widespread populations (van der Grift and Kuijsters 1998). However, the death of a few individuals of a rare or endangered species may further increase species extinction risk (van der Grift 1999).

Bird Mortality

From the assessment of bird carcasses found on railway tracks, it has been frequently deduced from monitoring studies that the cause of mortality was due to collisions (Peña and Llama 1997), although collisions with the catenary, electrocution and barotrauma induced by the train movement are also possible bird mortality causes related to the railways (SCV 1996).

The frequency of bird mortality by trains when compared to other taxonomic groups varies from 11% (USA), to 55% (Spain) and 57% (Netherlands) (Heske 2015; SCV 1996; van der Grift and Kuijsters 1998). In Chicago state, USA, most frequent victims were mallard duck (*Anas platyrhynchos*), common grackle (*Quicalus quiscula*) and sora (*Porzana carolina*), while songbirds were infrequent (Heske 2015). In the Netherlands, accidents concerned mostly swans (*Cygnus* spp.), ducks (family Anatidae) and coots (*Fulica* spp.). Gulls (family Laridae) and raptors such as hawks and owls were also reported frequently (van der Grift and Kuijsters 1998). In Guadarrama (Spain), Peña and Llama (1997) monitored for two years the bird mortality in 8 km of railways. The tawny owl (*Strix aluco*) was the most frequent bird victim (18%), followed by the carrion crow (*Corvus corone*; 16%) and the little owl (*Athene noctua*; 9.6%). Nearly half (46.8%) of this mortality occurred in the summer, while much lower values (13.8 and 16%) occurred in the winter and spring, respectively. Another report for Spanish railways (SCV 1996) also found that owls were the most common victims (22.5% of all birds). Barn owl (*Tyto alba*), tawny owl and little owl were the main victims, although there were also records of eagle owl (*Bubo bubo*), long-eared owl (*Asio otus*) and scops owl (*Otus scops*). As in road-related night mortality, train lights are likely responsible for the majority of owl kills. It was observed that little owls when perched on the train catenary, became disoriented with the approaching train, hence increasing the likelihood of being killed (Peña and Llama 1997; SCV 1996).

Birds of prey were also frequently registered as train casualties in Spanish railways being 19.2% of all birds killed (SCV 1996). The most recorded species were the buzzard (*Buteo buteo*), black kite (*Milvus migrans*) and griffon vulture (*Gyps fulvus*). One possible explanation is the attractiveness of perches along the trails and of railway verges as a hunting ground for birds of prey and owls (SCV 1996; van der Grift and Kuijsters 1998). Moreover, all three species scavenge regularly the trails for food carcasses, increasing their vulnerability to collisions (SCV 1996).

In some areas, train drivers observed several partridges (*Alectoris rufa*) using the gravel from the track ballast or within the train track on rainy days, being thus killed altogether (Peña and Llama 1997). This last behaviour of galliforme birds, seeking refuge on the train tracks, has also been referred by Havlin (1987) for the common pheasant (*Phasianus colchicus*).

As is often the case, for small or isolated populations the death of a few individuals represents a serious risk to population decline (van der Grift and Kuijsters 1998). In particular, birds of prey and owls are sensitive in this respect, even though some species may be considered common. Railway traffic is the second cause of mortality in the Netherlands for the buzzard and the kestrel (*Falco tinnunculus*), with 7.1 and 4.6%, respectively, of all dead birds found (van der Grift and Kuijsters 1998 and references therein). For the barn owl, railway traffic is a genuine danger, as 16% of mortality in Brittany (France) was caused by trains (SCV 1996). Also, for the little owl, 3.6% of deaths result from train accidents (van der Grift and Kuijsters 1998 and references therein).

Concerning the mortality of protected birds, the Spanish train personnel referred one fatal collision with one young imperial eagle (*Aquila adalberti*) in a rail section close to the nest, and collisions with several black vultures (*Aegypius monachus*) when feeding on carcasses on the rail (Peña and Llama 1997). Also, in the Swiss Alps, the collisions with trains or cars were the third cause of mortality among eagle owls, accounting for 30% of the anthropogenic mortality (Schaub et al. 2010).

Not all deaths in railways are caused by train collisions. Some of the mortality arises from the collision with the rail electric lines. This has been clearly observed in Spain for the buzzard, barn owl, eagle owl, lesser kestrel (*Falco naumanni*), sparrowhawk (*Accipiter nisus*), song thrush (*Turdus philomelos*), and starlings (*Sturnus* sp). But, still, the most frequent cause of death was train collisions (SCV 1996).

Amphibian and Reptile Mortality

As stated before, most railway studies focus only on the mortality of large species (van der Grift 1999). In addition, reports on train-collisions often underestimate the true number of collisions, as detectability and persistence rates of carcasses are rarely estimated (see Chap. 3, and van der Grift 1999), problems that affect mainly small body size species (SCV 1996). Thus, there is still an important difference between the abundance of amphibian and reptile mortality in roads and in railways. While there are plenty of data on mortality of these taxa on roads (Beebee 2013;

Hels and Buchwald 2001; Carvalho and Mira 2011), records for railways are much rarer (van der Grift and Kuijsters 1998). Unfortunately, there are no studies that compare detectability and persistence rates of carcasses between road and railway surveys, so the difference in numbers remains an open question.

Considering only the studies that aimed to survey train casualties of all species, the frequency of amphibians can reach up to 47% of all vertebrate records (Heske 2015), while the records of reptiles represent ca. 4% (Spain) and 6% (USA) of carcasses (Heske 2015; SCV 1996). In Netherlands, for example, no carcasses of amphibians or reptiles were ever found (van der Grift and Kuijsters 1998), while in Spain, only reptiles were recorded (SCV 1996).

In Chicago state, USA, amphibians were the most abundant taxonomic group recorded in train collisions (Heske 2015), mainly northern leopard frogs (*Lithobates pipiens*) and American toads (*Bufo americanus*), common species in the region. Mortalities of amphibians were particularly high after rain events, when these species are most active and are also frequently found dead on the roads (Heske 2015).

Budzic and Budzic (2014) conducted a survey specifically aimed to estimate amphibian mortality on 34 km of Polish railways. They found that three species were most affected by train accidents: the common toad (*Bufo bufo*), the common frog (*Rana temporaria*) and the green frog (*Pelophylax kl. esculentus*), and most of dead individuals (77%) were adult common toads. Although mortality rates can be high, all three species are considered common in Europe. In fact, two of the species affected by railway mortality (common toad and common frog), are among the most common European amphibians, for which there is also evidence of high road-mortality (Hels and Buchwald 2001; Matos et al. 2012; Orłowski 2007).

As in roads, the spring migration (or autumn migration, in some regions) seems to be the period of highest amphibian mortality on railway tracks, as 87% of all accidents occurred in that season (Budzic and Budzic 2014). Yet, railway mortality of amphibians seems to depend on animals' physical features (such as body size and limb length) and should be associated with the agility of the species (Budzic and Budzic 2014). While in the case of roads, the velocity of individuals is used as a proxy for agility (Hels and Buchwald 2001), in the case of railway tracks, agility relates mainly to the ability to overcome obstacles (Budzic and Budzic 2014; SCV 1996). Due to its physical features, the common toad was more likely to become trapped between rails, indicating that this species (and others with similar physical features) may be more vulnerable to railway mortality (Budzic and Budzic 2014). Some species of small size may be less vulnerable because they cannot cross the rail track (Heske 2015). However, they may be affected by railways at the level of gene flow due to barrier effects (Holderegger and Di Giulio 2010), which may be also a conservation problem.

Concerning reptile mortality, 13 snakes (4%) were recorded across the whole Spanish railway network between 1990 and 1995. Only two species were identified: the ladder snake (*Elaphe scalaris*) and the Montpellier snake (*Malpolon monspessulanus*), both common reptiles in the country (SCV 1996).

The railway bed may be lethal itself for smaller animals that can become trapped between the rails, where they may be susceptible to predation or physiological stress. This is the case of railway-induced mortality of Eastern box turtles (*Terrapene carolina*), in the USA, that often cannot escape when trapped between railway tracks. In a controlled experiment with 12 turtles, Kornilev et al. (2006) showed that most turtles clearly have the ability to escape at railway crossings, although mortality rates within the railway were still high. Notice, as well, that when between railway tracks, turtles can quickly reach critically high body temperatures, even on relatively mild days (Kornilev et al. 2006).

How Different Is Mortality in Railways and Roads?

Only a few studies have directly compared railway and road mortality, a comparison that is not easy because railway impacts are more difficult to detect (SCV 1996). Considering the diversity of taxonomic groups that are frequently killed on railways, it may be surprising that trains can cause such high mortality among some species (see ungulates, for example), because, compared to cars, trains are often less frequent, noisier, larger and, most of the times, travel at low to medium speeds (Heske 2015; Morelli et al. 2014). However, even freight trains can sometimes reach speeds close to 200 km/h and cannot stop quickly when encountering animals on the rails. This obviously leads to high mortality numbers on railways that could be avoided more easily on roads (Dorsey et al. 2015; Heske 2015). Accordingly, the number of dead bears and moose along railways often exceed death rates along roads (Belant 1995; Boscagli 1987; Modafferi and Becker 1997; Waller and Servheen 2005). On the other hand, for other species the railways may lead to lower mortality. For instance, the vibrations of approaching trains can be felt along the rails and this may give warning to some terrestrial vertebrates. This seems to be the case of snakes that may be warned of approaching trains by vibrations transmitted through the rails or the ballast (Heske 2015). In addition, while the traffic on roads can be simultaneously two-way, trains approach from one direction at a time, and the width of a road is greater than the width of a railway, decreasing the vulnerability of vertebrates that cross railways (Heske 2015). In a study that compared mortality rates between roads and railways, it was found that railways had notably lower mortalities of songbirds, small mammals, and turtles when compared to those of roads (Heske 2015), suggesting that diurnal and vagile species may be more efficient at avoiding trains than avoiding cars and trucks on busy two-way roads (Heske 2015).

Future Directions

In Spain, estimates of vertebrate mortality due to train collisions averages 36.5 vertebrates/km/year (not including high speed trains; SCV 1996). In high-speed railways, this estimate grows to, at least, 92 vertebrates/km/year (SCV 1996). Nevertheless, in general, most of the species recorded as train casualties are considered common within the adjacent natural areas (Budzic and Budzic 2014; Heske 2015; van der Grift and Kuijsters 1998). However, determining the impacts of high-speed railways is crucial because this type of railway is expanding rapidly across the world. We expect that the differences in noise levels, speed and the common fencing practices in high-speed lines promotes different impacts between these and the traditional rails (Dorsey et al. 2015).

Acknowledgements We would like to thank Infraestruturas de Portugal, and in particular João Morais Sarmento, Ana Cristina Martins, Cândida Osório de Castro and Graça Garcia, without which this book would not have been possible.

We also would like to thank Sasha Vasconcelos for her willingness to help with our constant questions on English.

Finally, we would like to thank Margaret Deignan at Springer for her invaluable help in putting together this book.

References

Beebee, T. J. C. (2013). Effects of road mortality and mitigation measures on amphibian populations: Amphibians and roads. *Conservation Biology, 27,* 657–668.

Belant, J. L. (1995). Moose collisions with vehicles and trains in Northeastern Minnesota. *Alces, 31,* 1–8.

Boscagli, G. (1987). Brown bear mortality in central Italy from 1970 to 1984. *International Conference on Bear Research and Management, 7,* 97–98.

Budzic, K. A., & Budzic, K. M. (2014). A preliminary report of amphibian mortality patterns on railways. *Acta Herpetologica, 9,* 103–107.

Carvalho, F., & Mira, A. (2011). Comparing annual vertebrate road kills over two time periods, 9 years apart: A case study in Mediterranean farmland. *European Journal Wildlife Research, 57,* 157–174.

Child, K. (1983). Railways and moose in the central interior of BC: A recurrent management problem. *Alces, 19,* 118–135.

Davenport, J., & Davenport, J. L. (Eds.). (2006). *The ecology of transportation: Managing mobility for the environment.* Dordrecht: Springer.

Deka, R. N., & Sarma, K. K. (2012). Pattern recognition based anti collision device optimized for elephant-train confrontation. *IRNet Transactions on Electrical and Electronics Engineering, 1* (2), 92–97.

Dorsey, B., Olsson, M., & Rew, L. J. (2015). Ecological effects of railways on wildlife. In R. van der Ree, D. J. Smith, & C. Grilo (Eds.), *Handbook of road ecology* (pp. 219–227). West Sussex: Wiley.

Eriksson, C. (2014). *Does tree removal along railroads in Sweden influence the risk of train accidents with moose and roe deer?* Dissertation, Second cycle, A2E. Grimsö och Uppsala: SLU, Dept. of Ecology, Grimsö Wildlife Research Station.

Forman, R. T. T., Sperling, D., Bissonette, J. A., Clevenger, A. P., Cutshall, C. D., Dale, V. H., et al. (2003). *Road ecology: Science and solutions.* Washington, DC: Island Press.

Gibeau, M. L., & Herrero, S. (1998). Roads, rails and Grizzly bears in the Bow River Valley, Alberta. In G. L. Evink, P. Garrett, D. Zeigler, & J. Berry (Eds.), *Proceedings of the international conference on wildlife ecology and transportation* (pp. 104–108). Florida Department of Transportation, Tallahassee, Florida, USA.

Gibeau, M. L., & Heuer, K. (1996). Effects of transportation corridors on large carnivores in the Bow River Valley, Alberta. In G. L. Evink, P. Garrett, D. Zeigler, & J. Berry (Eds.),

Proceedings of the international conference on wildlife ecology and transportation (pp. 67–79). Florida Department of Transportation, Tallahassee, Florida, USA.

Gundersen, H., & Andreassen, H. P. (1998). The risk of moose *Alces alces* collision: A predictive logistic model for moose-train accidents. *Wildlife Biology, 4,* 103–110.

Gundersen, H., Andreassen, H. P., & Storaas, T. (1998). Spatial and temporal correlates to Norwegian moose-train collisions. *Alces, 34,* 385–394.

Havlin, J. (1987). On the importance of railway lines for the life of avifauna in agrocoenoses. *Folia Zoologica, 36,* 345–358.

Hels, T., & Buchwald, E. (2001). The effect of road kills on amphibian populations. *Biological Conservation, 99,* 331–340.

Heske, E. J. (2015). Blood on the tracks: Track mortality and scavenging rate in urban nature preserves. *Urban Naturalist, 4,* 1–13.

Holderegger, R., & Di Giulio, M. (2010). The genetic effects of roads: A review of empirical evidence. *Basic and Applied Ecology, 11,* 522–531.

Huber, D., Kusak, J., & Frikovic, A. (1998). Traffic kills of brown bears in Gorski Kotar, Croatia. *Ursus, 10,* 167–171.

Huggard, D. J. (1993). Prey selectivity of wolves in Banff National Park. II. Age, sex and condition of elk. *Canadian Journal of Zoology, 71,* 140–147.

Huijser, M. P., Begley, J. S., & van der Grift, E. A. (2012). *Mortality and live observations of wildlife on and along the yellowhead highway and the railroad through Jasper National Park and Mount Robson Provincial Park, Canada.* Calgary, Canada: Salmo Consulting Inc., On behalf of Kinder Morgan Canada.

Jaren, V., Andersen, R., Ulleberg, M., Pedersen, P., & Wiseth, B. (1991). Moose-train collisions: The effects of vegetation removal with a cost-benefit analysis. *Alces, 27,* 93–99.

Kaczensky, P., Knauer, F., Krze, B., Jonozovic, M., Adamic, M., & Gossow, H. (2003). The impact of high speed, high volume traffic axes on brown bears in Slovenia. *Biological Conservation, 111,* 191–204.

Kornilev, Y., Price, S., & Dorcas, M. (2006). Between a rock and a hard place: Responses of eastern box turtles (*Terrapene carolina*) when trapped between railroad tracks. *Herpetological Reviews, 37,* 145–148.

Krofel, M., Jonozovič, M., & Jerina, K. (2012). Demography and mortality patterns of removed brown bears in a heavily exploited population. *Ursus, 23,* 91–103.

Kušta, T., Ježek, M., & Keken, Z. (2011). Mortality of large mammals on railway tracks. *Scientia Agriculturae Bohemica, 42,* 12–18.

Kušta, T., Holá, M., Keken, Z., Ježek, M., Zíka, T., & Hart, V. (2014). Deer on the railway line: Spatiotemporal trends in mortality patterns of roe deer. *Turkish Journal of Zoology, 38,* 479–485.

Matos, C., Sillero, N., & Argaña, E. (2012). Spatial analysis of amphibian road mortality levels in northern Portugal country roads. *Amphibia-Reptilia, 33,* 469–483.

Modafferi, R. D., & Becker, E. F. (1997). Survival of radiocollared adult moose in lower Susitna river valley, southcentral Alaska. *Journal of Wildlife Management, 61,* 540–549.

Morelli, F., Beim, M., Jerzak, L., Jones, D., & Tryjanowski, P. (2014). Can roads, railways and related structures have positive effects on birds? A review. *Transportation Research Part D, 30,* 21–31.

Munro, R. (1997). Assessing the impact of the Trans-Canada Highway and the Canadian Pacific Railway on bear movements and habitat use patterns in the Beaver Valley, British Columbia. In A. P. Clevenger & K. Wells (Eds.), *Proceedings of the second roads, rails and the environment workshop* (pp. 8–13). Parks Canada, Banff National Park, Alberta & Columbia Mountains Institute of Applied Ecology, Revelstoke, British Coliumbia, Canada.

Orłowski, G. (2007). Spatial distribution and seasonal pattern in road mortality of the common toad Bufo bufo in an agricultural landscape of south-western Poland. *Amphibia-Reptilia, 28,* 25–31.

Peña, O. L., & Llama, O. P. (1997). *Mortalidad de aves en un tramo de linea de ferrocarril* (32 pp.). Grupo Local SEO-Sierra de Guadarrama, Spain.

Rausch, R. A. (1958). *The problem of railroad-moose conflicts in the Susitna valley.* Federal aid in wildlife restoration completion reports (Vol. 12, no. 1). Alaska Resources Library & Information Services, Anchorage, Alaska, USA.

Santos, S. M., Carvalho, F., & Mira, A. (2011). How long do the dead survive on the Road? Carcass persistence probability and implications for road-kill monitoring surveys. *PLoS ONE, 6,* e25383.

S. C. V. (1996). *Mortalidad de vertebrados en líneas de ferrocarril.* Documentos Técnicos de Conservacion SCV 1, Sociedad Conservación Vertebrados, Madrid.

Schaub, M., Aebischer, A., Gimenez, O., Berger, S., & Arlettaz, R. (2010). Massive immigration balances high anthropogenic mortality in a stable eagle owl population: Lessons for conservation. *Biological Conservation, 143,* 1911–1918.

Seiler, A., & Helldin, J.-O. (2006). Mortality in wildlife due to transportation. In J. Davenport & J. L. Davenport (Eds.), *The ecology of transportation: Managing mobility for the environment* (pp. 165–189). Dordrecht: Springer.

Singh, A. K., Kumar, A., Mookerjee, A., & Menon, V. (2001). *A scientific approach to understanding and mitigating elephant mortality due to train accidents in Rajaji National Park.* An occasional report no 3 by Wildlife Trust of India and the International Fund for Animal Welfare of a Rapid Action Project on understanding and mitigating the problem of elephant mortality due to train hits.

van der Grift, E. A. (1999). Mammals and railroads: Impacts and management implications. *Lutra, 42,* 77–98.

van der Grift, E. A., & Kuijsters, H. M. J. (1998). Mitigation measures to reduce habitat fragmentation by railway lines in the Netherlands. In G. L. Evink, P. Garrett, D. Zeigler, & J. Berry (Eds.), *Proceedings of the international conference on wildlife ecology and transportation* (pp. 166–170). Florida Department of Transportation, Tallahassee, Florida, USA.

Waller, J. S., & Servheen, C. (2005). Effects of transportation infrastructure on grizzly bears in northwestern Montana. *Journal of Wildlife Management, 69,* 985–1000.

Wells, P., Woods, J. G., Bridgewater, G., & Morrison, H. (1999). *Wildlife mortalities on railways: Monitoring methods and mitigation strategies.* Revelstoke, British Columbia. Unpublished report.

Chapter 3
Methods to Monitor and Mitigate Wildlife Mortality in Railways

Filipe Carvalho, Sara M. Santos, António Mira and Rui Lourenço

Abstract Recording wildlife mortality on railways is challenging as they have narrow corridors and lower accessibility. To improve mitigation measures, surveys must be systematic and their frequency depending on the targeted species traits and biology. To obtain unbiased estimates in diverse contexts, the data should be corrected using mortality estimators. Mitigation measures must avoid that animals remain on the tracks, as trains cannot be instantly stopped. Box culverts, amphibian tunnels, and under- or overpasses allow a safe crossing, whereas exclusion fences, olfactory repellents, sound signals and sound barriers prevent the crossing of railways. Habitat management in railway verges improves the animal capability to evade trains.

Keywords Bias sources · Mitigation · Monitoring methods · Monitoring plan · Mortality estimators

Filipe Carvalho and Rui Lourenço have contributed equally to this chapter.

F. Carvalho (✉) · S.M. Santos · A. Mira
CIBIO/InBIO, Centro de Investigação em Biodiversidade e Recursos Genéticos,
Universidade de Évora, Pólo de Évora, Casa do Cordovil 2º Andar,
7000-890 Évora, Portugal
e-mail: filipescpcarvalho@gmail.com

F. Carvalho
Department of Zoology and Entomology, School of Biological and Environmental Sciences,
University of Fort Hare, Private Bag X1314, Alice 5700, South Africa

A. Mira
Unidade de Biologia da Conservação, Departamento de Biologia,
Universidade de Évora, Mitra, 7002-554 Évora, Portugal

R. Lourenço
LabOr – Laboratório de Ornitologia, ICAAM—Instituto de Ciências Agrárias e Ambientais
Mediterrânicas, Universidade de Évora, Pólo da Mitra, 7002-554 Évora, Portugal

© The Author(s) 2017
L. Borda-de-Água et al. (eds.), *Railway Ecology*,
DOI 10.1007/978-3-319-57496-7_3

Monitoring Wildlife Mortality in Railways: General Approaches

Railway mortality can negatively affect some species. It is, therefore, crucial to accurately monitor casualties that will reveal areas of high killing rates (i.e., hotspots). This will allow a correct assessment of environmental factors influencing higher mortality rates and the application of proper mitigation measures (Gunson et al. 2011). However, most of our understanding on railway impacts comes from a small number of studies in North America and Europe on a few species, primarily large mammals, such as moose and bears (e.g., Dorsey et al. 2015).

Why Is Monitoring Important?

Ideally, before implementing any linear infrastructure, it is advisable to assess its possible impacts (e.g., vegetation removal, soil movement, noise, light pollution, etc.) in adjacent areas. In addition, species richness and abundance must be recorded to establish a baseline to identify possible conservation concerns (e.g., extinction risk of rarer species) (van der Grift et al. 2013). This is especially relevant when the infrastructure is to be placed across a natural protected area, encompassing several ecosystems (Dorsey et al. 2015). Then, during and after construction, a monitoring plan aiming to measure the impact of collisions, electrocution, rail entrapment, use of culverts, and other aspects of the railway, is crucial for forecasting and understanding the behavioral responses of animals living in the vicinity (Iuell et al. 2003; van der Grift et al. 2013). Therefore, based on the results obtained, we endeavor to set up the best mitigation solutions for each situation.

So far, most research and monitoring studies have assessed the use and effectiveness of wildlife crossing structures (Hunt et al. 1987; Rodríguez et al. 1996). To counteract this, we focus on the monitoring of wildlife mortality, as it is the most visible direct impact of linear infrastructures (van der Grift et al. 2013). Some impacts of railways differ from those of roads for example, the casualties from electrocution, rail entrapment and wire strikes (Dorsey et al. 2015), which suggests the need for specific monitoring programs.

Designing a Wildlife Mortality Monitoring Plan for Railways

Monitoring is limited by practical considerations of cost and feasibility, so a survey of all species is unrealistic (Rytwinski et al. 2015; van der Grift et al. 2013). In fact, costs are one of the main reasons that the state of the habitats and species crossed by a railway are often not assessed. However, regardless of the cost considerations, the impacts of new projected linear infrastructures must always be assessed. Thus, measures should be taken before construction starts, and then during the construction phases to mitigate those impacts on wild species present: this includes paying attention to fences (adequate mesh size, buried, and without holes), fauna passes, and, mainly because of amphibians, the preservation of existing ponds (Iuell et al. 2003). Usually, a monitoring plan must follow four important steps: (1) the choice of the target species; (2) the selection of the spatio-temporal scale of the study; (3) the selection of methods to estimate mortality; and (4) the standardization of the variables influencing wildlife mortality, so that they are easily replicable and comparable (Roedenbeck et al. 2007; van der Grift et al. 2013). To achieve these goals, the co-operation between railway company workers, stakeholders and wildlife researchers at all stages of the monitoring plan (Iuell et al. 2003; Roedenbeck et al. 2007).

Selecting the Target Species

The selection of the species to be monitored depends on their traits (e.g., vagility, conservation status, or sensitivity to fragmentation), on features of the landscape crossed by the railway, and on the characteristics of the railway. For instance, if the railway has overhead electric lines and pylons special care should be taken with birds and bat species (Peña and Llama 1997; Rose and Baillie 1989; SCV 1996); if there is a risk of entrapment inside the rail lines, low vagile reptiles (e.g., turtles) and amphibians (e.g., toads) must be prioritized (Kornilev et al. 2006; Pelletier et al. 2006).

Species that show strong responses toward linear infrastructures and traffic flow, such as carnivores and ungulates, are often selected as target species. These animals are good models for evaluating the factors influencing mortality, because they have large individual territories and large daily and seasonal movements that increase the probability of crossing and of being hit by a train (Iuell et al. 2003; van der grift et al. 2013). Some habitat specialists, like forest-dwelling ones (e.g., the tawny owl *Strix aluco*), should also be considered in this group as they often use specific corridors that cross railways, thus increasing their risk of collision.

Relying on larger species (or species with high visibility) may increase the detection probability because recording wildlife mortality on railways is often harder than on roads, as railways have lower accessibility, narrower corridors, lack of lateral dirt roads, and a highly variable topographic profile (Dorsey 2011; Wells et al. 1999). In addition, choosing abundant species will likely increase sample sizes, and thus the power of the analyses and some species might serve as a proxy for rarer ones that allows obtaining crucial data that otherwise will be not possible (e.g., Roedenbeck et al. 2007).

Selecting the Spatio-Temporal Scale

The rationale for selecting the spatio-temporal scales depends on the species to be monitored (Roedenbeck et al. 2007; van der Grift et al. 2013). Large animals with large home ranges and daily movements require the survey of a railway stretch large enough (e.g., 30 km) to reflect their spatial requirements (Iuell et al. 2003; Seiler and Helldin 2006). By contrast, if we aim to study a local amphibian population during seasonal migrations, a 1 km railway stretch can be reasonable (Hels and Buchwald 2001). Another important issue is the habitat preference. If the target species is forest-dwelling, we must focus on stretches crossing forests (Iuell et al. 2003). For more general species, the railway stretch selection should cover and reflect a hierarchical distribution and abundance of the existing habitats (van der Grift et al. 2013) or, alternatively, we may choose different railway stretches, each representing a different habitat (Roedenbeck et al. 2007). If a species shows clear seasonal movements with a high probability of railway crossings at certain periods of the year (e.g., amphibian breeding season migrations), then wildlife mortality studies should focus on those periods.

Estimating the Number of Casualties on Railways

After assessing the baseline situation in a given region, the following guidelines are important to achieve an appropriate monitoring plan. In electrified railways, besides the mortality due to train-collisions (almost all vertebrate groups), we also expect mortality due to overhead electric line collisions (birds and bats) and electrocution at cables and pylons (mainly birds). Once the sources of mortality are established, we should adopt the most practical prospecting methods (e.g., on foot, on foot with a search dog, or using a motor vehicle) according to the railway features (width, train speed and volume, vegetation on verges, topography, etc.) (Dorsey et al. 2015; SCV 1996).

Counting dead animals along railways is more challenging than on roads because railways often cross remote areas and their accessibility is often difficult (tunnels, steep topography, etc.) (Dorsey et al. 2015; Wells et al. 1999). Therefore, most studies report counts obtained by the transportation agency personnel, such as, train drivers and maintenance workers, who often lack wildlife experience, leading to inaccurate identifications and underestimation of the mortality (Wells et al. 1999).

An important issue is the sampling effort required in each situation to effectively detect patterns of causalities (Costa et al. 2015; Santos et al. 2011, 2015). As recommended for roads, railway surveys should be carried out early in the morning (to reduce scavenging, but see also sources of bias below), preferably by two experienced observers walking at specified railway stretches, one on each side of the rail (Peña and Llama 1997), and covering a 10 m sight strip whenever logistically possible. The use of a vehicle could be a better choice if parallel dirty roads exist (at least on one side), and in cases where the surveyed stretch is too long

(>10 km), but their detectability as proved to be lower (Garrah et al. 2015). Once a carcass is found, and depending on its state of decay, several variables should be recorded: species, age, sex, GPS location, time of day, weather conditions, position on the railway (verges, between lines, rock ballast, etc.), and surrounding habitat (Santos et al. 2011; Wells et al. 1999).

Surveys should be systematic, but their temporal frequency depends on species traits and biology. For small animals such as passerines, small mammals or bats, daily surveys are needed to reduce bias on estimates due to lower carcass persistence time (Santos et al. 2011, 2015). Santos et al. (2011) showed in a road study that sampling at intervals greater than one day results in the loss of 60% of the casualties for small animals (<500 g), and reaches 73 and 85% for lizards and bats, respectively. By contrast, they found that larger animals (>500 g, e.g., birds of prey and carnivores) persist for more than two days on average, which can be used as a baseline for their monitoring (Santos et al. 2011). In order to obtain a good balance among costs and survey frequency, the priority should be given to the animals' activity peaks, including those during dispersal, migration, and hunting seasons (Costa et al. 2015; Stevens and Dennis 2013). After the beginning of the railway operations, and in order to get a good evaluation of mortality estimates, monitoring should cover a minimum period of 3 years, so that animals adapt to the new environment (Iuell et al. 2003).

On electrified railways the risk of bird and bat mortality increases due to collisions with the overhead electric lines. Additionally, the electrocution of birds can occur at pylons and wherever the cable isolation has flaws (Kušta et al. 2011; Peña and Llama 1997). Because overhead electric lines are placed at the sides of the rails, animals killed by electrocution may be projected further away than those suffering collisions with trains, thus decreasing their detectability. Accordingly, to obtain accurate estimates on mortality when there are electric components (overhead electric lines and pylons), surveys should be done daily, and if possible, supported by search dogs, as these dogs have been shown to be more efficient than humans at detecting carcasses under power lines (73 vs. 20%) and at wind farms (Mathews et al. 2013).

Review of the Methods Used on Wildlife Mortality Surveys

Most studies addressing mortality on railways rely on incidental reports provided by the railway staff (van der Grift and Kuijsters 1998). Few have described their methods exhaustively (Seiler and Helldin 2006), and most do not report the monitoring frequency (e.g., Cserkész and Farkas 2015; van der Grift and Kuijsters 1998), or the methodology used (Jaren et al. 1991). Typically, studies used one of the following survey methods:

- *Surveys by rail companies' workers and non-expert citizens*

These surveys are characterized by fortuitous observations. Reports of collisions are often done by train drivers, and occasionally by maintenance crew personnel, who visit the railways to maintain the rails, electric components, or fences (Huijser et al. 2012), thus occasionally recording casualties but without a fixed periodicity (Gundersen et al. 1998; Kušta et al. 2014; SCV 1996). Recently, to minimize bias and identify hotspots properly, some studies have used reports of train drivers obtained systematically, with information on both the species and location of the accident (Dorsey 2011; Kušta et al. 2011, 2015). However, these studies are limited to large species, those that impede train operations and are easily observed (e.g., Singh et al. 2001). Some studies have depended on inquiries to railway staff about mortality data (Huijser et al. 2012), while others use anecdotal data from inhabitants, forest workers, naturalists, hunters, etc. (Singh et al. 2001).

- *Surveys by sidewalks along the railways*

Sidewalk surveys were mainly used in systematic studies by experienced technicians to detect mortality hotspots (Heske 2015; Peña and Llama 1997; Wells et al. 1999). Usually, these surveys aim to monitor and quantify all taxonomic groups (from small to large animals) independently of the mortality source (train collisions, entrapment in the track lines or electrocution) (Peña and Llama 1997; Wells et al. 1999).

- *Surveys by video vigilance*

In Alaska (USA), Rea et al. (2010) used movies from the *YouTube* video-sharing website, made mainly by employees of train companies and some anonymous people, to record the behavior of moose and other ungulates in the presence of trains. The use of video vigilance devices attached to trains, or cameras set on specific railway stretches, allowed continuous and clear recording of the animals' behavior toward trains (Kinley et al. 2003; Babińska-Werka et al. 2015). Video vigilance is particularly valuable in the winter in regions where snow cover and/or steep embankments impede the use of other methods (Rea et al. 2010). However, this method is expensive and thus has a limited observational range, and it cannot capture the entire vertebrate community in the railway vicinity.

Standardization of Monitoring

Studies should have standardized survey methods across different geographic areas, and combine several methods in order to reduce bias and contribute to better estimates of mortality on railways (Iuell et al. 2003; van der Grift et al. 2013). For instance, railway mortality should be presented as an index, reflecting the number of casualties/km/year, and these results should include information on sampling effort and periodicity (Peña and Llama 1997; SCV 1996). In order to achieve consistency in the use of similar methods over temporal and spatial scales so that comparisons can be made, cooperation is required, and specialized personnel should be trained (Roedenbeck et al. 2007; Rytwinski et al. 2015).

How to Describe and Evaluate Hotspots of Mortality

Most methods developed to identify mortality hotspots have been developed for roads (Gomes et al. 2009; Malo et al. 2004). One of the methods includes the assumption that the expected number of kills per segment (km) follows a Poisson distribution, and hotspots correspond to segments where the number of casualties is higher than the upper 95% confidence limit of the mean (Malo et al. 2004; Santos et al. 2015). The kernel method, identifies clusters of casualties by using a moving function to weight points of mortality within the influence of the function by their proximity to the location where density is being calculated (Ramp et al. 2005). The nearest neighbour hierarchical clustering identifies groups of points based on "nearest-neighbor-method" criteria (Gomes et al. 2009). Finally, the Getis-Ord Gi* statistic identifies hotspots by adding the number of casualties associated with a given segment of the road to the casualties of its neighboring segments, and compares that value with an overall expected distribution (Garrah et al. 2015). Integrating the information on rail sectors with high mortality rates with GIS mapping tools leads to the identification of the locations where mitigation measures (drift fences, rail passages, traffic regulation, etc.) should be applied (Costa et al. 2015). After obtaining the location of mortality hotspots, it is essential to verify the accuracy of the estimates, as overestimation can result in false hotspots (i.e., areas wrongly identified as having high mortality rates; Santos et al. 2015).

Sources of Bias in Wildlife Mortality Estimates

The reduction of mortality is one of the main aims of mitigation measures, hence high mortality rates and their locations need to be as accurate as possible (Guinard et al. 2012). As stated, even in systematic surveys, the number of carcasses found largely underestimates mortality because (1) only a subset of animals killed stay within the area that is searched by the observer; (2) many carcasses are removed by scavengers, or decomposed until the survey occurs; and (3) some dead animals remain undetected because of the observers failure (Korner-Nievergelt et al. 2015). Thus, to obtain unbiased estimates, the numbers obtained during surveys should be corrected by taking into account: (1) the proportion of animals killed in the area searched; (2) carcass persistence probability; and (3) searcher efficiency. In order to take into account these sources of bias, several mortality estimators were developed in wind farm studies (Korner-Nievergelt et al. 2015) and some have been applied to road casualties (Gerow et al. 2010; Teixeira et al. 2013), but rarely to railways, where these issues also apply.

The proportion of casualties (dead and/or injured that have not moved away) recorded at the search area can be obtained from the size and spatial distribution of the total area that can be searched, the spatial distribution of the carcasses, and the proportion of injured animals (still alive) that manage to move away. However, the spatial distribution of the carcasses depends on the type of obstacle they collided with, the size of the animal, and the wind speed at the time of collision (Korner-Nievergelt et al. 2015). Calculating the proportion of dead animals detected in the search area relative to the total area is rarely done because of the difficulties in estimating the variables involved (Teixeira et al. 2013).

Carcass persistence is the time a carcass stays on the rail or ground before it is removed by scavengers or has been severely decomposed, becoming undetectable to the observers. It depends mainly on the carcass size, the abundance and activity of scavengers, and temperature and humidity (Guinard et al. 2012; Santos et al. 2011).

Searcher efficiency, or detectability, is the probability that an observer actually finds a carcass in the search area. Detectability depends mostly on the survey method used, the experience and motivation of the observer, characteristics of the ground (vegetation density, ballast color), time of day, weather conditions, and type and size of the carcass. The method used for monitoring (e.g., by vehicle or on foot) greatly influences the detectability rates. For roads, Hels and Buchwald (2001) concluded that surveys by car detected between 7 and 67% of the amphibian carcasses that had been detected by surveys on foot and, similarly, Teixeira et al. (2013) reported a reduction in the detection rates from 1 to 27% for small and large animals, respectively. The observer's experience largely determines the probability of detectability: nevertheless, after several hours of work (>3 h), there is a "saturation effect" that affects the observer performance. Motivation is also important, because a motivated observer (e.g., working on a thesis) may stop more often to collect data and see details that otherwise would have gone undetected (PVMC 2003). On railways, the color of the rock ballast is one of the main factors influencing carcass detectability. Generally, smooth terrain and light colors of rock increase detectability as does clear verges on both sides of the rail. Additionally, if the vegetation beyond the verge is sparse and short (or even absent), detectability increases. Another crucial aspect is the accessibility of the rail line, as well as the topography on its sides. Weather conditions also influence detection by altering visibility—for instance, detectability decreases on rainy or foggy days (e.g., Mathews et al. 2013). Intrinsic characteristics of the species, such as, the size, shape and color of a species, can compromise their identification (e.g., small carcasses with cryptic colors are more difficult to detect). These sources of error can be taken into account by mathematical estimators to recalculate a detectability function that is then used to correct the observed mortality estimates (Bernardino et al. 2013; Korner-Nievergelt et al. 2015), but this is not yet common practice in railway mortality studies.

How to Mitigate Wildlife Mortality in Railways?

Over the last few decades, research and investment on mitigation measures to reduce wildlife mortality on roads have increased greatly (Jackson and Griffin 2000; Glista et al. 2009; Polak et al. 2014; Kociolek et al. 2015; Ward et al. 2015). In railways the efforts to increase wildlife safety have not been as considerable (van der Grift 1999; Dorsey et al. 2015), and these efforts have been concentrated mostly on mitigating the impacts on existing railways, ignoring the expansion of the high-speed railway network (Dorsey et al. 2015). An important difference between roads and railways is that in the latter, the speed and trajectory of a train cannot be changed to avoid collisions; therefore, mitigation measures must rely almost entirely on preventing the animals from entering or remaining on the train tracks.

Crossing Structures

The structures that facilitate wildlife crossing of railways can be part of the original engineering design of the infrastructure; these include culverts or bridges over roads or rivers, even when that was not their original purpose. At other times, these structures can become wildlife crossing locations with just minor adaptations, as happens with dry ledges. Alternatively, structures can be built with the specific goal of enabling movements across the linear infrastructure, usually designated as wildlife underpasses or overpasses.

Although crossing structures contribute to mitigating both mortality and barrier effects of linear infrastructures (Dorsey et al. 2015), their main role has been focused on barrier effects, ensuring connectivity through the landscapes crossed by railways and roads (Glista et al. 2009; Jackson and Griffin 2000; van der Ree et al. 2008). In this book, the application of crossing structures as a mitigation measure in railways is described in greater detail in Chap. 4. Here, we give a short description of the most frequently used crossing structures, and how they may help reduce wildlife mortality on railways.

Pipe culverts are small structures designed to let water flow under the railway, being regularly flooded during the rainy seasons. Pipe culverts form part of the design of most linear infrastructures and are sometimes used by small animals (Glista et al. 2009; Jackson and Griffin 2000). The efficiency of pipe culverts in reducing mortality would increase if it were associated with railway exclusion methods (see section below).

Box culverts are also designed for water drainage, and being larger than pipe culverts, often remain dry except in periods of heavy run-off. Box culverts are generally better as crossing structures than pipe culverts (Glista et al. 2009).

Culverts can be adapted before or after construction to facilitate their use by wildlife (Clevenger and Waltho 1999; Jackson and Griffin 2000; Rodríguez et al. 1996), namely by (1) including dry ledges, (2) modifying habitats at entrances,

(3) adding drift fences, (4) avoiding culverts with steep slopes and large hydraulic jumps or steps; and (5) placing culverts in dry habitats (i.e., dry drainage culverts) and not only along streams. Flooded or very steep culverts may contribute little as wildlife crossing structures. Modifying culverts for wildlife crossing can represent one of the most economical measures to mitigate mortality on railways (Clevenger and Waltho 1999).

Amphibian tunnels are sometimes used in roads to facilitate crossing in areas where amphibians concentrate their movements (Glista et al. 2009), and they can be easily adapted to railways. Similarly, some tunnels can be designed for reptile crossing—**reptile tunnels**. Turtles and other small animals may often get caught between the rails, and a simple measure to prevent their becoming trapped can be the excavation of the rock ballast between pairs of railway sleepers, thus allowing animals to cross below the tracks (Dorsey et al. 2015; Pelletier et al. 2006).

Wildlife underpasses facilitate animal movement under linear infrastructures, and are generally located where railways cross watercourses and roads. However, underpasses can be specifically designed to be used by animals (Glista et al. 2009; Jackson and Griffin 2000). Therefore, underpasses show considerable differences in size, and provide variable crossing facilities. Underpasses can be very large (i.e., viaducts, expanded bridges), where railways cross large watercourses and extensive valleys, in which it is assumed that passage for wildlife is limited, or they can be relatively small when, for example, they are only meant to allow the access of local vehicles between agricultural fields. Small underpasses may provide limited use for wildlife crossing if they are associated with roads with considerable traffic, or when they are completely flooded by the watercourse.

Wildlife overpasses, also called "ecoducts," are structures mainly designed for large animal crossings, such as ungulates and large carnivores, and have often been used as a mitigation measure in large highways (Jackson and Griffin 2000). Overpasses with strips of natural vegetation are referred to as "green bridges," while the term "landscape connectors" is used for very wide overpasses designed to maintain landscape connectivity (Forman et al. 2003). Overpasses are often less confining than underpasses, facilitating the movement of a greater number of species, and they maintain ambient conditions more easily throughout the year (Glista et al. 2009; Jackson and Griffin 2000). The main drawback is their high cost of construction.

Crossing structures designed specifically for wildlife use should take into account the following characteristics to improve their effectiveness (Glista et al. 2009). However, as different species will favor different structure design, the best options must be planned for each particular case:

1. *Location.* The place where crossing structures are implemented is probably the single, most important factor for their effectiveness; thus, structures should be implemented where animal movements are more likely (Ando 2003; Jackson and Griffin 2000; Rodríguez et al. 1996; Yanes et al. 1995).
2. *Dimensions.* There is no reference size for passage structures, depending greatly on the target species. However, passages should have a relatively large diameter

and openness (relative width-length of underpasses) and remain dry (at least partially) for most of the year (Baofa et al. 2006; Jackson and Griffin 2000; Yanes et al. 1995). Some species, such as rabbits, seem to prefer underpasses where it is possible to see the opposite end (Rosell et al. 1997).

3. *Substrate, moisture, temperature, and light.* These characteristics of crossing structures will influence their use by animals (Glista et al. 2009; Jackson and Griffin 2000; Rosell et al. 1997). The set of adequate conditions can be species-specific, but in general animals prefer crossing structures with a natural substrate, no temperature differences, and natural light (avoiding dark or artificially lit structures). Amphibians in particular require moist conditions.

4. *Approaching habitat.* The habitat near passing structures should be attractive to animals; in particular, the presence of covers, or their absence, may determine the use of crossing structures by some species (Baofa et al. 2006; Jackson and Griffin 2000; Rodríguez et al. 1996; Yanes et al. 1995).

5. *Guiding structures.* Fences and barrier walls guiding wildlife to crossing structures maximize their effectiveness (Jackson and Griffin 2000). For example, in Florida, the implementation of a barrier wall in conjunction with a culvert system reduced the road mortality of vertebrate species (excluding hylid treefrogs) by 94% (Dodd et al. 2004).

6. *Disturbance.* The use of crossing structures by humans and vehicles should be reduced as much as possible (Baofa et al. 2006; Clevenger and Waltho 2000). Noisy underpasses are likely to be avoided by more sensitive species (Jackson and Griffin 2000).

7. *Interspecific interactions.* The regular use of passages by predators may discourage their use by potential prey species (Clevenger and Waltho 1999, 2000). Habitats frequently used by predators of the target species should then be avoided—namely, the preferred hunting grounds and core areas of the home range of predators.

Although crossing structures can strongly contribute to reducing the mortality of non-flying animals, they are less effective in preventing collisions of birds and bats with trains, as these animals fly above the railway (Tremblay and St. Clair 2009). The exception to this are the large railway bridges and viaducts going over watercourses and valleys, in which the large size of the crossing structure facilitates the movement of birds and bats beneath the railway (Tremblay and St. Clair 2009).

Structures that Restrict Wildlife Access to Railways

One of the most effective measures for reducing wildlife mortality in railways is the implementation of structures that prevent crossing or, in the case of flying animals, forcing the crossing above the trains and overhead wires (Dorsey et al. 2015; Glista et al. 2009; van der Grift 1999). However, the application of measures that restrict movements will inevitably increase barrier effects, unless accompanied by adequate crossing structures (Dorsey et al. 2015; Jackson and Griffin 2000; van der Grift 1999). Railways have already been recognized as potential ecological barriers, namely for large mammals (Ito et al. 2013, and see also Chap. 14), a similar and cumulative effect with that of roads (see Chap. 4). Therefore, the generalized use of barriers along railways should be discouraged (Dorsey et al. 2015; Ito et al. 2013; Jackson and Griffin 2000). Barriers should be erected only along the stretches with high incidence of wildlife-train collisions, i.e., mortality hotspots.

Exclusion fences are currently considered to be the most effective means to restrict wildlife access to railways (Ito et al. 2013; van der Grift 1999) being, probably, the best cost-effective measure to mitigate wildlife mortality in the long run (Dorsey et al. 2015). However, fences may be less effective for species capable of climbing, jumping over, or digging under them (Jackson and Griffin 2000). When fencing is used, it is crucial to provide escapes to avoid animals becoming trapped between fences on both sides of the railway (Jackson and Griffin 2000). One-way gates or returning ramps should be regularly installed in places where animal crossing is more intense. Short retaining walls can be efficient barriers for reptiles, amphibians and small mammals (Jackson and Griffin 2000). While leaving a small passage close to the ground (like raising the lower wire of fences) may reduce the barrier effect for smaller and more agile species, it still prevents the access of livestock to the railway (Ito et al. 2013).

Olfactory repellents consist of chemicals applied on structures or on the vegetation along the railway (e.g., Andreassen et al. 2005; Kušta et al. 2015). The aim of olfactory repellents is to drive animals away from the structure or to keep them more alert, and thus more prone to evade approaching trains. There are several types of olfactory repellents used for scent-marking, namely synthetic products of predator substances (Lutz 1994; Andreassen et al. 2005). In Norway, Andreassen et al. (2005) sprayed a repellant on trees and bamboo canes at 5 m intervals along the railway, and found that it had an effect on reducing moose casualties, although with variable efficiency. In the Czech Republic, the application of odor repellents reduced overall animal mortality on roads and railways, but it was not effective for reptiles or amphibians (Kušta et al. 2015). In addition, repellants seem to be less effective at low temperatures (Castiov 1999; Kušta et al. 2015).

Sound signalling and **sound-barriers** can represent a promising alternative mitigation measure, especially if they do not greatly limit movements across railways. Sound signalling consists of warning animals of approaching trains (Babińska-Werka et al. 2015), while sound barriers are mostly intended to keep animals off the railway. Stationary systems can be mounted in critical areas, using a motion-activated sensor triggering an audio (and sometimes also visual) stimulus to frighten the animals. Alternatively, stationary systems can be activated only when there is a train coming. Electronic systems can also be installed on the front of the trains. Displayed sounds can be audible or ultrasonic to humans depending on the target species. For instance, trains equipped with ultrasonic wildlife warning contributed to the reduction of moose-train collisions along a railway in Canada (Muzzi and Bisset 1990). Babińska-Werka et al. (2015) studied the effectiveness of an acoustic wildlife warning device in reducing casualties along railways. The sound sequence in their devices combined: (1) the alarm call of the jay *Garrulus glandarius*; (2) the sound of a frightened brown hare *Lepus europaeus*; (3) a dog *Canis familiaris* growling and barking; (4) a wolf *Canis lupus* howling; (5) squeal of a wild boar *Sus scrofa*; and (6) warning sounds of roe deer *Capreolus capreolus* (Babińska-Werka et al. 2015). These natural warning calls should promote the alerted behavior of wild animals thus increasing their escape time from oncoming trains. The sound warning devices may be placed along critical railway stretches (e.g., in Babińska-Werka et al. 2015 they were separated by 70 m), being activated by a signal sent by the automatic railway system in advance of a train (30 s to 3 min). The use of sound signalling increased the proportion of wildlife escaping from the tracks, with individuals reacting faster and showing no evidence of habituation to the warning signals (Babińska-Werka et al. 2015). Multiple horn blasts have been used by train drivers to effectively flush animals (mostly ungulates) from rails (Helldin et al. 2011).

Physical barriers such as trees, diversion poles, flight diverters, or noise barriers, may contribute to the reduction of mortality, especially among birds (Bard et al. 2002; Jacobson 2005; Kociolek et al. 2015; Zuberogoitia et al. 2015) and bats (Ward et al. 2015). Pole barriers may represent a relatively inexpensive mitigation measure, as they can be effective in diverting the flight of medium- or large-sized birds above the poles (Zuberogoitia et al. 2015). The pole barriers used by Zuberogoitia et al. (2015) consisted of: (1) gray PVC poles 2 m high and 8 cm wide regularly separated by 1 or 2 m, and some had shredded pieces of coloured paper (white or orange) attached to the top of the pole, or (ii) tree trunks (20–26 cm diameter and 350 cm height) separated by 1 m.

Trees and other hedgerow vegetation can work as a barrier, preventing access to railways, especially to large animals, and forcing birds and bats to fly above the trains. However, this habitat can also become a refuge or foraging site for some species (e.g., small birds and mammals), thus potentially working as an ecological trap by attracting wildlife to the proximity of the railway.

Lighting and reflectors have been mainly tested on roads as wildlife deterrents. This mitigation measure is used mostly for nocturnal species, and may reflect the lights of vehicles or, alternatively, flash signals before an oncoming train. Lights and reflectors may be combined with sound signalling to provide a faster response. Nevertheless, lights should not be too intense in order to avoid blinding the animals, whose reaction will probably be to stay still, thus remaining on the tracks.

Habitat and Wildlife Population Management

Lower numbers of wildlife species near railways can be achieved by controlling populations (e.g., selective hunting, trapping), or by habitat modification. Changes in habitat structure along railway verges may also increase animals' capability to detect and evade the train.

Habitat management includes vegetation mowing or pruning at railway verges, or the removal of specific fruiting plants (Andreassen et al. 2005; Eriksson 2014; Jacobson 2005). Vegetation removal (forest clearing) was successfully employed along a railway in Norway, achieving a reduction in about half the number of moose casualties (Andreassen et al. 2005; Jaren et al. 1991). It combined a decreased attractiveness of verges for wildlife, hence leading to animals spending less time foraging close to railways, and higher visibility, providing a shorter reaction time by animals. However, Helldin et al. (2011) and Eriksson (2014) found the opposite effect, suggesting that tree-clearing may increase moose and roe deer train collisions in Sweden, as areas cleared after mowing offer attractive foraging opportunities for some species. In another study using Before-After-Control-Impact (BACI), tree clearance had no effect on the frequency of wildlife collisions (Eriksson 2014). Helldin et al. (2011) showed that vegetation management requires regular maintenance; otherwise, mortality may return to previous levels after plant growth. These authors mention that tree-clearance can reduce collisions if applied frequently, but it can increase collisions if applied at intervals greater than 3–4 years. Finally, it is relevant to mention that habitat management (vegetation removal) can limit the movements of some species, hence increasing barrier effects; for example, small vertebrates are generally reluctant to cross large open spaces due to greater exposure to predators (Hunt et al. 1987; Yanes et al. 1995).

Population control of a particular species may sometimes be used to reduce its numbers near railways. This method should only be applied on very common species, or those that can compromise human safety due to collisions. This method has been used to prevent collision with vehicles on roads (Glista et al. 2009), but its use in railways may not be as necessary since most animals will not affect trains' movements.

Supplemental feeding stations placed far from railways may influence animal movement, keeping animals away from linear infrastructures and thus contributing to reducing mortality (Andreassen et al. 2005; Wood and Wolfe 1988).

Reducing Train Speed

Greater speeds are undoubtedly associated with a greater risk of wildlife mortality on roads as well as on, railways (e.g., Cserkész and Farkas 2015; Frikovic et al. 1987). Thus, train speed moderation, at least at critical points (mortality hotspots) and during periods of higher crossing movements (e.g., migration), should contribute to reducing animal mortality, as slower trains have fewer collisions (Becker and Grauvogel 1991; Belant 1995).

Collision and Electrocution in Overhead Wires

Flying animals may collide or be electrocuted in the overhead wires that provide electrical energy to trains—the catenary (SCV 1996). Thus, specific mitigation measures should be taken whenever high levels of mortality are expected or identified. These measures are similar to those used to prevent collision and electrocution in powerlines (Bevanger 1994; Barrientos et al. 2011). Additionally, the tubular poles often used to support the overhead wires can cause death if uncapped, especially among cavity nesting birds (Malo et al. 2016). The compulsory use of capped poles can easily avoid this problem, since it prevents birds from falling inside the poles.

Evaluating the Effectiveness of Mortality Mitigation Measures

Along with adequate planning for railway corridors, which accounts for the expected impacts on wildlife, another fundamental and final step to reduce the negative effects of railways is a thorough evaluation of the effectiveness of the mitigation measures (Grilo et al. 2010; Glista et al. 2009; van der Grift et al. 2013). Poor mitigation measures, which do little to reduce wildlife mortality, represent a waste of valuable resources. Moreover, some structures may reduce animal movement, disrupt landscape connectivity, or have negative implications for erosion and grazing.

Guidelines for evaluating the effectiveness of road mitigation measures have been proposed for roads (van der Grift et al. 2013; Rytwinski et al. 2015; van der Grift and van der Ree 2015), but since no similar research has been done specifically for railways, one can consider the guidelines for roads as a starting point.

Step 1 Identify the species requiring mitigation measures and determine the specific goals for them.

Step 2 Select the species (from the targeted species list) that will be used for evaluation.

Step 3 Select the best measures of interest, i.e., those that are most closely related to the aim of the mitigation (e.g., casualty rate, number of crossings, population trend).

Step 4 Select the adequate study design, including a spatial and temporal sampling strategy. Whenever possible, the best study design is a replicated BACI at mitigation sites (where measures were taken) and control sites (with similar characteristics but where no mitigation measures were applied).

Step 5 Determine the best sampling scheme, including the number of sites, the frequency of visits, the monitoring methods, and the number of replicates.

Step 6   Select the appropriate sites, both mitigated and control. This includes choosing the best spatial scale for the evaluation study.

Step 7   Select the best covariates to measure (e.g., railway characteristics, type of fences, type of culvert, presence of noise barriers, human disturbance, presence of vegetation on verges, train speed, surrounding landscape characteristics).

Step 8   Select the most suitable survey methods, namely preferring methods that monitor several species simultaneously, and choosing ways to reduce bias.

Step 9   Determine the costs and feasibility of the evaluation study, and act in agreement by implementing the necessary adaptations.

It is crucial to do a research-based evaluation of the success of mitigation measures, which should be of a broad scope that includes wildlife mortality and movements, landscape constraints, and safety. Finally, as indirect ecological consequences of many mitigation measures are poorly understood and often neglected, they should be taken into account in the overall evaluation process of their effectiveness.

Acknowledgments S.M. Santos and R. Lourenço were supported by post-doctoral Grants of the Fundação para a Ciência e Tecnologia (FCT; SFRH/BPD/70124/2010 and SFRH/BPD/78241/2011, respectively).

References

Ando, C. (2003). The relationship between deer-train collisions and daily activity of the sika deer, *Cervus nippon. Mammal Study, 28,* 135–143.

Andreassen, H. P., Gundersen, H., & Storaasthe, T. (2005). The effect of scent-marking, forest clearing and supplemental feeding on moose-train collisions. *Journal of Wildlife Management, 69,* 1125–1132.

Babińska-Werka, J., Krauze-Gryz, D., Wasilewski, M., & Jasińska, K. (2015). Effectiveness of an acoustic wildlife warning device using natural calls to reduce the risk of train collisions with animals. *Transportation Research D, 38,* 6–14.

Baofa, Y., Huyin, H., Yili, Z., Le, Z., & Wanhong, W. (2006). Influence of the Qinghai-Tibetan railway and highway on the activities of wild animals. *Acta Ecologica Sinica, 26,* 3917–3923.

Bard, A. M., Smith, H. T., Egensteiner, E. D., Mulholland, R., Harber, T. V., Heath, G. W., et al. (2002). A simple structural method to reduce road-kills of royal terns at bridge sites. *Wildlife Society Bulletin, 30,* 603–605.

Barrientos, R., Alonso, J. C., Ponce, C., & Palacín, C. (2011). Meta-analysis of the effectiveness of marked wire in reducing avian collisions with power lines. *Conservation Biology, 25,* 893–903.

Becker, E. F., & Grauvogel, C. A. (1991). Relationship of reduced train speed on moose-train collisions in Alaska. *Alces, 27,* 161–168.

Belant, J. L. (1995). Moose collisions with vehicles and trains in Northeastern Minnesota. *Alces, 31,* 1–8.

Bernardino, J., Bispo, R., Costa, H., & Mascarenhas, M. (2013). Estimating bird and bat fatality at wind farms: A practical overview of estimators, their assumptions and limitations. *New Zeland Journal of Zoology, 40,* 63–74.

Bevanger, K. (1994). Bird interactions with utility structures: Collision and electrocution, causes and mitigating measures. *Ibis, 136,* 412–425.

Castiov, F. (1999). Testing potential repellents for mitigation of vehicle-induced mortality of wild ungulates in Ontario. Ph.D. thesis Dissertation, Laurentian University, Sudbury.

Clevenger, A. P., & Waltho, N. (1999). Dry drainage culvert use and design considerations for small- and medium-sized mammal movement across a major transportation corridor. In G. L. Evink, P. Garrett, & D. Zeigler (Eds.), *Proceedings of the third international conference on wildlife ecology and transportation* (pp. 263–277). Tallahassee, FL: Florida Department of Transportation.

Clevenger, A. P., & Waltho, N. (2000). Factors influencing the effectiveness of wildlife underpasses in Banff National Park, Alberta, Canada. *Conservation Biology, 14,* 47–56.

Costa, A. S., Ascensão, F., & Bager, A. (2015). Mixed sampling protocols improve the cost-effectiveness of roadkill surveys. *Biodiversity and Conservation, 24,* 2953–2965.

Cserkész, T., & Farkas, J. (2015). Annual trends in the number of wildlife-vehicle collisions on the main linear transport corridors (highway and railway) of Hungary. *North-Western Journal of Zoology, 11,* 41–50.

Dodd, C. K., Jr., Barichivich, W. J., & Smith, L. L. (2004). Effectiveness of a barrier wall and culverts in reducing wildlife mortality on a heavily traveled highway in Florida. *Biological Conservation, 118,* 619–631.

Dorsey, B. (2011). Factors affecting bear and ungulate mortalities along the Canadian Pacific Railroad through Banff and Yoho National Parks. Master's thesis, Montana State University.

Dorsey, B., Olsson, M., & Rew, L. J. (2015). Ecological effects of railways on wildlife. In R. van der Ree, D. J. Smith, & C. Grilo (Eds.), *Handbook of road ecology* (pp. 219–227). West Sussex: Wiley.

Eriksson, C. (2014). Does tree removal along railroads in Sweden influence the risk of train accidents with moose and roe deer? Dissertation, Second cycle, A2E. Grimsö och Uppsala: SLU, Department of Ecology, Grimsö Wildlife Research Station.

Forman, R. T. T., Sperling, D., Bissonette, J. A., Clevenger, A. P., Cutshall, C. D., Dale, V. H., et al. (2003). *Road ecology: Science and solutions.* Washington, DC: Island Press.

Frikovic, A., Ruff, R. L., Cicnjak, L., & Huber, D. (1987). Brown bear mortality during 1946–85 in Gorski Kotar, Yugoslavia. In *International conference on bear research and management* (Vol. 7, pp. 87–92).

Garrah, E., Danby, R. K., Eberhardt, E., Cunnington, G. M., & Mitchell, S. (2015). Hot spots and hot times: Wildlife road mortality in a regional conservation corridor. *Environmental Management, 56,* 874–889.

Gerow, K., Kline, N. C., Swann, D. E., & Pokorny, M. (2010). Estimating annual vertebrate mortality on roads at Saguaro National Park, Arizona. *Human and Wildlife Interactions, 4,* 283–292.

Glista, D. J., DeVault, T. L., & DeWoody, J. A. (2009). A review of mitigation measures for reducing wildlife mortality on roadways. *Landscape and Urban Planning, 91,* 1–7.

Gomes, L., Grilo, C., Silva, C., & Mira, A. (2009). Identification methods and deterministic factors of owl roadkill hotspot locations in Mediterranean landscapes. *Ecological Research, 24,* 355–370.

Grilo, C., Bisonette, J. A., & Cramer, P. C. (2010). Mitigation measures to reduce impacts on biodiversity. In S. R. Jones (Ed.), *Highways: Construction, management, and maintenance* (pp. 73–114). Hauppauge: Nova Science Publishers.

Guinard, E., Julliard, R., & Bardraud, C. (2012). Motorways and bird traffic casualties: Carcass surveys and scavenging bias. *Biological Conservation, 147,* 40–51.

Gundersen, H., Andreassen, H. P., & Storaas, T. (1998). Spatial and temporal correlates to Norwegian moose-train collisions. *Alces, 34,* 385–394.

Gunson, K. E., Mountrakis, G., & Quackenbush, L. J. (2011). Spatial wildlife-vehicle collision models: A review of current work and its application to transportation mitigation projects. *Journal of Environmental Management, 92,* 1074–1082.

Helldin, J. O., Seiler, A., Olsson, M., & Norin, H. (2011). *Klövviltolyckor på järnväg: kunskapsläge, problemanalys och åtgärdsförslag. Ungulate-train collisions in Sweden— review, GIS-analyses and train-drivers experiences* (in Swedish). Sweden: Trafikverket.

Hels, T., & Buchwald, E. (2001). The effect of road kills on amphibian populations. *Biological Conservation, 99,* 331–340.

Heske, E. J. (2015). Blood on the tracks: Track mortality and scavenging rate in urban nature preserves. *Urban Naturalist, 4,* 1–13.

Huijser, M. P., Begley, J. S., & van der Grift, E. A. (2012). *Mortality and live observations of wildlife on and along the Yellowhead Highway and the Railroad through Jasper National Park and Mount Robson Provincial Park, Canada.* Salmo Consulting Inc., On behalf of Kinder Morgan Canada, Calgary, Canada.

Hunt, A., Dickens, H. J., & Whelan, R. J. (1987). Movement of mammals through tunnels under railway lines. *Australian Zoology, 24,* 89–93.

Ito, T. Y., Lhagvasuren, B., Tsunekawa, A., Shinoda, M., Takatsuki, S., Buuveibaatar, B., et al. (2013). Fragmentation of the habitat of wild ungulates by anthropogenic barriers in Mongolia. *PLoS ONE, 8,* e56995.

Iuell, B., Bekker, G. J., Cuperus, R., Dufek, J., Fry, G., Hicks, C., et al. (2003). *Wildlife and traffic: A European handbook for identifying conflicts and designing solutions.* Utrecht: KNNV Natural History Publishers.

Jackson, S. D., & Griffin, C. R. (2000). A strategy for mitigating highway impacts on wildlife. In T. A. Messmer & B. West (Eds.), *Wildlife and highways: Seeking solutions to an ecological and socio-economic dilemma* (pp. 143–159). Bethesda: The Wildlife Society.

Jacobson, S. L. (2005). *Mitigation measures for highway-caused impacts to birds.* PSW-GTR-191. USDA Forest Service General Technical Report.

Jaren, V., Andersen, R., Ulleberg, M., Pedersen, P., & Wiseth, B. (1991). Moose-train collisions: The effects of vegetation removal with a cost-benefit analysis. *Alces, 27,* 93–99.

Kinley, T. A., Page, H. N., & Newhouse, N. J. (2003). *Use of infrared camera video footage from a wildlife protection system to assess collision-risk behavior by deer in Kootenay National Park, British Columbia.* Invermere, BC: Sylvan Consulting Ltd.

Kociolek, A., Grilo, C., & Jacobson, S. (2015). Flight doesn't solve everything: Mitigation of road impacts on birds. In R. van der Ree, D. J. Smith, & C. Grilo (Eds.), *Handbook of road ecology* (pp. 281–289). West Sussex: Wiley.

Korner-Nievergelt, F., Behr, O., Brinkmann, R., Etterson, M. A., Huso, M. M. P., Dalthorp, D., et al. (2015). Mortality estimation from carcass searches using the R-package carcass—A tutorial. *Wildlife Biology, 21,* 30–43.

Kornilev, Y., Price, S., & Dorcas, M. (2006). Between a rock and a hard place: Responses of eastern box turtles (*Terrapene carolina*) when trapped between railroad tracks. *Herpetological Reviews, 37,* 145–148.

Kušta, T., Holá, M., Keken, Z., Ježek, M., Zíka, T., & Hart, V. (2014). Deer on the railway line: Spatiotemporal trends in mortality patterns of roe deer. *Turkish Journal of Zoology, 38,* 479–485.

Kušta, T., Ježek, M., & Keken, Z. (2011). Mortality of large mammals on railway tracks. *Scientia Agriculturae Bohemica, 42,* 12–18.

Kušta, T., Keken, Z., Ježek, M., & Kůta, Z. (2015). Effectiveness and costs of odor repellents in wildlife-vehicle collisions: A case study in Central Bohemia, Czech Republic. *Transportation Research Part D, 38,* 1–5.

Lutz, W. (1994). Trial results of the use of a "Duftzaun®" (scent fence) to prevent game losses due to traffic accidents. *Zeitschrift für Jagdwissenschaft, 40,* 91–108.

Malo, J. E., García de la Morena, E. L., Hervás, I., Mata, C., & Herranz, J. (2016). Uncapped tubular poles along high-speed railway lines act as pitfall traps for cavity nesting birds. *European Journal of Wildlife Research, 62,* 483–489.

Malo, J. E., Suarez, F., & Diez, A. (2004). Can we mitigate animal-vehicle accidents using predictive models? *Journal of Applied Ecology, 41,* 701–710.

Mathews, F., Swindells, M., Goodhead, R., August, T. A., Hardman, P., Linton, D. M., et al. (2013). Effectiveness of search dogs compared with human observers in locating bat carcasses at wind-turbine sites: A blinded randomized trial. *Wildlife Society Bulletin, 37,* 34–40.

Muzzi, P. D., & Bisset, A. R. (1990). Effectiveness of ultrasonic wildlife warning devices to reduce moose fatalities along railway corridors. *Alces, 26,* 37–43.

Pelletier, S. K., Carlson, L., Nein, D., & Roy, R. D. (2006). Railroad crossing structures for spotted turtles: Massachusetts Bay Transportation Authority–Greenbush rail line wildlife crossing demonstration project. In C. L. Irwin, P. Garrett, & K. P. McDermott (Eds.), *Proceedings of the 2005 international conference on ecology and transportation* (pp. 414–425). Raleigh, NC: Center for Transportation and the Environment, North Carolina State University.

Peña, O. L., & Llama, O. P. (1997). *Mortalidad de aves en un tramo de línea de ferrocarril* (32 p). Grupo Local SEO-Sierra de Guadarrama, Spain.

Polak, T., Rhodes, J. R., Jones, D., & Possingham, H. P. (2014). Optimal planning for mitigating the impacts of roads on wildlife. *Journal of Applied Ecology, 51,* 726–734.

PVMC. (2003). *Mortalidad de vertebrados en carreteras. Proyecto provisional de seguimiento de la mortalidad de vertebrados en carreteras.* Coordinadora de Organizaciones de Defensa Ambiental. Documentos Técnicos de Conservación S.C.V., 4.

Ramp, D., Caldwell, J., Edwards, K., Warton, D., & Croft, D. (2005). Modelling of wildlife fatality hotspots along the Snowy Mountain Highway in New South Wales, Australia. *Biological Conservation, 126,* 474–490.

Rea, R. V., Child, K. N., & Aitken, D. A. (2010). Youtube™ insights into moose-train interactions. *Alces, 46,* 183–187.

Rodríguez, A., Crema, G., & Delibes, M. (1996). Use of non-wildlife passages across a high speed railway by terrestrial vertebrates. *Journal of Applied Ecology, 33,* 1527–1540.

Roedenbeck, I. A., Fahrig, L., Findlay, C. S., Houlahan, J. E., Jaeger, J. A. G., Klar, N., et al. (2007). The Rauischholzhausen agenda for road ecology. *Ecology and Society, 12,* 11.

Rose, P., & Baillie, S. (1989). *The effects of collisions with overhead lines on British birds: An analysis of ringing recoveries.* Research Report, 42. British Trust for Ornithology.

Rosell, C., Parpal, J., Campeny, R., Jove, S., Pasquina, A., & Velasco, J. M. (1997). Mitigation of barrier effect of linear infrastructures on wildlife. In K. Canters (Ed.), *Habitat fragmentation & infrastructure* (pp. 367–372). Proceedings of the International Conference on Habitat Fragmentation, Infrastructure and the Role of Ecological Engineering. Delft, The Netherlands: Ministry of Transport, Public Works and Water Management.

Rytwinski, T., van der Ree, R., Cunnington, G. M., Fahrig, L., Findlay, C. S., Houlahan, J., et al. (2015). Experimental study designs to improve the evaluation of road mitigation measures for wildlife. *Journal of Environmental Management, 154,* 48–64.

Santos, S. M., Carvalho, F., & Mira, A. (2011). How long do the dead survive on the Road? Carcass persistence probability and implications for road-kill monitoring surveys. *PLoS ONE, 6,* e25383.

Santos, S. M., Marques, J. T., Lourenço, A., Medinas, D., Barbosa, A. M., Beja, P., et al. (2015). Sampling effects on the identification of roadkill hotspots: implications for survey design. *Journal of Environmental Management, 162,* 87–95.

S. C. V. (1996). *Mortalidad de vertebrados en líneas de ferrocarril.* Documentos Técnicos de Conservacion SCV 1, Sociedad Conservación Vertebrados, Madrid.

Seiler, A., & Helldin, J.-O. (2006). Mortality in wildlife due to transportation. In J. Davenport & J. L. Davenport (Eds.), *The ecology of transportation: Managing mobility for the environment* (pp. 165–189). Dordrecht: Springer.

Singh, A. K., Kumar, A., Mookerjee, A., & Menon, V. (2001). *A scientific approach to understanding and mitigating elephant mortality due to train accidents in Rajaji National Park.* An Occasional Report no 3 by Wildlife Trust of India and the International Fund for Animal Welfare of a Rapid Action Project on understanding and mitigating the problem of elephant mortality due to train hits.

Stevens, B. S., & Dennis, B. (2013). Wildlife mortality from infrastructure collisions: Statistical modelling of count data from carcass surveys. *Ecology, 94,* 2087–2096.

Teixeira, F. Z., Coelho, A. V. P., Esperandio, I. B., & Kindel, A. (2013). Vertebrate road mortality estimates: Effects of sampling methods and carcass removal. *Biological Conservation, 157,* 317–323.

Tremblay, M. A., & St. Clair, C. C. (2009). Factors affecting the permeability of transportation and riparian corridors to the movements of songbirds in an urban landscape. *Journal of Applied Ecology, 46,* 1314–1322.

van der Grift, E. A. (1999). Mammals and railroads: Impacts and management implications. *Lutra, 42,* 77–98.

van der Grift, E. A., & Kuijsters, H. M. J. (1998). Mitigation measures to reduce habitat fragmentation by railway lines in the Netherlands. In G. L. Evink, P. Garrett, D. Zeigler, & J. Berry (Eds.), *Proceedings of the international conference on wildlife ecology and transportation* (pp. 166–170). Tallahassee, FL: Florida Department of Transportation.

van der Grift, E. A., & van der Ree, R. (2015). Guidelines for evaluation use of wildlife crossing structures. In R. van der Ree, D. J. Smith, & C. Grilo (Eds.), *Handbook of road ecology* (pp. 119–128). West Sussex: Wiley.

van der Grift, E. A., van der Ree, R., Fahrig, L., Findlay, S., Houlahan, J., Jaeger, J. A. G., et al. (2013). Evaluating the effectiveness of road mitigation measures. *Biodiversity and Conservation, 22,* 425–448.

van der Ree, R., Clarkson, D. T., Holland, K., Gulle, N., & Budden, M. (2008). *Review of mitigation measures used to deal with the issue of habitat fragmentation by major linear infrastructure.* Report for Department of Environment, Water, Heritage and the Arts (DEWHA), Contract No. 025/2006, Published by DEWHA.

Ward, A. I., Dendy, J., & Cowan, D. P. (2015). Mitigating impacts of roads on wildlife: An agenda for the conservation of priority European protected species in Great Britain. *European Journal of Wildlife Research, 61,* 199–211.

Wells, P., Woods, J. G., Bridgewater, G., & Morrison, H. (1999). *Wildlife mortalities on railways: Monitoring methods and mitigation strategies,* Revelstoke, BC. Unpublished report.

Wood, P., & Wolfe, M. L. (1988). Intercept feeding as a means of reducing deer-vehicle collisions. *Wildlife Society Bulletin, 16,* 376–380.

Yanes, M., Velasco, J. M., & Suaréz, F. (1995). Permeability of roads and railways to vertebrates: The importance of culverts. *Biological Conservation, 7,* 217–222.

Zuberogoitia, I., del Real, J., Torres, J. J., Rodríguez, L., Alonso, M., de Alba, V., et al. (2015). Testing pole barriers as feasible mitigation measure to avoid bird vehicle collisions (BVC). *Ecological Engineering, 83,* 144–151.

Chapter 4
Railways as Barriers for Wildlife: Current Knowledge

Rafael Barrientos and Luís Borda-de-Água

Abstract In this chapter we provide practical suggestions, together with examples, to identify, monitor and mitigate railway barrier effects on wildlife, as this is considered one of the railways' greatest impacts. Railways can be both physical and behavioral barriers to wildlife movement, as well as disturbance to populations living close to them. Also, mortality is recognized as an important contribution to the barrier effect. However, the consequences of habitat loss, and fragmentation due to railways alone remain largely unexplored. Barrier effects have mainly been mitigated with wildlife passes, with the effectiveness of this tool being one of the most-studied topics in Railway Ecology. Methods formerly employed to monitor pass usage, such as track beds or video-surveillance, are now being replaced by molecular ones. Among the latter methods, genetic fingerprinting allows individual-based approaches, opening the door to population-scale studies. In fact, genetic sampling allows for the assessment of functional connectivity, which is closely linked to successful reproduction and population viability, variables not necessarily coupled with crossing rates. There is strong evidence that railway verges offer new habitats for generalist species and for opportunistic individuals, a point that deserves to be experimentally explored in order to find wildlife-friendly policies. Preventing animals from crossing (e.g., by fencing), should be reserved for collision hotspots, as it increases barrier effects. Instead, it has been shown that warning signals or pole barriers effectively reduce collisions without increasing barrier effects. In this respect, we argue that computer simulations are a promising field to investigate potential impact scenarios. Finally, we present a protocol to guide planners and managers when assessing barrier effects, with emphasis on monitoring and mitigation strategies.

R. Barrientos (✉) · L. Borda-de-Água
CIBIO/InBIO, Centro de Investigação em Biodiversidade e Recursos Genéticos, Universidade do Porto, Campus Agrário de Vairão, Rua Padre Armando Quintas, 4485-661 Vairão, Portugal
e-mail: barrientos@cibio.up.pt

R. Barrientos · L. Borda-de-Água
CEABN/InBIO, Centro de Ecologia Aplicada "Professor Baeta Neves", Instituto Superior de Agronomia, Universidade de Lisboa, Tapada da Ajuda, 1349-017 Lisboa, Portugal

© The Author(s) 2017
L. Borda-de-Água et al. (eds.), *Railway Ecology*,
DOI 10.1007/978-3-319-57496-7_4

Keywords Barrier effect · Connectivity · Habitat fragmentation · Permeability · Wildlife pass

Introduction

We know today that linear infrastructures are one of the largest threats to biodiversity worldwide, including habitat loss and fragmentation (Forman et al. 2003; Benítez-López et al. 2010; van der Ree et al. 2015). As mentioned in the introductory chapter, most of what we know regarding this comes from studies on roads however, as some of these impacts are shared by roads and railways, many of the management tools developed for the former can be applied to the latter. Nevertheless, new approaches are needed in some cases, because some impacts are railway-specific (Dorsey et al. 2015).

This chapter provides some conceptual insight into, and summarizes in a single document, what is known about barrier effects caused by railways on wildlife populations, which is considered by some to be one of the greatest ecological impacts of these infrastructures (Iuell et al. 2003; Rodríguez et al. 2008). This review is organized as follows: We begin by identifying the factors contributing to railway barrier effects and we then describe the approaches used to study these effects, and the effectiveness of the measures implemented to mitigate them. Finally, we suggest management guidelines for railway companies, and present a protocol to minimize these barrier effects.

We searched the ISI Web of Science and Zoological Record databases for railway ecology studies. For all searches, our search terms were combinations of the following words: barrier effect, collision, fragmentation, habitat loss, impact, permeability, radio-tracking, railroad, railway, train, underpass, wildlife pass. We looked for publications in English, Spanish, Portuguese, and French. We also searched Google Scholar™ for additional studies not published in peer-reviewed journals (gray literature) but citing peer-reviewed papers, such as Ph.D. dissertations, reports, or conference papers.

Barrier Effects

The various sources of railway barrier effects are closely related, and sometimes it is not easy to clearly separate them. For instance, noisy traffic can cause both habitat loss (if wildlife refuses to use the area adjacent to the railway right-of-way), and it can create a behavioral barrier if this noise implies a perceived risk. We identified four broad types of impacts causing barrier effects.

Physical and Behavioral Barriers

Barriers can be physical, when a species cannot trespass the railway, or behavioral, when the species may be physically able to cross the barrier but does not do so because of unfavorable ambient conditions or perceived risk.

Physical barrier constraints mainly affect species of small size with reduced mobility, such as herptiles. For instance, Kornilev et al. (2006) reported that eastern box turtles (*Terrapene carolina*) in the USA enter a railway at crossings with roads (where surfaces are at the same level), but then become trapped between the rails as they are unable to climb over them, and finally die due to thermal stress. A study of Hermann's tortoises (*Testudo hermanni*) in Romania found that the impossibility of overtaking obstacles (e.g., ditches with angles of over 60°) led to an increase in the distribution of railway-kills at the end of the ditches (Iosif 2012).

Bumblebees (*Bombus impatiens* and *B. affinis*) in the USA (Bhattacharya et al. 2003) provide an example of a species to which the railway is a behavioral barrier. These species are reluctant to cross railways (and roads) because of their high fidelity to their foraging site. In experiments, individuals of these species could come back to their patches of origin after being translocated, or could leave their patch when their food was removed, but these movements were rarely in-control (non-translocated) individuals (Bhattacharya et al. 2003). Bhattacharya et al. (2003) observed that foraging bees turned back when they reached the edge of a patch (i.e., bisected by a railway); thus, rather than physically impassable barriers to bumblebees, railways acted as barriers because they are likely to be strong landmarks. On the other hand, most of the translocated gatekeeper butterflies (*Pyronia tithonus*) crossed the French High Speed Rail (hereinafter "HSR") to return to their capture plot, as this species shows a strong homing behavior (Vandevelde et al. 2012; see also Chap. 16). Mongolian gazelles (*Procapra gutturosa*) are able to cross fenced railways in Mongolia (Ito et al. 2008), but they usually do not do so (Ito et al. 2005, 2013, and see Chap. 14). Therefore, these fenced railways represented a barrier effect affecting population dynamics as it caused disruptions in long-distance gazelle migrations (Ito et al. 2005, 2008, 2013). Stopping migration routes prevents gazelles from reaching their traditional food-rich winter quarters, potentially increasing their winter mortality due to starvation (Ito et al. 2005, 2008, 2013). This conservation problem is compounded by the current climate change, as drought events reducing vegetation productivity could require gazelles to migrate longer distances, instead of the current average of 600 km (Olson et al. 2009). To date, railways do not seem to have been a barrier for gazelle gene flow, although this assertion should be explored with more suitable markers (Okada et al. 2012; see below).

Although it has been less studied, the conservation threat of the critically endangered saiga antelope (*Saiga tatarica*) in Kazakhstan—as a consequence of migration disruption due to fenced railways—seems to be similar to that of Mongolian gazelles (Olson 2013; see also Olson and van der Ree 2015). In a study carried out in Sweden, Kammonen (2015) found that a motorway and a railway running in parallel acted as barriers for two bat species (whiskered bat *Myotis mystacinus* and Brandt's bat *M. brandtii*) in a forest-dominated area. Although this author did not differentiate between these two infrastructures, she found that bats did not directly cross them; rather they used either the green bridge or the underpass both to cross it and to forage (Kammonen 2015).

Disturbance

Traffic noise, vibrations, chemical pollution, and human presence can impact animal populations living close to railways, contributing to the barrier effects (van der Grift 1999; Dorsey et al. 2015; see a more complete discussion in the Chap. 6). These impacts can be divided into those related to siting and construction, and those related to the operation of the railway line (De Santo and Smith 1993; see also Chap. 12). The former has received little attention, likely because they are considered short-term, but they can nevertheless affect local animal populations. For instance, although construction was halted during the migration period of the human-wary Tibetan antelopes (*Pantholops hodgsonii*) in China, the animals did not use wildlife passes because of the presence of machinery and debris (Xia et al. 2007). However, this species finally adapted to the presence of the railway and used wildlife passes once the was train operative (Yang and Xia 2008).

Among the long-term disturbances related to barrier effects, one of the most significant seems to be the noise produced by trains, because railways' right-of-way has little vegetation to absorb the sound, as it is frequently mowed. Studies to date have been carried out with simple census-based approaches, and have provided contradictory findings. For instance, whereas in the Netherlands, Waterman et al. (2002) found a reduction in the density of meadow birds close to railways, Wiącek et al. (2015), counting forest birds at 30, 280 and 530 m from the tracks, found higher species richness close to the railway in a Polish forest crossed by a railway. In fact, guilds like insectivorous or those ecotone-specialists were more common close to railways, and species with low-frequency calls were not averse to it (Wiącek et al. 2015). In a study in a Brazilian Atlantic forest, Cerboncini et al. (2016) did not find an effect of traffic noise (up to 120 dB, higher close to the track) on small mammals, probably because of the infrequent train passage.

Several field studies did not explore the potential causes of the abundance patterns observed and the distance to railway tracks. For instance, Li et al. (2010) found higher ground-dwelling bird richness and abundance in China close to the railway when sampled at 0, 300, 600, 900 and 1,200 m. In a study on the rodent community in the same railway, Qian et al. (2009) found no effect on species composition or densities at 50, 200 and 500 m from the railway in non-grazing areas, but grazing-disturbed areas showed different species ratios (Qian et al. 2009). The density of urban foxes (*Vulpes vulpes*) in the UK was similar in squares with and without railways (Trewhella and Harris 1990); however, in areas with poor habitat quality, fox dens were commonly placed in railway banks, most likely because they acted as refuges (Trewhella and Harris 1990).

Mortality

Chapters 2 and 3 deal exclusively with the impact of railways on wildlife mortality; however, because we consider mortality to be a barrier effect, and in order to ensure that current chapter is self-contained, we now summarize the main results.

Train-related mortality can directly prevent connectivity among sub-populations, or reduce their reproductive success, if individuals seeking mates die or if their off-spring are railway-killed. Wildlife-train collisions (hereinafter "WTC") are the most commonly documented source of animal mortality, although electrocution or col-lisions with wires can also occur (Rodríguez et al. 2008; Dorsey et al. 2015). A recently documented source of mortality is that of cavity nester birds in uncapped catenary poles from HSR in Spain, as these tubular poles act as pitfall traps for birds, because poles have smooth internal walls, not allowing birds that enter them to fall down to fly out (Malo et al. 2016). As the drainage hole at the base of the pole is too narrow to allow trapped birds to escape, the problem could be readily solved by simply capping the tops of all types of tubular poles (Malo et al. 2016). A non-exclusive mitigation alternative would be to widen drainage holes.

Most of our understanding about railway impacts on wildlife comes from studies focused on large mammal railway-kills, as they are the most conspicuous, and can cause accidents, delays to the trains' operation, significant damage to trains due to their large size, and, overall, significant financial losses (Dorsey et al. 2015, and also Chap. 17). Among train-related factors influencing the rates of WTC, traffic flow is the most important one, with the highest mortality occurring, in fact, in lines with moderate traffic flow because higher traffic volume deters animals from attempting to cross (Dorsey et al. 2015). For instance, in the USA, the killing by trains of black bears (*Ursus americanus*) attempting to cross areas with low vegetation cover through train bridges in lines with moderate traffic has been documented (van Why and Chamberlain 2003). The existence of lapses of time without traffic encourages the bears to cross the valley by the bridge, and these are railway-killed when a train arrives and they cannot jump off the railway bed (van Why and Chamberlain 2003).

Animal behavior also determines their mortality rates. The isotope analyses using brown bear (*U. arctos*) hair in Canada showed that some individuals fed on carcasses from train-killed animals, or on plants growing in railway verges (Hopkins et al. 2014; see also Wells et al. 1999; see also Chap. 9). Similarly, granivore species in Canada consumed grain spilled by wagons (Wells et al. 1999), and in Norway moose (*Alces alces*) took advantage of the availability of branches resulting from logging activities in the right-of-way (Gundersen et al. 1998). On the contrary, despite Cerboncini et al. (2016) trapped more small mammals close to the tracks in Brazil, their body condition was similar to that of animals trapped far from the railway. Thus, it seems that this guild of Atlantic forest specialists was not taking advantage of grain falling from trains (Cerboncini et al. 2016). Railways can also facilitate the movements of some species, as wildlife uses these homogeneous surfaces for travelling faster. This was the case of moose and wolves (*Canis lupus*) in Canada, and moose and roe deer (*Capreolus capreolus*) in Sweden, when snow was very deep (Child 1983; Paquet and Callaghan 1996; Eriksson 2014), of brown bears in Slovenia, when the terrain was steep (Kaczensky et al. 2003).

One frequently neglected issue is to what extent WTCs could affect population dynamics. In some species, this impact seems small, but in others it could be important. This impact can also vary at the population level, as was the case of different moose populations in the USA, with WTCs ranging from less than 1%

(Belant 1995) to 70% (Schwartz and Bartley 1991) of the estimated population size, or 5% of adult radio-collared animals (n = 204; Modafferi and Becker 1997). For brown bears, it varied from 5% of radio-tracked animals in the USA (n = 43; Waller and Servheen 2005) to 18% in Slovenia (n = 17; Kaczensky et al. 2003). The mortality due to collisions with trains affected 5% (n = 21) of the radio-tagged eagle owls (*Bubo bubo*) in Switzerland, with WTCs being less important than electrocution and cable or car collisions (Schaub et al. 2010). However, all these figures provide little information without the corresponding population viability analyses (PVAs). In this sense, the latter paper estimated an annual 31% of population growth if the entire anthropogenic mortality of eagle owls was eliminated, but without information on the effects of removing WTCs alone (Schaub et al. 2010).

Habitat Loss and Fragmentation

Habitat loss takes place when railway construction leads to the reduction of the available habitat, since the transformed railway bed is unsuitable for several species. Habitat fragmentation is often, but not necessarily, mediated by habitat loss. During fragmentation, large, continuous fragments are divided resulting in smaller, often isolated, patches that may not be able to maintain viable populations in the long run (Fahrig 2003). Whereas general information on habitat fragmentation is abundant (see Fahrig 2003 for a review), studies exclusively focused on railway-related fragmentation are non-existent, because researchers did not differentiate between railway- and road-related fragmentation, assessing these two different infrastructures as a whole (e.g., Jaeger et al. 2007; Girvetz et al. 2008; Bruschi et al. 2015).

When a population's territory is bisected by a railway, part of its habitat is lost, and the remainder may be degraded, usually via cascade effects. The latter is what is happening to the woodland caribous (*Rangifer tarandus*) in Canada, as the construction of railways and other linear infrastructures facilitated the access of wolves to remote areas where there are still populations of this ungulate, being their viability threatened (James and Stuart-Smith 2000; Whittington et al. 2011).

Habitat changes also take place in railway corridors, as their verges commonly differ from the surrounding landscape, but are homogeneous along the railway network. These changes can be exploited by generalist species or by opportunistic individuals, using them as shelters or corridors. They can be used by invasive species as well (for more details on the latter, see Chap. 5). Some authors have suggested that the creation of new habitats by mowing the right-of-way, and the presence of associated structures like powerlines and their pylons, provide new opportunities for several species to breed or hunt (see Morelli et al. 2014 regarding birds). For instance, Vandevelde et al. (2014) (see Chap. 16) found that in France, in intensive agricultural landscapes, where linear semi-natural elements like hedgerows tend to disappear, bat species that forage in more open habitats benefited from railway verges. For Polish butterflies, railways not only acted as corridors, but also sheltered greater species richness than forest clearings or degraded meadows

(Kalarus and Bąkowski 2015). The wide range of environmental conditions occurring in tracks, led to the presence of a large number of nectar plant species, allowing the existence of many butterfly species, from those selecting dry and warm microhabitats to forest specialists (Kalarus and Bąkowski 2015). Similar results have been found for other pollinators due to the high diversity of bee forage flora, although their diversity is higher in lines with intermediate traffic volume, and differs between microhabitats within the embankments (Wrzesień et al. 2016). On the contrary, Cerboncini et al. (2016) found no effects of railway edge on micro-climate in a Brazilian Atlantic forest, probably because railway track was narrow and the forest was well developed. Finally, especially in more impacted landscapes, there is an opportunity to integrate old tracks once they are abandoned into the regional conservation schema. They can act, for instance, as habitat corridors among protected areas, as many of their new uses—like rail-cycle or hiking—are wildlife-friendly activities. However, much more research is needed on the conservation potential of abandoned tracks as well as on cost-effective maintenance methods, like the weeding by domestic animals used in France (Orthlieb 2016). Also in France, Kerbiriou et al. (2015) found a strong increase in the population of hibernating pipistrelle bats (*Pipistrellus pipistrellus*) in a railway tunnel as a result of the end of the exploitation of the railway line, remarking on the idea that abandoned railway structures can have second-life fulfilling conservation purposes.

Methods to Estimate Barrier Effects

Some of the methods we describe below were first used in road ecology studies (Smith and van der Ree 2015), but all of them are useful for railway ecology studies as well.

Direct Methods

(1) *Wildlife-train collisions data.* This is the simplest method to estimate barrier effects due to wildlife mortality. A reduction of WTCs after applying mitigation measures has been commonly argued to be a measure of the effectiveness of management policies (e.g., Andreassen et al. 2005; Kušta et al. 2015), although WTC data without PVAs may be a poor surrogate of the impact of the railway.

(2) *Track beds.* These consist of a layer of fine sand, marble dust, or clay powder of 3–30 mm thick, spread across the entire pass (usually underpass or culvert), and smoothed with a brush. It should be fine enough to detect the tracks of small vertebrates such as mice or amphibians, or even macroinvertebrates. The pass must be reviewed every 1–2 days and, if necessary, the material must be removed and extra material added (Yanes et al. 1995; Rodríguez et al. 1996; Baofa et al. 2006). This method, combined with strips of soot-coated paper, as well as

trapping and indirect evidence, such as scat identification, was used in Australia for the first time by Hunt et al. (1987). This technique has been repeatedly improved, as it was the first widely used method to confirm culvert/wildlife pass usage. Due to its low costs, it is more cost-effective for short-term surveys than modern alternatives, like video-surveillance (Ford et al. 2009; Mateus et al. 2011). However, the use of track pads is limited to optimal conditions, as the material employed can become useless by rain or livestock passage (Rodríguez et al. 1996, 1997; Mateus et al. 2011), and there can be track misidentification and under-estimation of crossings due to track overlapping (Ford et al. 2009). Tracks in snowy landscapes can help to estimate the qualitative crossing of certain sections (Olsson et al. 2010), although limitations of this method regarding misidentifi-cation and track overlapping are similar to those from sandy beds.

(3) *Video-surveillance*. Modern technology allows the monitoring of pass usage thanks to cameras activated by infrared motion detectors at the pass entrance (Ford et al. 2009; Mateus et al. 2011). This method is constrained by the sen-sitiveness of the camera monitor sensor (Ford et al. 2009), which is especially limiting in large passes and with small animals (Mateus et al. 2011), because video cameras cover small areas and only animals close to the sensor are recorded (Ford et al. 2009; García-Sánchez et al. 2010; Mateus et al. 2011). Thus, a logical next step to evaluate the use of wildlife passes has been the development of wireless sensor networks (García-Sánchez et al. 2010). These are low-cost devices that, by using a camera at the entrance of the pass and an infrared motion sensor network deployed in the surrounding area, enable the recording of reactions of animals approaching the wildlife pass and their eventual crossings (García-Sánchez et al. 2010).

(4) *Capture-mark-recapture* (hereinafter "*CMR*"). These are fine-scale methods as they are individual-based, but they are intrusive, time- and budget-consuming, and require safety measures both for the animals and the researchers working around the transport infrastructures; all this leads to small- to modest-sized samples (Simmons et al. 2010). Tagging must be adapted to the size and the ecology of the target species. CMR with numbered plastic tags allowed Bhattacharya et al. (2003) to monitor bumblebee movement across a railway, with a recapture rate of 31% (n = 367). In France, Vandevelde et al. (2012) (see Chap. 16) used a thin-point permanent pen to mark gatekeeper butterflies with a recapture rate of 30% (n = 149). Alternatively, passive integrated transponders have been proven to be useful to monitor wildlife passages. Once the animal crosses, an antenna connected to a decoder unit installed in the pass records the individual, time, and date of crossing (recapture rate = 50%, n = 6; Soanes et al. 2013).

Radio-tracking-based projects share some of the advantages (fine-scale, individual-based) and disadvantages (intrusive, time- and budget-consuming) of the previous methods, although the increasing effectiveness (e.g., satellite-based telemetry) and decreasing price currently make radio-tracking suitable for a wide variety of organisms (Simmons et al. 2010). Furthermore, the devices are becoming miniaturized to the point of being a feasible alternative for some invertebrates (e.g.,

Hedin and Ranius 2002). Radio-tracking can also provide detailed information, not only on individual crossings, but also on the full territory use by animals relative to the railway location, therefore allowing the impact of the linear infrastructure on the movements of individuals to be estimated (Clevenger and Sawaya 2010).

Non-invasive genetic sampling (hereinafter "NGS") methods have been used mainly to measure the genetic sub-structure of a population bisected by a railway (see below). However, NGS also enables individual identification based on microsatellite analysis (Balkenhol and Waits 2009; Clevenger and Sawaya 2010; Simmons et al. 2010), the so-called "fingerprinting" or "DNA profiling," which is a kind of capture-mark-recapture method. For instance, Clevenger and Sawaya (2010) showed that the passive hair-collection methods based on barbed wire and/or adhesive strings, followed by microsatellite analyses, was an effective technique for monitoring wildlife pass use at an individual level for cougars (*Puma concolor*), and black and brown bears in Canada. Furthermore, as costs for genetic analyses are become lower, the sample sizes have increased in recent studies (Simmons et al. 2010). The main limitation of this method is that it is only suitable for large animals whose remains (hairs, feathers, scats) can be found in the field in sufficient quantity to extract DNA from them. In addition, fingerprinting requires a relatively high number of microsatellites. If species-specific microsatellites have not been developed in the target species, they have to be specifically developed, increasing both time and costs. However, microsatellites already developed for related species can be tested, as sometimes they amplify the DNA from the target species as well.

Indirect Methods

(1) *Census at both sides of a railway*. This is the most simplistic approach, either to assess population densities (e.g., Waterman et al. 2002; Li et al. 2010; Wiącek et al. 2015), or to calculate diversity indices of community structures (e.g., Qian et al. 2009). However, it is also the most limited approach to identifying causal factors or population dynamic-related processes. Failure to control for potentially confounding variables makes that detected patterns cannot be clearly associated with railway impact.

(2) *Genetic-based assessment of functional connectivity*. Even more important than confirming crossing is to assessing the functional connectivity (or the barrier effect)—that is, to detect whether individuals reproduce on both sides of the railway. It is worth noting that moderate to low crossing rates may not necessarily imply functional connectivity (Riley et al. 2006). This type of information is logistically difficult to obtain using other than genetic methods (Clevenger and Sawaya 2010; Simmons et al. 2010).

Because the impact of railways is relatively recent, highly variable markers such as microsatellites are the most suitable method for estimating demographic and population genetic effects. Balkenhol and Waits (2009) found that 76% of the 33 reviewed studies employed these markers in road ecology studies (see below for railway studies). Indeed, authors who used mitochondrial analyses failed to find genetic structuring related to railway-related barrier effects, as happened with the Mongolian gazelle (Okada et al. 2012).

The easiest design consists of sampling individuals at both sides of a railway to infer whether this acts as a barrier driving population differentiation. This approach was used by Gerlach and Musolf (2000) to study the genetic sub-structuring between bank vole (*Clethrionomys glareolus*) populations bisected by a railway in Germany and Switzerland. The authors found that a 40-year-old railway did not contribute to genetic substructuring in bank voles. In their study in the USA with the marbled salamander (*Ambystoma opacum*) in the USA, Bartoszek and Greenwald (2009) found that the populations from two ponds just separated by a railway, although potentially connected by a culvert, were genetically differentiated, although some gene flow was still occurring. On the contrary, the Qinghai-Tibetan railway seems not to be a barrier structuring the toad-headed lizard (*Phrynocephalus vlangalii*) populations, as samples from both sides of the railway were genetically similar, whereas those sampled at 20 km away were different, as expected due to the distance (Hu et al. 2012).

Genetics should be complemented with landscape analyses to control, among other things, for the relationship between genetic and geographic distances. Reh and Seitz (1990), using a sample design that included several sites and enzyme analyses, found that railways contributed to the isolation, and thus the inbreeding, of common frogs (*Rana temporaria*) in Germany. Yang et al. (2011) used landscape analysis and genetics based on microsatellites to identify the factors influencing the differentiation among Przewalski's gazelle (*P. przewalskii*) populations in China. Prunier et al. (2014) used a microsatellite individual-based sampling scheme combined with computer simulations to determine whether HSR in west-central France was old enough to cause genetic discontinuities in the alpine newt (*Ichthyosaura alpestris*). The latter authors did not detect any barrier effects, which could be due both to the relatively recent existence of the railway (29 years), or to the highly nomadic behavior of this amphibian. Also, the small size of newts could allow them to move under the rails, minimizing their risk of being railway-killed. The few smooth newts (*Lissotriton vulgaris*), a species of similar size, that were found railway-killed in Poland, were found at pedestrian crossings, where newts are forced to move over instead of under the rails (Kaczmarski and Kaczmarek 2016). On the other hand, the simultaneous use of genetic approaches and landscape analyses has also shown that linear infrastructures can, in some cases, increase connectivity. For instance, the use of microsatellites enabled Fenderson et al. (2014) to identify railways, powerlines, or even road sides as dispersal facilitators of New England cottontails (*Sylvilagus transitionalis*), but their approach was limited as they did not evaluate the relative importance of each of these infrastructures to the observed increase in connectivity. Such an increase in the landscape connectivity can be detrimental in some cases, such as that of invasive species, a theme that is discussed at length in Chap. 5.

(3) *Computer Simulations*. These tools have the potential to play a major role in understanding the impact of railways in wild animal populations and, accordingly, in planning new railway networks or developing mitigation measures. However, to the best of our knowledge, there are only a few simulation studies that specifically target the impact of the fragmentation caused by railways.

Simulations can be used before and after railway construction. Before the construction phase, an impact assessment is desirable to compare alternatives. In particular, it is important to avoid cutting through areas of great natural value, but if this is unavoidable, then it is necessary to identify the sectors most affected in order to implement mitigation measures. In such situations, a region-wide focus that includes future projections is the most suitable approach. For these purposes, graph theory is being increasingly used in conservation biology, as graph models provide simplified representations of ecological networks with flexible data requirements (Urban et al. 2009). For instance, Clauzel et al. (2013) (see also Chap. 13) combined graph-based analysis and species distribution models to assess the impact of a railway line on the future distribution of the European tree frogs (*Hyla arborea*) in France. This study was able to identify—among potential routes—the railway line with the lowest impact on the species distribution.

Mateo-Sánchez et al. (2014) conducted computer simulations to assess the degree of connectivity of two populations of the endangered brown bear in north-western Spain. They used a multi-scale habitat model to predict the presence of bears as a function of habitat suitability, combined with a factorial least-cost path density analysis. With this model, the authors identified possible corridors that could connect the two populations and the locations that should be prioritized in order to ameliorate the permeability of the local railways (and roads). In a study to identify the most suitable corridor in a future railway line in Sweden, Karlson et al. (2016) integrated models with ecological and geological information by using spatial multi-criteria analysis techniques to generate a set of potential railway corridors, followed by the application of the lowest cost path analysis in order to find the corridor with the best environmental performance within the set.

Much of what has been learned from simulations applied to the impact of roads (e.g., Roger et al. 2011; Borda-de-Água et al. 2011) can also be used in railway ecology. Among the techniques used, we highlight the individuals-based models, (hereafter "IBM") (e.g., Lacy 2000; Jaeger and Fahrig 2004; Kramer-Schadt et al. 2004; Grimm et al. 2006). While more traditional simulation approaches use variables to study the collective behavior of certain entities (for instance, an entire population could be characterized by a single variable describing the total number of individuals), IBMs explicitly simulate all individuals as separate entities, each with its own set of characteristics, and interacting among them and with the environment. The main advantage of

IBMs is that we are no longer constrained by the difficulties of obtaining analytical or numerical solutions using differential calculus, or are dependent on the mathematical tractability of complicated systems of differential (or integral) equations (Railsback and Grimm 2011). With IBMs, one can model a wide variety of behaviors, thus increasing the realism of the models—although often at the expense of time consuming simulations.

The most effective tools for reducing barrier effects, such as overpasses or viaducts, increase construction costs considerably (Smith et al. 2015). Therefore, either before or after railway construction, simulations at the landscape level can be used to identify the location and type of the mitigation measures to be implemented. For instance, Gundersen and Andreassen (1998) included both train- and environment-related variables to model the occurrence of WTCs during seasonal moose migration to valley bottoms, when WTC risk is highest (Gundersen et al. 1998). and they tested the model predictability with a subset of data not used for model development, a method that can be used to infer future WTCs (Gundersen and Andreassen 1998).

Effectiveness of Mitigation Measures

Avoiding Crossing

Several studies have tested measures, like fencing, to avoid animal crossings of railways. These can reduce WTCs but, on the other hand, they can increase barrier effects (e.g., Ito et al. 2005, 2008, 2013; see also Chap. 14). Thus, they should only be implemented in areas of high concentration of WTCs, and combined with wildlife passes to maintain railway permeability (van der Grift 1999). Building exclusion fences seems the most effective (van der Grift 1999; Ito et al. 2013), and is even the most cost-effective measure, in the long run (Dorsey et al. 2015). The application of odor repellents reduced animal mortality in a study in the Czech Republic, but with contrasting results among taxa (Kušta et al. 2015). This technique was less effective at low temperatures, when repellents froze (Kušta et al. 2015; see also Castiov 1999). Indeed, Andreassen et al. (2005) found odour repellents to have highly variable efficiency in reducing moose WTCs in Norway. Instead, these authors found a more consistent decrease in WTCs after the placement of feeding stations to keep animals away from railways, and forest clearing in the vicinity of the railway (Andreassen et al. 2005).

Interesting alternatives are those devices that aim to reduce WTCs without having barrier effects (see Chap. 17). For example, trains equipped with ultrasonic warning devices killed fewer moose in Canada than those without (Muzzi and Bisset 1990). More recently, Babińska-Werka et al. (2015) reported the development of a device in Poland that uses alarm calls from several wild animals in advance (30 s to 3 min) of an oncoming train that allows animals near the railway to react and escape in a natural way. The proportion of wildlife escaping from the tracks was higher, and individuals reacted faster, when the device was switched on and, importantly, animals did not show evidence of habituation to the warning signals (Babińska-Werka et al. 2015).

Planting trees (Tremblay and St. Clair 2009) or erecting pole barriers (Zuberogoitia et al. 2015) in the railway corridor can reduce the WTCs of flying animals. However, trees could attract animals as well, for perching or feeding, so pole barriers are preferred since they have fewer side effects. An experiment with medium- to large-sized birds in Spain showed that most birds shifted or raised the flight when approaching the pole barrier (Zuberogoitia et al. 2015).

Habitat Management

Vegetation mowing at railway verges was successfully applied to reduce moose WTCs in Norway, and it could have had three complementary benefits: first, reducing the attractiveness of the verges for animals, therefore reducing foraging close to railways (Jaren et al. 1991; Andreassen et al. 2005); second, reducing the time spent by animals close to the railway, as they could perceive the clearing as dangerous (Jaren et al. 1991); and, third, allow them 'see and be seen' rule, as provided for both the train driver and the animal, with a greater amount of time to react to each other and avoid a collision (Jaren et al. 1991). This technique reduced the number of moose WTCs by half (Jaren et al. 1991; Andreassen et al. 2005). On the contrary, Eriksson (2014) originally hypothesized that tree-clearing could be the factor behind the increase in moose and roe deer train collisions in Sweden, as early successional stages created after mowing provided attractive foraging opportunities for ungulates, but she found that it had no effect on the increase of WTCs in her "Before-After-Control-Impact", the so-called BACI design (Eriksson 2014). Finally, it is worth mentioning that vegetation removal may increase barrier effects for small vertebrates, as these do not cross open spaces due to their associated high predation risks (Hunt et al. 1987; Yanes et al. 1995).

Crossing Structures

Animals use both non-wildlife passes (i.e., those placed and designed for purposes other than to allow wildlife crossing, like drainage culverts), or wildlife passes specifically designed on the basis of the target species traits (small tunnels for amphibians or small mammals; underpasses, overpasses, ecoducts or green bridges for large mammals) (Smith et al. 2015). Large passes mimicking natural habitat are more expensive, but they are also the most effective technique for reducing barrier effects, commonly suitable for most species, including the most demanding ones, like large carnivores and ungulates (Iuell et al. 2003; Clevenger and Waltho 2005; Smith et al. 2015). In some cases, structures that were not originally designed as wildlife passes have been adapted to better allow animal crossings. For instance, culverts were modified by the addition of a bench to facilitate wildlife crossing when the culvert is wet (Iuell et al. 2003). In some cases, the adaptation is as easy as removing the gravel below pairs of sleepers to create a gap to allow small vertebrates, like spotted turtles (*Clemmys guttata*) in the USA, to cross under the sleepers, to where they were funnelled by a fence (Pelletier et al. 2006). Culverts have been found to be used by animals to bypass railways in Australia (Hunt et al. 1987) and in Spain (Yanes et al. 1995; Rodríguez et al. 1996, 1997). However, culvert dimensions or the surrounding habitats influence their use by vertebrates (Hunt et al. 1987; Yanes et al. 1995; Rodríguez et al. 1996, 1997). For instance, small culverts were used by small mammals, but they were unsuitable for ungulates (Rodríguez et al. 1996) and the addition of natural vegetation and refuges such as stones increased crossing rates for small animals (Hunt et al. 1987; Yanes et al. 1995). Not unexpectedly, longer passes have lower crossing rates for several taxa (Hunt et al. 1987; Yanes et al. 1995).

Notice, however, that all the examples of pass monitoring reported, at the most, the intensity of crossing, not the functional connectivity—two variables that are not necessarily coupled (Riley et al. 2006). Thus the implementation of these mitigation measures should be complemented with genetic analyses at the population level to assess whether they contribute to the effective reduction of barrier effects (Riley et al. 2006).

Management Guidelines

In Fig. 4.1 we present in a schematic way the steps to be followed in an "Ideal Protocol to Mitigate Railway Barrier Effects". These include:

Forecasting Impacts

To know the wildlife status in the whole region, it is necessary to assess the impact of potential routes, to understand the target-species ecology, as WTCs are usually correlated with animal abundance (e.g., D'Amico et al. 2015), and the latter can temporally and geographically change along the biological cycle (e.g., moose in Canada or Norway; Child 1983; Gundersen et al. 1998; or sika deer *Cervus nippon* in Japan; Ando 2003). Impacts can be predicted by NGS (Balkenhol and Waits 2009) or by censuses (Species Distribution Models, Clauzel et al. 2013). In addition, individual assignment tests and graph theory could be combined in landscape analyses to identify connectivity zones that should be preserved, and computer simulations could be run to evaluate population dynamics under several barrier effect levels (Balkenhol and Waits 2009; Clauzel et al. 2013). Thus, by combining these approaches, planners will be able to select the alternatives with the lowest barrier effects on wildlife.

IDEAL PROTOCOL FOR MITIGATING
RAILWAY BARRIER EFFECTS

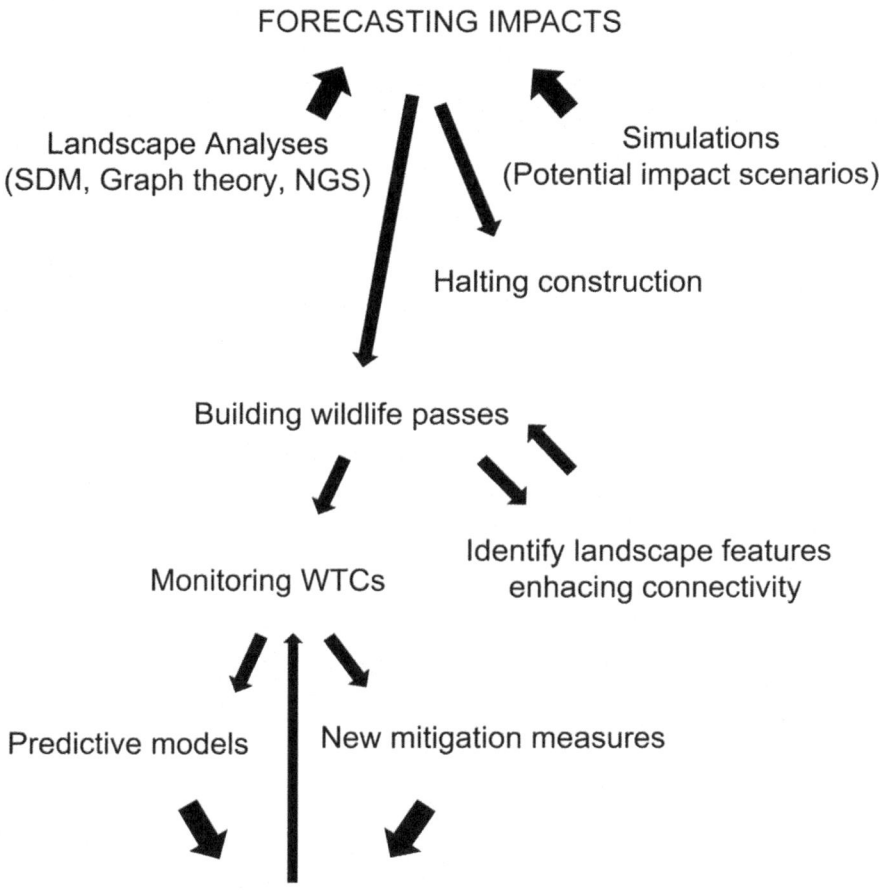

FORECASTING IMPACTS

Landscape Analyses
(SDM, Graph theory, NGS)

Simulations
(Potential impact scenarios)

Halting construction

Building wildlife passes

Monitoring WTCs

Identify landscape features
enhacing connectivity

Predictive models

New mitigation measures

Effectiveness evaluation

Fig. 4.1 Ideal protocol for mitigating railway barrier effects

Halting Construction

Disturbances should be minimized during strategic stages of the life cycle of the
target species, as when construction of the railway was halted to allow the migration
of the Tibetan antelopes in China (Xia et al. 2007) or the reproduction of wetland
birds in Portugal (see Chap. 12).

Wildlife Passes Combined with Funnelling Fencing

This approach should be considered only in those railway sections previously identified with landscape analysis studies to be particularly suitable to maintain connectivity, since their construction is expensive. The option of elevating the railway on pile-supported structures is usually more feasible than it is for similar highway segments (De Santo and Smith 1993). Demanding species, such as large animals, prefer to cross under bridges rather than through underpasses (Rodríguez et al. 1996; Baofa et al. 2006; Yang and Xia 2008) but, on the other hand, bridges can represent a risk for flying species as they tend to cross above them (Tremblay and St. Clair 2009). Thus, bridges should be flanked by pole barriers to ensure safe passage well above moving traffic (Zuberogoitia et al. 2015). Finally, functional connectivity should be evaluated with molecular methods and BACI designs implemented to know the effectiveness of mitigation measures (Balkenhol and Waits 2009; Corlatti et al. 2009; Clevenger and Sawaya 2010; Simmons et al. 2010; Soanes et al. 2013).

Identify Landscape Features Enhancing Connectivity

The identification of those features that enhance landscape connectivity and their adoption when designing and operating railways should be promoted. For instance, as mowed railway rights-of-way have been found to restore population connectivity the for New England cottontail, an early successional habitat specialist of conservation concern, a management recommendation to reduce the isolation of their populations is not to allow vegetation development beyond shrub stratum (Fenderson et al. 2014).

Monitoring of WTCs to Identify Hotspots

A periodic schema should be implemented to monitor the railway line to identify sections with high collision rates, i.e., hotspots, and simulation models should be built to forecast WTCs (Gundersen and Andreassen 1998). In identified hotspots, anti-collision measures, such as warning signals (Babińska-Werka et al. 2015; see also Chap. 17) or pole barriers, (Zuberogoitia et al. 2015) should be placed. Their effectiveness should be monitored with BACI designs.

Conclusions

Railways can cause barrier effects in several ways. Some, like mortality, have been widely studied for some charismatic species, but their effects for others are poorly known. Studies on barrier effects due to behavioral responses to disturbances or the effects of habitat loss and fragmentation due to railway implementation are scarce. The monitoring of the effectiveness of mitigation measures should incorporate more recent approaches, such as genetic tools. It would allow, for instance, to broaden the current scope based on the qualitative use of wildlife passes to a more interesting functional connectivity-based framework. The use of computer simulations has advantages not fully applied at present, but useful both before railway construction and during its operation.

Acknowledgements We thank Graça Garcia and Clara Grilo for carefully reading an early version of this chapter. This research was supported by FEDER funds through the Operational Programme for Competitiveness Factors—COMPETE, by National Funds through FCT—the Foundation for Science and Technology under UID/BIA/50027/2013, POCI-01-0145-FEDER-006821, and by the Infraestruturas de Portugal Biodiversity Chair. Rafael Barrientos acknowledges financial support by the Infraestruturas de Portugal Biodiversity Chair—CIBIO—Research Center in Biodiversity and Genetic Resources.

References

Ando, C. (2003). The relationship between deer-train collisions and daily activity of the sika deer, *Cervus nippon*. *Mammal Study, 28*, 135–143.

Andreassen, H. P., Gundersen, H., & Storaasthe, T. (2005). The effect of scent-marking, forest clearing and supplemental feeding on moose-train collisions. *Journal of Wildlife Management, 69*, 1125–1132.

Babińska-Werka, J., Krauze-Gryz, D., Wasilewski, M., & Jasińska, K. (2015). Effectiveness of an acoustic wildlife warning device using natural calls to reduce the risk of train collisions with animals. *Transportation Research Part D, 38*, 6–14.

Balkenhol, N., & Waits, L. P. (2009). Molecular road ecology: Exploring the potential of genetics for investigating transportation impacts on wildlife. *Molecular Ecology, 18*, 4151–4164.

Baofa, Y., Huyin, H., Yili, Z., Le, Z., & Wanhong, W. (2006). Influence of the Qinghai-Tibetan railway and highway on the activities of wild animals. *Acta Ecologica Sinica, 26*, 3917–3923.

Bartoszek, J., & Greenwald, K. R. (2009). A population divided: Railroad tracks as barriers to gene flow in an isolated population of marbled salamanders (*Ambystoma opacum*). *Herpetological Conservation and Biology, 4*, 191–197.

Belant, J. L. (1995). Moose collisions with vehicles and trains in Northeastern Minnesota. *Alces, 31*, 45–52.

Benítez-López, A., Alkemade, R., & Verweij, P. A. (2010). The impacts of roads and other infrastructure on mammal and bird populations: A meta-analysis. *Biological Conservation, 143*, 1307–1316.

Bhattacharya, M., Primack, R. B., & Gerwein, J. (2003). Are roads and railroads barriers to bumblebee movement in a temperate suburban conservation area? *Biological Conservation, 109*, 37–45.

Borda-de-Água, L., Navarro, L., Gavinhos, C., & Pereira, H. M. (2011). Spatio-temporal impacts of roads on the persistence of populations: Analytic and numerical approaches. *Landscape Ecology, 26,* 253–265.

Bruschi, D., Garcia, D. A., Gugliermetti, F., & Cumo, F. (2015). Characterizing the fragmentation level of Italian's National Parks due to transportation infrastructures. *Transportation Research Part D, 36,* 18–28.

Castiov, F. (1999). *Testing potential repellents for mitigation of vehicle-induced mortality of wild ungulates in Ontario.* Doctoral dissertation. Sudbury: Laurentian University.

Cerboncini, R. A. S., Roper, J. J., & Passos, F. C. (2016). Edge effects without habitat fragmentation? Small mammals and a railway in the Atlantic Forest of southern Brazil. *Oryx, 50,* 460–467.

Child, K. (1983). Railways and moose in the central interior of BC: A recurrent management problem. *Alces, 19,* 118–135.

Clauzel, C., Girardet, X., & Foltête, J.-C. (2013). Impact assessment of a high-speed railway line on species distribution: Application to the European tree frog (*Hyla arborea*) in Franche-Comté. *Journal of Environmental Management, 127,* 125–134.

Clevenger, A. P., & Sawaya, M. A. (2010). Piloting a non-invasive genetic sampling method for evaluating population-level benefits of wildlife crossing structures. *Ecology and Society, 15,* 7.

Clevenger, A. P., & Waltho, N. (2005). Performance indices to identify attributes of highway crossing structures facilitating movement of large mammals. *Biological Conservation, 121,* 453–464.

Corlatti, L., Hackländer, K., & Frey-Roos, F. (2009). Ability of wildlife overpasses to provide connectivity and prevent genetic isolation. *Conservation Biology, 23,* 548–556.

D'Amico, M., Román, J., de los Reyes, L., & Revilla, E. (2015). Vertebrate road-kill patterns in Mediterranean habitats: Who, when and where. *Biological Conservation, 191,* 234–242.

De Santo, R. S., & Smith, D. G. (1993). Environmental auditing: An introduction to issues of habitat fragmentation relative to transportation corridors with special reference to high-speed rail (HSR). *Environmental Management, 17,* 111–114.

Dorsey, B., Olsson, M., & Rew, L. J. (2015). Ecological effects of railways on wildlife. In R. van der Ree, D. J. Smith, & C. Grilo (Eds.), *Handbook of road ecology* (pp. 219–227). West Sussex: Wiley.

Eriksson, C. (2014). *Does tree removal along railroads in Sweden influence the risk of train accidents with moose and roe deer?* Second cycle, A2E. Uppsala: Uppsala University.

Fahrig, L. (2003). Effects of habitat fragmentation on biodiversity. *Annual Review of Ecology Evolution and Systematic, 34,* 487–515.

Fenderson, L. E., Kovach, A. I., Litvaitis, J. A., O'Brien, K. M., Boland, K. M., & Jakuba, W. J. (2014). A multiscale analysis of gene flow for the New England cottontail, an imperiled habitat specialist in a fragmented landscape. *Ecology and Evolution, 4,* 1853–1875.

Ford, A. T., Clevenger, A. P., & Bennett, A. (2009). Comparison of methods of monitoring wildlife crossing-structures on highways. *Journal of Wildlife Management, 73,* 1213–1222.

Forman, R. T. T., Sperling, D., Bissonette, J. A., Clevenger, A. P., Cutshall, C. D., Dale, V. H., et al. (2003). *Road ecology. Science and solutions.* Washington: Island Press.

García-Sánchez, A. J., García-Sánchez, F., Losilla, F., Kulakowski, P., García-Haro, J., Rodríguez, A., et al. (2010). Wireless sensor network deployment for monitoring wildlife passages. *Sensors, 10,* 7236–7262.

Gerlach, G., & Musolf, K. (2000). fragmentation of landscape as a cause for genetic subdivision in bank voles. *Conservation Biology, 14,* 1066–1074.

Girvetz, E. H., James, H., Thorne, J. H., Berry, A. M., & Jaeger, J. A. G. (2008). Integration of landscape fragmentation analysis into regional planning: A statewide multi-scale case study from California, USA. *Landscape and Urban Planning, 86,* 205–218.

Grimm, V., Berger, U., Bastiansen, F., Eliassen, S., Ginot, V., Giske, J., et al. (2006). A standard protocol for describing individual-based and agent-based models. *Ecological Modelling, 198,* 115–126.

Gundersen, H., & Andreassen, H. P. (1998). The risk of moose *Alces alces* collision: A predictive logistic model for moose-train accidents. *Wildlife Biology, 4*, 103–110.

Gundersen, H., Andreassen, H. P., & Storaas, T. (1998). Spatial and temporal correlates to Norwegian moose-train collisions. *Alces, 34*, 385–394.

Hedin, J., & Ranius, T. (2002). Using radio telemetry to study dispersal of the beetle *Osmoderma eremita*, an inhabitant of tree hollows. *Computers and Electronic in Agriculture, 35*, 171–180.

Hopkins, J. B., III, Whittington, J., Clevenger, A. P., Sawaya, M. A., & St. Clair, C. C. (2014). Stable isotopes reveal rail-associated behavior in a threatened carnivore. *Isotopes in Environmental and Health Studies, 50*, 322–331.

Hu, D., Fu, J., Zou, F., & Qi, Y. (2012). Impact of the Qinghai-Tibet railway on population genetic structure of the toad-headed lizard, *Phrynocephalus vlangalii*. *Asian Herpetological Research, 3*, 280–287.

Hunt, A., Dickens, H. J., & Whelan, R. J. (1987). Movement of mammals through tunnels under railway lines. *Australian Zoologist, 24*, 89–93.

Iosif, R. (2012). Railroad-associated mortality hot spots for a population of Romanian Hermann's tortoise (*Testudo hermanni boettgeri*): A gravity model for railroad-segment analysis. *Procedia Environmental Sciences, 14*, 123–131.

Ito, T. Y., Lhagvasuren, B., Tsunekawa, A., Shinoda, M., Takatsuki, S., Buuveibaatar, B., et al. (2013). Fragmentation of the habitat of wild ungulates by anthropogenic barriers in Mongolia. *PLoS ONE, 8*, e56995.

Ito, T. Y., Miura, N., Lhagvasuren, B., Enkhbileg, D., Takatsuki, S., Tsunekawa, A., et al. (2005). Preliminary evidence of a barrier effect of a railroad on the migration of Mongolian gazelles. *Conservation Biology, 19*, 945–948.

Ito, T. Y., Okada, A., Buuveibaatar, B., Lhagvasuren, B., Takatsuki, S., & Tsunekawa, A. (2008). One-sided barrier effect of an international railroad on Mongolian gazelles. *Journal of Wildlife Management, 72*, 940–943.

Iuell, B., Bekker, G. J., Cuperus, R., Dufek, J., Fry, G., Hicks, C., et al. (2003). *Wildlife and traffic: A European handbook for identifying conflicts and designing solutions*. Utrech: KNNV Publishers.

Jaeger, J. A. G., & Fahrig, L. (2004). Effects of road fencing on population persistence. *Conservation Biology, 18*, 1651–1657.

Jaeger, J. A. G., Schwarz-von Raumer, H.-G., Esswein, H., Müller, M., & Schmidt-Lüttmann, M. (2007). Time series of landscape fragmentation caused by transportation infrastructure and urban development: A case study from Baden-Württemberg, Germany. *Ecology and Society, 12*, 22.

James, A. R., & Stuart-Smith, A. K. (2000). Distribution of caribou and wolves in relation to linear corridors. *Journal of Wildlife Management, 64*, 154–159.

Jaren, V., Andersen, R., Ulleberg, M., Pedersen, P. H., & Wiseth, B. (1991). Moose-train collisions: The effects of vegetation removal with a cost-benefit analysis. *Alces, 27*, 93–99.

Kaczensky, P., Knauer, F., Krze, B., Jonozovic, M., Adamic, M., & Gossow, M. (2003). The impact of high speed, high volume traffic axes on brown bears in Slovenia. *Biological Conservation, 111*, 191–204.

Kaczmarski, M., & Kaczmarek, J. M. (2016). Heavy traffic, low mortality—Tram tracks as terrestrial habitat of newts. *Acta Herpetologica, 11*, 227–231.

Kalarus, K., & Bąkowski, M. (2015). Railway tracks can have great value for butterflies as a new alternative habitat. *Italian Journal of Zoology, 82*, 565–572.

Kammonen, J. (2015) *Foraging behaviour of Myotis mystacinus and M. brandtii in relation to a big road and railway in south-central Sweden*. Bachelor's thesis. Uppsala: Uppsala University.

Karlson, M., Karlsson, C. S. J., Mörtberg, U., Olofsson, B., & Balfors, B. (2016). Design and evaluation of railway corridors based on spatial ecological and geological criteria. *Transportation Research Part D, 46*, 207–228.

Kerbiriou, C., Julien, J. F., Monsarrat, S., Lustrat, P., Haquart, A., & Robert, A. (2015). Information on population trends and biological constraints from bat counts in roost cavities: A

22-year case study of a pipistrelle bats (*Pipistrellus pipistrellus* Schreber) hibernaculum. *Wildlife Research, 42,* 35–43.

Kornilev, Y. V., Price, S. J., & Dorcas, M. E. (2006). Between a rock and a hard place: Responses of eastern box turtles (*Terrapene carolina*) when trapped between railroad tracks. *Herpetological Review, 37,* 145–148.

Kramer-Schadt, S., Revilla, E., Wiegand, T., & Breitenmoser, U. (2004). Fragmented landscapes, road mortality and patch connectivity: Modelling influences on the dispersal of Eurasian lynx. *Journal of Applied Ecology, 41,* 711–723.

Kušta, T., Keken, Z., Ježek, M., & Kůta, Z. (2015). Effectiveness and costs of odor repellents in wildlife–vehicle collisions: A case study in Central Bohemia, Czech Republic. *Transportation Research Part D, 38,* 1–5.

Lacy, R. C. (2000). Considering threats to the viability of small populations using individual-based models. *Ecological Bulletins, 48,* 39–51.

Li, Z., Ge, Ch., Li, J., Li, Y., Xu, A., Zhou, K., et al. (2010). Ground-dwelling birds near the Qinghai-Tibet highway and railway. *Transportation Research Part D, 15,* 525–528.

Malo, J. E., García de la Morena, E. L., Hervás, I., Mata, C., & Herranz, J. (2016). Uncapped tubular poles along high-speed railway lines act as pitfall traps for cavity nesting birds. *European Journal of Wildlife Research, 62,* 483–489.

Mateo-Sánchez, M. C., Cushman, S. A., & Saura, S. (2014). Connecting endangered brown bear subpopulations in the Cantabrian Range (north-western Spain). *Animal Conservation, 17,* 430–440.

Mateus, A. R. A., Grilo, C., & Santos-Reis, M. (2011). Surveying drainage culvert use by carnivores: Sampling design and cost–benefit analyzes of track-pads vs. video-surveillance methods. *Environmental Monitoring and Assessment, 181,* 101–109.

Modafferi, R. D., & Becker, E. F. (1997). Survival of radiocollared adult moose in lower Susitna river valley, southcentral Alaska. *Journal of Wildlife Management, 61,* 540–549.

Morelli, F., Beim, M., Jerzak, L., Jones, D., & Tryjanowski, P. (2014). Can roads, railways and related structures have positive effects on birds? A review. *Transportation Research Part D, 30,* 21–31.

Muzzi, P. D., & Bisset, A. R. (1990). The effectiveness of ultrasonic wildlife warning devices to reduce moose fatalities along railway corridors. *Alces, 26,* 37–43.

Okada, A., Ito, T. Y., Buuveibaatar, B., Lhagvasuren, B., & Tsunekawa, A. (2012). Genetic structure of Mongolian gazelle (*Procapra gutturosa*): The effect of railroad and demographic change. *Mongolian Journal of Biological Sciences, 10,* 59–66.

Olson, K. A. (2013). *Saiga crossing options: Guidelines and recommendations to mitigate barrier effects of border fencing and railroad corridors on saiga antelope in Kazakhstan.* Smithsonian Conservation Biology Institute. http://www.cms.int/sites/default/files/publication/Kirk_Olson_Saiga_Crossing_Options_English.pdf. Accessed November 11, 2016.

Olson, K. A., Mueller, T., Leimgruber, P., Nicolson, C., Fuller, T. K., Bolortsetseg, S., et al. (2009). Fences impede long-distance Mongolian Gazelle (*Procapra gutturosa*) movements in drought-stricken landscapes. *Mongolian Journal of Biological Sciences, 7,* 45–50.

Olson, K. A., & van der Ree, R. (2015). Railways, roads and fences across Kazakhistan and Mongolia threaten the survival of wide-ranging wildlife. In R. van der Ree, D. J. Smith, & C. Grilo (Eds.), *Handbook of road ecology* (pp. 472–478). West Sussex: Wiley.

Olsson, M., Augustsson, E., Seiler, A., Widén, P., & Helldin, J.-O. (2010). The barrier effect of twin tracked, non fenced railroads in Sweden. In *Proceedings of the 2010 IENE International Conference on Ecology and Transportation* (p. 76). Velence: Hungarian Academy of Sciences.

Orthlieb, X. (2016). Find a second life for old regional railway tracks in France allying transport perspective for the future and protection of the environment. In *Proceedings of the 2016 IENE International Conference on Ecology and Transportation* (p. 157). Lyon: Cerema and Fondation pour la Recherche sur la Biodiversité.

Paquet, P., & Callaghan, C. (1996). Effects of linear developments on winter movements of gray wolves in the Bow River Valley of Banff National Park, Alberta. In G. L. Evink, P. Garrett, D.

Zeigler, & J. Berry (Eds.), *Proceedings of the transportation related wildlife mortality seminar* (pp. 46–66). Tallahassee: Florida Department of Transportation.

Pelletier, S. K., Carlson, L., Nein, D., & Roy, R. D. (2006). Railroad crossing structures for spotted turtles: Massachusetts Bay Transportation Authority—Greenbush rail line wildlife crossing demonstration project. In C. L. Irwin, P. Garrett, & K. P. McDermott (Eds.), *Proceedings of the 2005 International conference on ecology and transportation* (pp. 414–425). Raleigh: North Carolina State University.

Prunier, J. G., Kaufmann, B., Léna, J.-P., Fenet, S., Pompanon, F., & Joly, P. (2014). A 40-year-old divided highway does not prevent gene flow in the alpine newt *Ichthyosaura alpestris*. *Conservation Genetics, 15*, 453–468.

Qian, Z., Lin, X., Jun, M., Pan-Wen, W., & Qi-Sen, Y. (2009). Effects of the Qinghai-Tibet Railway on the community structure of rodents in Qaidam desert region. *Acta Ecologica Sinica, 29*, 267–271.

Railsback, S. F., & Grimm, V. (2011). *Agent-based and individual-based modeling: A practical introduction*. Princeton: Princeton University Press.

Reh, W., & Seitz, A. (1990). The influence of land use on the genetic structure of populations of the common frog *Rana temporaria*. *Biological Conservation, 54*, 239–249.

Riley, S. P. D., Pollinger, J. P., Sauvajot, R. M., York, E. C., Bromley, C., Fuller, T. K., et al. (2006). A southern California freeway is a physical and social barrier to gene flow in carnivores. *Molecular Ecology, 15*, 1733–1741.

Rodríguez, A., Crema, G., & Delibes, M. (1996). Use of non-wildlife passages across a high speed railway by terrestrial vertebrates. *Journal of Applied Ecology, 33*, 1527–1540.

Rodríguez, A., Crema, G., & Delibes, M. (1997). Factors affecting crossing of red foxes and wildcats through non-wildlife passages across a high-speed railway. *Ecography, 20*, 287–294.

Rodríguez, J. J., García de la Morena, E., & González, D. (2008). *Estudio de las medidas correctoras para reducir las colisiones de aves con ferrocarriles de alta velocidad*. Madrid: Centro de Estudios y Experimentación, Ministerio de Fomento.

Roger, E., Laffan, S. W., & Ramp, D. (2011). Road impacts a tipping point for wildlife populations in threatened landscapes. *Population Ecology, 53*, 215–227.

Schaub, M., Aebischer, A., Gimenez, O., Berger, S., & Arlettaz, R. (2010). Massive immigration balances high anthropogenic mortality in a stable eagle owl population: Lessons for conservation. *Biological Conservation, 143*, 1911–1918.

Schwartz, C., & Bartley, B. (1991). Reducing incidental moose mortality: Considerations for management. *Alces, 27*, 93–99.

Simmons, J. M., Sunnucks, P., Taylor, A. C., & van der Ree, R. (2010). Beyond roadkill, radiotracking, recapture and *FST*—A review of some genetic methods to improve understanding of the influence of roads on wildlife. *Ecology and Society, 15*, 9.

Smith, D. J., & van der Ree, R. (2015). Field methods to evaluate the impacts of roads on wildlife. In R. van der Ree, D. J. Smith, & C. Grilo (Eds.), *Handbook of road ecology* (pp. 82–95). West Sussex: Wiley.

Smith, D. J., van der Ree, D., & Rosell, C. (2015). Wildlife crossing structures: An effective strategy to restore or maintain wildlife connectivity across roads. In R. van der Ree, D. J. Smith, & C. Grilo (Eds.), *Handbook of road ecology* (pp. 172–183). West Sussex: Wiley.

Soanes, K., Lobo, M. C., Vesk, P. A., McCarthy, M. A., Moore, J. L., & van der Ree, R. (2013). Movement re-established but not restored: Inferring the effectiveness of road-crossing mitigation for a gliding mammal by monitoring use. *Biological Conservation, 159*, 434–441.

Tremblay, M. A., & St. Clair, C. C. (2009). Factors affecting the permeability of transportation and riparian corridors to the movements of songbirds in an urban landscape. *Journal of Applied Ecology, 46*, 1314–1322.

Trewhella, W. J., & Harris, S. (1990). The effect of railway lines on urban fox (*Vulpes vulpes*) numbers and dispersal movements. *Journal of Zoology, 221*, 321–326.

Urban, D. L., Minor, E. S., Treml, E. A., & Schick, R. S. (2009). Graph models of habitat mosaics. *Ecology Letters, 12*, 260–273.

van der Grift, E. A. (1999). Mammals and railroads: Impacts and management implications. *Lutra, 42*, 77–98.

van der Ree, R., Smith, D. J., & Grilo, C. (2015). *Handbook of road ecology*. West Sussex: Wiley.

van Why, K. R., & Chamberlain, M. J. (2003). Mortality of black bears, *Ursus americanus*, associated with elevated train trestles. *Canadian Field Naturalist, 117*, 113–115.

Vandevelde, J.-C., Bouhours, A., Julien, J.-F., Couvet, D., & Kerbiriou, C. (2014). Activity of European common bats along railway verges. *Ecological Engineering, 64*, 49–56.

Vandevelde, J.-C., Penone, C., & Julliard, R. (2012). High-speed railways are not barriers to *Pyronia tithonus* butterfly movements. *Journal of Insect Conservation, 16*, 801–803.

Waller, J. S., & Servheen, C. (2005). Effects of transportation infrastructure on grizzly bears in northwestern Montana. *Journal of Wildlife Management, 69*, 985–1000.

Waterman, E., Tulp, I., Reijnen, R., Krijgsveld, K., & Braak, C. (2002). Disturbance of meadow birds by railway noise in The Netherlands. *Geluid, 1*, 2–3.

Wells, P., Woods, J. G., Bridgewater, G., & Morrison, H. (1999). Wildlife mortalities on railways; Monitoring methods and mitigation strategies. In G. Evink, P. Garrett, & D. Zeigler (Eds.), *Proceedings of the third international conference on wildlife ecology and transportation* (pp. 237–246). Tallahassee: Florida Department of Transportation.

Whittington, J., Hebblewhite, M., DeCesare, N. J., Neufeld, L., Bradley, M., Wilmshurst, J., et al. (2011). Caribou encounters with wolves increase near roads and trails: A time-to-event approach. *Journal of Applied Ecology, 48*, 1535–1542.

Wiącek, J., Polak, M., Filipiuk, M., Kucharczyk, M., & Bohatkiewicz, J. (2015). Do birds avoid railroads as has been found for roads? *Environmental Management, 56*, 643–652.

Wrzesień, M., Jachuła, J., & Denisow, B. (2016). Railway embankments—Refuge areas for food flora, and pollinators in agricultural landscape. *Journal of Apicultural Science, 60*, 97–110.

Xia, L., Yang, Q., Li, Z., Wu, Y., & Feng, Z. (2007). The effect of the Qinghai-Tibet railway on the migration of Tibetan antelope *Pantholops hodgsonii* in Hoh-xil National Nature Reserve, China. *Oryx, 41*, 352–357.

Yanes, M., Velasco, J. M., & Suaréz, F. (1995). Permeability of roads and railways to vertebrates: The importance of culverts. *Biological Conservation, 71*, 217–222.

Yang, J., Jiang, Z., Zeng, Y., Turghan, M., Fang, H., & Li, C. (2011). Effect of anthropogenic landscape features on population genetic differentiation of Przewalski's Gazelle: Main role of human settlement. *PLoS ONE, 6*, e20144.

Yang, Q., & Xia, L. (2008). Tibetan wildlife is getting used to the railway. *Nature, 452*, 810–811.

Zuberogoitia, I., del Real, J., Torres, J. J., Rodríguez, L., Alonso, M., de Alba, V., et al. (2015). Testing pole barriers as feasible mitigation measure to avoid bird vehicle collisions (BVC). *Ecological Engineering, 83*, 144–151.

Chapter 5
Aliens on the Move: Transportation Networks and Non-native Species

Fernando Ascensão and César Capinha

Abstract Biological invasions are a major component of global environmental change, threatening biodiversity and human well-being. These invasions have their origin in the human-mediated transportation of species beyond natural distribution ranges, a process that has increased by orders of magnitude in recent decades as a result of accelerating rates of international trade, travel, and transport. In this chapter, we address the role that overland transportation corridors, particularly railways, have in the transport of non-native species. We focus specifically on the role of rail vehicles in dispersing stowaway species, i.e. species that are moved inadvertently and that are not specific to the commodities being transported; we also focus on the natural dispersal and establishment of non-native species along railway edges. We place these processes in the context of biological invasions as a global phenomenon and provide examples from the literature. We also list general management recommendations for biological invasions highlighting the particularities associated with their management in railway transport systems. Following previous studies, we briefly outline four possible management approaches: (1) "Do nothing;" (2) "Manage propagule supply;" (3) "Manage railway environments;" and (4) "Act over the invasive populations directly". These approaches are not mutually exclusive, and they range from an expectation that natural processes (e.g. ecological succession) will drive the invaders out of the ecosystems, to the application of

Fernando Ascensão and César Capinha have contributed equally to this chapter.

F. Ascensão
CIBIO/InBIO, Centro de Investigação em Biodiversidade e Recursos Genéticos,
Universidade do Porto, Campus Agrário de Vairão, 4485-661 Vairão, Portugal

F. Ascensão
CEABN/InBIO, Centro de Ecologia Aplicada "Professor Baeta Neves",
Instituto Superior de Agronomia, Universidade de Lisboa,
Tapada da Ajuda, 1349-017 Lisboa, Portugal

C. Capinha (✉)
Global Health and Tropical Medicine Centre (GHTM), Instituto de Higiene e Medicina
Tropical (IHMT), Universidade Nova de Lisboa, Lisboa, Portugal
e-mail: ccapinha@cibo.up.pt

© The Author(s) 2017
L. Borda-de-Água et al. (eds.), *Railway Ecology*,
DOI 10.1007/978-3-319-57496-7_5

65

measures to extirpate the invaders directly (e.g. manual removal). We highlight that best practices for the management of invaders in railway-related systems may be difficult to generalize and that they may have to be considered on a case-by-case basis. We end by stressing that research on railways in the context of biological invasions remains scarce, and that fundamental knowledge for understanding the relative importance of this transport system in the dispersal of species and on how this process should be dealt with remains largely lacking.

Keywords Biological invasions · Stowaway species · Verges · Invasibility

Introduction

Nowadays non-native species are widespread around the world. These species, which have been moved beyond the limits of their natural ranges by human activities, include not only most farm animals and plants, forestry species and pets, but also many unwanted organisms such as mosquitoes, weeds, fungus and bacteria. Although most of these species perish soon after arriving at the new area, or once humans cease to care for them, a few do form wild populations and become part of the ecosystems (Williams and Newfield 2002). These species, often called as "invasive," are now one of the most important causes of global change, being responsible for the decline and extinction of native species, economic losses, and human health problems (Simberloff et al. 2013).

Humans have been moving species since pre historic times (Di Castri 1989), but the number of biological invasions and the magnitude of their impacts in the last few decades is unprecedented (Ricciardi 2007). This is to a great extent the result of the recent large scale expansion of the transportation network and of the increasing exchange of commodities to a nearly global coverage (Hulme 2009). Because of this, most of the earth's surface is now within reach of non-native species. Moreover, the volume and diversity of the cargo and number of passengers now being transported is also much greater than in any period in the past (Hulme 2009). This allows for a greater diversity of species being introduced in new areas as well as a greater number of their propagules, which increases both the number of potential invaders and their ability to establish (Lockwood et al. 2009). Finally, the latest technical advances in transportation allow species to move much faster. Even the least suspicious species, e.g. those that are more environmentally sensitive, may become invasive in remote regions of the original distribution (Wilson et al. 2009). In summary, as the extent, volume, and efficiency of the transportation of people and freight increase, the burden of biological invasions should also increase (Bradley et al. 2012).

Hulme et al. (2008) distinguish three main mechanisms by which human-activities cause the dispersal of non-native species: (1) through importation as a commodity or with a commodity; (2) as stowaways, i.e. through a direct influence of a transportation vehicle; and (3) by means of natural dispersal along artificial infrastructures, such as water canals, roads, and, especially relevant for our concerns, railways. The first mechanism arises directly from trading activities, where the traded commodity may itself be the non-native species. In these cases, the "importation" of the species is deliberate, because some of its attributes are desired in the area of destination. The reasons behind the importation of species are varied, and include farming, forestry, livestock, ornamental plants and pets, laboratory testing or biocontrol. In some situations, the goal is purposely the formation of wild populations, as is the case of gaming and fishing or biocontrol agents. In other cases, the imported species are to be stored in enclosed environments, but they often escape from captivity. For instance, the American mink (*Neovison vison*) has invaded many European countries due to accidental escape or deliberate release from fur farms (Vidal-Figueroa and Delibes 1987; Bonesi and Palazon 2007). It is worrisome that a great number of problematic invaders is still actively marketed today, including freshwater macroinvertebrates and fish (Capinha et al. 2013; Consuegra et al. 2011) and plants (Humair et al. 2015).

Trading activities may also lead to the arrival of non-native species as an accidental "by-product" of a commodity, such as a commensal, a parasite or a disease. These species can remain undetected for long periods of time and may benefit from measures towards the establishment to their hosts in the wild. An illustrative example of such by-products of a traded commodity is the crayfish plague (*Aphanomyces astaci*) in Europe. This fungus-like disease is hosted by several North American crayfish species that were introduced in European wetlands in order to boost the wild stocks of this food item. However, unlike for North American species that co-evolved with the disease, the crayfish plague is deadly to European species and has already caused numerous extinctions of local populations (Capinha et al. 2013).

Stowaway species can be associated with trading activities or any other activity that involves a transportation vehicle. In other words, a non-native stowaway species is not specific to a particular commodity, and it can be any organism that at some point is displaced by a vehicle or its load. This occurs more often with species that are difficult to detect, such as those that are small or stealthy. Known examples of these stowaway species include land snails attached to trains or their cargo (Peltanová et al. 2011), plankton in ship's ballast waters (Hulme et al. 2008), or seeds in soil attached to automobiles (Hodkinson and Thompson 1997). Centers of human and commodity transportation and nearby areas (e.g. seaports and railroad stations and yards) often provide the first records of non-native stowaways (e.g. Noma et al. 2010) and can host diverse communities of invasive species (Drake and Lodge 2004).

Finally, transport infrastructures may also act as "corridors" for the natural dispersal of non-native biodiversity. These infrastructures facilitate the dispersal of non-native organisms by allowing their movement across physical and environmental barriers (e.g. a mountain range now crossed by a tunnel), or by supplying suitable habitat for expanding invasive populations. Concerning the latter case, a few characteristics of the areas managed by transportation companies (e.g. road and railway verges) are considered beneficial to the establishment of non-native species, particularly the regular occurrence of disturbance that gives rise to "vacant" niches and low biological diversity, reducing the number of potential competitors for space and resources (Catford et al. 2012).

In this chapter, we focus on the role of terrestrial transportation systems, particularly railways, in dispersing stowaway fauna and flora and in facilitating the natural dispersal of non-native species, i.e. the second and third mechanisms identified above. We start by providing a contextual overview of the impacts of invasive species in a global context, in order to better familiarize readers with the significance of the problem. We then focus on the role of railway traffic in transporting non-native stowaway species and describe some of the best-known examples. Finally, we discuss and provide examples of natural dispersal of non-native species along transportation corridors and conclude by discussing some of the management actions that could be taken to help reducing the spread of those species in railways.

Why Care about Invasive Species? An Overview of Their Global-Scale Relevance

In natural environments, invasive species compete, predate and hybridize with native species, and alter community structure and ecosystem processes, ultimately leading to irreversible changes on the diversity and distribution of life on earth (Simberloff et al. 2013; Capinha et al. 2015). Examples of mass extinctions precipitated by the introduction of non-native species include nearly every native bird species on the Pacific island of Guam after the arrival of the invasive brown tree snake (*Boiga irregularis*) (Wiles et al. 2003), or the extinction of more than 100 terrestrial gastropods due to the introduction of the predatory rosy wolf snail (*Euglandina rosea*) in tropical oceanic islands worldwide (Régnier et al. 2009). The Nile perch (*Lates niloticus*) is another paradigmatic example of the negative impacts of invasive species on native species. In the 1950s, this predatory freshwater fish was intentionally introduced in Lake Victoria, Africa, to boost the lake's fish stocks, which were becoming severely overfished. In the decades after its introduction, the Nile perch density grew massively leading to the extinction of nearly 200 endemic species of cichlid fishes (Craig 1992).

The economic costs of invasive species can be striking. It is estimated that invasive species can cost many billions of dollars in the USA and in Europe alone (Pimentel et al. 2005; Davis 2009; Hulme 2009; Marbuah et al. 2014). For example, Bradshaw et al. (2016) recently compiled a comprehensive database of economic costs of invasive insects. Taking all reported goods and service estimates, according to the authors' study, invasive insects cost a minimum of US$70.0 billion per year globally and the associated health costs exceed US$6.9 billion per year. These values mainly reflect observable damages, such as those caused on other economically important species, e.g. the cinnamon fungus (*Phytophthora cinnamomi*) on the sweet chestnut (*Castanea sativa* in Europe, and *Castanea dentata* in North America) (Vettraino et al. 2005), or on man-made infrastructures and equipment, e.g. the zebra mussel (*Dreissena polymorpha*) in water treatment works (Elliott et al. 2005). However, there are also many indirect damages that are difficult to quantify because they would require a deeper understanding of the ecosystem-economy relationship (Bradshaw et al. 2016). Furthermore, many invasive species are also vectors of human diseases, such as malaria, plague, typhus or yellow fever, and their transportation may result in outbreaks of these infections in previously unsuspected areas (Lounibos 2010; Capinha et al. 2014). Also, invasive species may cause ecological or landscape changes that have negative implications for human safety, these can assume multiple forms, such as the promotion of pathogen eruptions (Vanderploeg et al. 2001), or an increase in the vulnerability of landscapes to natural hazards such as fires (Berry et al. 2011).

Importantly, many of the future impacts of non-native species may still remain unknown. For instance, the increased profusion of invasive species may render a cascading effect on the vulnerability of ecosystems, i.e. by making these even more susceptible to future impacts. Climate change is a further source of concern. Changes in climatic patterns altering the geography of the areas that can be invaded may put additional pressures on native biodiversity. Understanding how these processes will interact in the future to determine the impacts of biological invasions is challenging. In fact, many future invasions may have already been set in motion, i.e. many non-native species are currently in a lag-phase, i.e. with little or no increase in species occurrence, to be followed by an increase-phase in which species occurrence and invasiveness rises rapidly before becoming invaders (Aikio et al. 2010; Essl et al. 2011).

Without increased efforts to manage non-native species in transportation infrastructures, including in vehicles, transported cargo and verges, the number of invasive species will likely continue to grow steadily (Keller et al. 2011). Hence, more effective policies to reduce the transport and release of non-native species, and to manage those already established, should become a priority (Pimentel et al. 2005; Keller et al. 2011).

Non-native Hitchhikers: Transportation Vehicles as Vectors of Stowaway Species

The surroundings of transportation infrastructures (e.g. verges and embankments) often host a high diversity of non-native species (Gelbard and Belnap 2003; Hansen and Clevenger 2005), in many cases due to their transportation as stowaways in vehicles. Species can be accidentally moved by a vehicle in many different ways, e.g. snails and slugs clinging to a train, insects flying inside a vehicle, plant seeds in passengers' boots, or any organism that makes frequent use of cargo yards and that is loaded unintentionally loaded with the cargo. Nevertheless, despite the range of possibilities, the movement of stowaway fauna and flora is poorly documented for terrestrial transportation systems. This is in contrast with aquatic transportation, especially maritime, for which there is a large body of research, particularly regarding the movement of species in ballast waters (Drake and Lodge 2004; Seebens et al. 2016). We speculate that overland transportation of biological stowaways is likely to be less frequent than those by aquatic vehicles; however, we also suspect that the lack of studies for the terrestrial counterpart does not reflect the true contribution of this process for non-native species dispersal. Below we describe a few documented cases where terrestrial transportation had a relevant role on the expansion of biological invaders, with special emphasis on trains.

The spotted knapweed (*Centaurea stoebe*) is an example of a non-native species dispersed by trains. This plant, native to south east and central Europe, arrived in North America in the late 1800s and by the year 2000 it was already found in most contiguous American states (Sheley et al. 1998). The mechanisms that enabled such a fast dispersal were unclear for a long time, but a recent reconstruction of the patterns of invasion of this plant showed a close agreement between the extent of colonized areas and its velocity dispersal on one side, and the spatial coverage and development of the railway network in the USA on the other (Broennimann et al. 2014). For instance, the wave of invasion was much faster and wider in the eastern states, where a denser and older network exists. In favor of the important role of stowaway transportation of this species in trains, particularly for some long-distance dispersal events, is the plants' known ability to become attached to the undercarriages of vehicles (Sheley et al. 1998).

Trains are known to have been a vehicle of dispersal also for ragwort plant species. One of such cases concerns the South African ragwort (*Senecio inaequidens*), a species introduced in Europe from South Africa in the first half of the twentieth century, and that is now found from Norway in the north to Italy in the south, and from Bulgaria in the east to Spain in the west (Heger and Böhmer 2006). The achene-type fruit of this species is able to stick to trains in movement or to their transported commodities, a characteristic that contributed to the rapid dispersal of the species along railways systems (Heger and Böhmer 2005). Another plant of this group known to "hitchhike" on trains is the Oxford ragwort (*Senecio squalidus*), a hybrid that escaped cultivation from the Oxford Botanic Garden. At the end of the eighteenth century it was established in some parts of Oxford (Harris 2002; Heger and Böhmer 2005), and it currently invades most of Britain. George Druce, an English botanist, described the dispersal of this plant as follows (1927b, p. 241, in Harris 2002): " ... the vortex of air following the express train carries the fruits in its wake. I have seen them enter a railway-carriage window near Oxford and remain suspended in the air in the compartment until they found an exit at Tilehurst [about 40 km from Oxford]".

A few records of animal species being transported as stowaways in trains can also be found in the scientific literature and the media. Perhaps the most recurrent cases refer to urban pest species, such as rats and mice (Li et al. 2007), but there are also references to ants (Elton 1958), beetles (White 1973), spiders (Nentwig and Kobelt 2010) and even armadillos (Hofmann 2009). However, in these cases the contribution of train-mediated transportation to the overall process of dispersal remains poorly studied.

Verges as Habitat and Corridors for Non-native Species

Vegetated verges generally border transportation infrastructures, in particular railways and roads. These verges are regularly mowed in order to ensure traffic safety, resulting in open and well-lighted areas. In places where the transportation corridor is wider, intense mowing is applied only to the zone closest to traffic, while the farthest zone is managed less intensively, leading to the creation of a well-developed vegetation structure. These two zones of management regime provide habitat not only for native species but also for many non-native organisms. In Portugal, for example, many kilometers of railway and road verges are dominated by silver wattle (*Acacia dealbata*), one of the most widespread and damaging invasive plants in the country (Sheppard et al. 2006; Vicente et al. 2011). This small tree is native to Australia, and was introduced in Europe in the 1820s. In addition to its great natural dispersal ability, silver wattle inhibits undergrowth species from growing, due to allelopathy, i.e. the ability to produce biochemical agents that the growth, survival, and reproduction of other organisms. Many other examples are referenced in the literature on the presence and sometimes dominance of non-native weedy species in transportation verges (Ernst 1998; Parendes and Jones 2000; Tikka et al. 2001; Gelbard and Belnap 2003; Albrecht et al. 2011; McAvoy et al. 2012; Penone et al. 2012; Suárez-Esteban et al. 2016). Even maritime or wetland species may spread their populations into inland areas along railway and road verges, as found in Finland (Suominen 1970), England (Scott and Davison 1982), or the USA (Wilcox 1989).

By facilitating dispersal, railways and roads may lead to homogeneous communities, sometimes dominated by invasive species. For example, Hansen and Clevenger (2005) measured the frequency of several non-native plant species along transects from 0 to 150 m from the edge of railways and highways in grasslands and forests, as well as at control sites away from corridors. These authors found that both transportation corridors had a higher frequency of non-native species than the respective control sites. Also, grasslands had a higher frequency of non-native species than forested habitats, but this frequency did not differ between the highways and the railways. Other studies have described a decline in the presence of invasive species as a function of the distance to the transportation corridor (see Gelbard and Belnap 2003). Interestingly, the penetration of non-native species in areas adjacent to the transportation corridor is likely to vary according to the dominant land cover. In the study mentioned above, Hansen and Clevenger (2005) discovered that the frequency of non-native species in grasslands along railways and highways was higher than at control sites up to 150 m from the corridor's edge, whereas in forested habitats the higher frequency of non-native species was only evident up to 10 m away from the corridor's edge. A similar result was found in forested areas in the Chequamegon National Forest (USA), where invasive species were most prevalent within 15 m of roads but were uncommon in the interior of the forest (Watkins et al. 2003). Hence, it appears that the dispersal of non-native species in transportation corridors may have a lower impact in forested landscapes than in open areas, such as grasslands.

The importance that verge areas represent for the spreading and fixation of non-native species is therefore clear. In fact, despite their relatively narrow width, verges can be long, creating continuous strips that may extend for many kilometers and occupy considerable areas. Knowing that the global railway and road lengths are, respectively, approximately 1.1 and 64.3 million km (CIA 2016), and using a conservative value of mean verge width of two meters on each side, we realize that more than 262,000 kilometers of the world are occupied by verges. For comparison, this is equivalent to approximately 33% of the terrestrial protected areas of Natura 2000 Network (EU 2016). Therefore, the management of the vegetation of verges is of utmost importance not only for traffic safety, but also for the control of non-native species.

Management of Non-native Species in Transportation Corridors

As discussed, transportation corridors can function as habitats and venues for the dispersal of non-native species, hence they are an important element to consider when managing and preventing the threats caused by biological invasions. On the other hand, verges can also help maintain conservation values and the connectivity among landscapes, particularly in areas that are heavily modified by human activities (see Chap. 4). Because verges provide habitat areas and corridors for both native and non-native species, it creates management challenges, such as how to maintain or increase the conservation value of transportation verges, while preventing non-native animal and plant species from spreading throughout the network and its surroundings. A delicate balance between restricting the arrival and dispersal of non-native species and maintaining or restoring the conservation value of verges is thus needed. This requires specific management actions, as we discuss below.

Identifying the factors that facilitate the arrival of propagules of non-native species and their dispersal along the corridors may render it possible to manage verges in ways that limit the expansion of an invasive species (Fagan et al. 2002; With 2002). Despite the difficulties in identifying such key factors for all organisms, some general management practices are likely to prevent the dispersal of non-native species in most circumstances.

Perhaps the best management option is to avoid setting up the conditions for non-native species dispersal and establishment when a corridor is under construction, being upgraded or under maintenance. Such activities may imply baring soil, clearing of natural vegetation, or drainage, resulting in considerable disturbance of natural communities. In turn, these disturbances underlie ecological processes that often facilitate the colonization by invasive species, as they "remove" any pre-existing advantage of native over non-natives species that could be present in the area. Hence, reducing the disturbance of the existing communities as much as possible, may be itself a good management practice.

Propagule pressure is also important. Christen and Matlack (2006) suggested that undisturbed areas are generally less invaded precisely because they may receive fewer non-native propagules. Avoiding the introduction of non-native propagules prevents not only the establishment of invasive species, but also offers much greater economic benefits than the management of invasive populations after their establishment (Keller et al. 2007). Equally important, management options available prior to invasion are more numerous and include legislation or quarantine rules (Keller et al. 2011; Buckley and Catford 2016). In this context, the inspection of cargo and their containers is of great importance, particularly for those having an international origin. For example, a recent inspection of international cargo entering the USA by rail enabled the identification of dozens of undeclared organisms, among which were invasive insects, noxious weeds and vectors of human diseases (https://goo.gl/fJ1uMo). Following legal rulings, some of the inspected cargo was re-exported to its origin in order to prevent the dispersal of the unwanted pests.

In many cases, however, prevention is not possible as the non-native species are already established in the landscape. In such cases, management actions should aim at containing or eradicating these species. However, the broad range of invasive species and the different ways that humans value the colonized ecosystems mean that few generalizations can be provided for management and policy guidelines. In other words, appropriate management and policies for invasive species is highly context dependent (Keller et al. 2011). However, based on the review provided by Catford et al. (2012), one can consider four not mutually exclusive main approaches for managing transportation corridor verges when attempting to control or eradicate invasive species: (1) "Do nothing;" (2) "Control at the introduction level by managing the propagule supply of non-native and native species;" (3) "Manage environmental conditions;" and (4) "Manage invasive species populations."

1. *Do nothing.* This is an option when invaders are well established and are successional colonists. Several studies have shown that the proportion of invasive species can decrease with time since disturbance–not only weedy plants (Bellingham et al. 2005), but also larger plants (Dewine and Cooper 2008). Hence, active management by removal of invasive species immediately following disturbance may be unnecessary and counterproductive. For example, tamarisk species (*Tamarix ramosissima, T. chinensis, T. gallica* and hybrids) have invaded riparian zones throughout western North America, southern Africa, Argentina, Hawaii, and Australia, demanding expensive control efforts. However, as a relatively recent addition to North American plant communities (the 1920s–1960s was the period of main invasion), the competitive and successional processes are still ongoing. In fact, Dewine and Cooper (2008) demonstrated that box elder (*Acer negundo*), a native species found in canyons throughout western North America, is a superior competitor to tamarisk and is capable of becoming established under dense tamarisk canopies, overtopping and eventually killing the tamarisks. Thus, superior shade tolerance appears to

be the mechanism for the successional replacement of tamarisk by box elder. The authors suggest that the preservation of box elder and other native tree populations is probably a better and cheaper means of tamarisk control than traditional control techniques.

As highlighted by Catford et al. (2012), however, there are several arguments against the "do nothing" approach. Firstly, ecosystem functions may be significantly altered by early successional invasive plants colonizing soon after disturbance (Peltzer et al. 2009). Secondly, some invasive species can outcompete functionally similar native species and therefore persist and dominate over long periods after disturbance (Christian and Wilson 1999). Thirdly, non-native species may establish themselves in highly disturbed areas first, and subsequently adapt and colonize nearby areas with different environmental conditions (Clark and Johnston 2011). Finally, most invasive species are not early successional species that will be replaced over time, some of them are even long-lived K-strategists that are highly competitive and able to invade even undisturbed areas (Wilsey et al. 2009).

2. *Manage the propagule supply.* By reducing the propagule pressure of non-native organisms and increasing that of native species, one can increase the dominance of the former. For example, one control action against silver wattle is the use of prescribed fire to favor the germination of the seed bank, therefore reducing it by destroying part of the seeds or by stimulating the germination of the remainders (http://invasoras.pt). Applying this or other measures to reduce the seed bank soon after disturbance activities may strongly reduce the propagule pressure and prevent or reduce the success of invasion. On the other hand, native species suitable for direct sowing should also be selected based on their traits and ability to establish and persist under the specific conditions of the transportation verges. This active selection towards native colonists may benefit not only local biodiversity but also regional agricultural areas. Blackmore and Goulson (2014) found that sowing native wildflowers in road verges significantly increased the abundance of native flowering plants and that of pollinator bumblebees (*Bombus* spp.) and hoverflies (Syrphidae), with benefits to agricultural crops. Furthermore, larvae of hoverflies prey on aphids, arthropod pest species that are responsible for enormous crop damages worldwide, every year. Hence, even easy to–implement interventions in verges may result in considerable environmental and economic gains.

3. *Manage the environmental conditions.* In some cases, it may be more effective to target the causes that boost the disturbance and facilitate the spread of non-native species than to attempt to manage their populations directly. For example, some weeds are particularly adapted to severely eroded verges. Rather than applying herbicide to control such weeds, it may be more effective to stabilize the soil (Catford et al. 2012). This option requires excellent knowledge of the ecological requirements of each invader and it may imply conducting experimental ecological work in order to evaluate the effectiveness of distinct management actions. Nevertheless, for some of the species the necessary

information already exists and is available in the various on-line databases that profile problematic invaders, such as the Invasive Species Compendium (http://sites.cabi.org/isc/), Invasoras (http://invasoras.pt) or the 'Global Invasive Species Database' (http://www.iucngisd.org/).

4. *Manage invasive species populations.* For some highly problematic invasive species it may be more effective to focus on management techniques that directly target their populations. Traditional control techniques, such as the use of prescribed fire or herbicides, as well as mechanical removal may help to alleviate the competitive effects of some species and limit further spread. However, care should be taken to minimize impacts on native species. If the non-native species, particularly plants, have unrestricted dispersal but infrequent propagule arrival, mapping and removing the individual colonist patches soon after they arise, may achieve effective control. This is most important, as such patches often pose a high risk of becoming sources of subsequent dispersal (Moody and Mack 1988). Conversely, if the invasive species has limited dispersal ability, an effective control may be achieved by creating spatial discontinuities on vegetation types at an extent greater than the seed dispersal range, thereby breaking the continuity of the habitat to the invader (Christen and Matlack 2006).

Currently, railroad companies routinely clear-cut and/or spray with herbicide all vegetation that grows too close to the tracks. However, although this method eradicates most of the vegetation (including native species), it also favors the dominance of species that respond favorably to clear-cutting or resist herbicides. Actually, in some cases, roadside herbicide treatments are known to reduce the cover of some non-native species favoring others (Gelbard and Belnap 2003). Hence, investing exclusively in the direct management of invasive populations should be accompanied by a careful evaluation of trade-offs and probably the best option is often a combination of various management possibilities.

Where to begin the control of invasive species is a major question when managing railway verges. One factor that apparently influences the propagule pressure is the disturbance level of the transportation infrastructure. For example, it is known that older roads typically have higher cumulative levels of traffic and maintenance than younger roads, which might result in an increase in non-native species occurrence near old roads simply due to higher disturbance and therefore of propagule pressure. This was found for earthworms (Cameron and Bayne 2009) and the invasive common reed (*Phragmites australis*) (Jodoin et al. 2008). Likewise, plant communities adjacent to roads that receive heavy traffic might be expected to be invaded more than those adjacent to infrequently used and unpaved roads (Parendes and Jones 2000; Gelbard and Belnap 2003). Hence, the degree of perturbation of the transportation corridor, namely the traffic level, could be used as an indicator of the invasion level of verges in railways, and to identify which corridor should be targeted for management.

During the project phase of new railway corridors, engineers should consider whether some routes might aid the dispersal of non-native species more than others. For example, routes expanding railways that are highly colonized by problematic invaders should be regularly monitored and managed to avoid the spread of propagules. Additionally, transportation networks can also be set to benefit people while minimizing their environmental and social costs. Integrating transportation verges in landscape connectivity management plans, including ecological corridors, would therefore better enable achieving the conservation goals, in particular to ensure a sustainable co-existence between transportation networks with the conservation of biodiversity.

Conclusions

Invasive species are responsible for many negative impacts on biodiversity and human welfare. What is worrisome is that as the transportation of people and goods around the world increases, so does the number of biological invasions. It is thus increasingly important to identify the role that each transportation system has in dispersing species beyond the limits of their natural ranges and to develop procedures by which this process can be reduced. Railways are responsible for several important invasion events but, in comparison to other transportation systems, their overall contribution to biological invasions remains poorly understood. Likewise, the guidelines for preventing or managing the transportation of unwanted organisms in trains or along railways are poorly synthesized and consist mainly of a hand-full of general principles that can be applied to overland transportation systems in general. Thus it is thus clear that more research must be devoted to this topic. Particularly relevant contributions include the identification or "profiling" of the species that go as stowaways in trains or their cargo, and the identification of the characteristics of railways verges that contribute to the natural dispersal and establishment of invaders over native species. Such knowledge remains vital for assessing the relative importance of railways systems in biological invasions and to helping prioritize which measures to take in order to reduce the human dispersal of unwanted species.

Acknowledgements Fernando Ascensão acknowledges financial support by the Infraestruturas de Portugal Biodiversity Chair—CIBIO—Research Center in Biodiversity and Genetic Resources. César Capinha acknowledges financial support from the Portuguese Foundation for Science and Technology (FCT/MCTES) and POPH/FSE (EC) trough grant SFRH/BPD/84422/2012.

References

Aikio, S., Duncan, R. P., & Hulme, P. E. (2010). Lag-phases in alien plant invasions: Separating the facts from the artefacts. *Oikos, 119,* 370–378.

Albrecht, H., Eder, E., Langbehn, T., & Tschiersch, C. (2011). The soil seed bank and its relationship to the established vegetation in urban wastelands. *Landscape and Urban Planning, 100,* 87–97.

Bellingham, P. J., Peltzer, D. A., & Walker, L. R. (2005). Contrasting impacts of a native and an invasive exotic shrub on flood-plain succession. *Journal of Vegetation Science, 16,* 135–142.

Berry, Z. C., Wevill, K., & Curran, T. J. (2011). The invasive weed *Lantana camara* increases fire risk in dry rainforest by altering fuel beds. *Weed Research, 51,* 525–533.

Blackmore, L. M., & Goulson, D. (2014). Evaluating the effectiveness of wildflower seed mixes for boosting floral diversity and bumblebee and hoverfly abundance in urban areas. *Insect Conservation and Diversity, 7,* 480–484.

Bonesi, L., & Palazon, S. (2007). The American mink in Europe: Status, impacts, and control. *Biological Conservation, 134,* 470–483.

Bradley, B., Blumenthal, D. M., Early, R., Grosholz, E. D., Lawler, J. J., Miller, L. P., et al. (2012). Global change, global trade, and the next wave of plant invasions. *Frontiers in Ecology and the Environment, 10,* 20–28.

Bradshaw, C. J. A., Leroy, B., Bellard, C., Roiz, D., Albert, C., Fournier, A., et al. (2016). Massive yet grossly underestimated global costs of invasive insects. *Nature Communications, 7,* 12986.

Broennimann, O., Mráz, P., Petitpierre, B., Guisan, A., & Müller-Schärer, H. (2014). Contrasting spatio-temporal climatic niche dynamics during the eastern and western invasions of spotted knapweed in North America. *Journal of Biogeography, 41,* 1126–1136.

Buckley, Y. M., & Catford, J. (2016). Does the biogeographic origin of species matter? Ecological effects of native and non-native species and the use of origin to guide management. *Journal of Ecology, 104,* 4–17.

Cameron, E. K., & Bayne, E. M. (2009). Road age and its importance in earthworm invasion of northern boreal forests. *Journal of Applied Ecology, 46,* 28–36.

Capinha, C., Essl, F., Seebens, H., Moser, D., & Pereira, H. M. (2015). The dispersal of alien species redefines biogeography in the Anthropocene. *Science, 348,* 1248–1251.

Capinha, C., Larson, E. R., Tricarico, E., Olden, J. D., & Gherardi, F. (2013). Effects of climate change, invasive species, and disease on the distribution of native European crayfishes. *Conservation Biology, 27,* 731–740.

Capinha, C., Rocha, J., & Sousa, C. A. (2014). Macroclimate determines the global range limit of Aedes aegypti. *EcoHealth, 11,* 420–428.

Catford, J. A., Daehler, C. C., Murphy, H. T., Sheppard, A. W., Hardesty, B. D., Westcott, D., et al. (2012). The intermediate disturbance hypothesis and plant invasions: Implications for species richness and management. *Perspectives in Plant Ecology, Evolution and Systematics, 14,* 231–241.

Christen, D., & Matlack, G. (2006). The role of roadsides in plant invasions: A demographic approach. *Conservation Biology, 20,* 385–391.

Christian, J. M., & Wilson, S. D. (1999). Long-term ecosystem impacts of an introduced grass in the northern great plains. *Ecology, 80,* 2397–2407.

CIA (2016). *The World Factbook—railways.* https://www.cia.gov/library/publications/the-world-factbook/rankorder/2121rank.html

Clark, G. F., & Johnston, E. L. (2011). Temporal change in the diversity-invasibility relationship in the presence of a disturbance regime. *Ecology Letters, 14,* 52–57.

Consuegra, S., Phillips, N., Gajardo, G., & de Leaniz, C. G. (2011). Winning the invasion roulette: Escapes from fish farms increase admixture and facilitate establishment of non-native rainbow trout. *Evolutionary Applications, 4,* 660–671.

Craig, J. F. (1992). Human-induced changes in the composition of fish communities in the African Great Lakes. *Reviews in Fish Biology and Fisheries, 2,* 93–124.

Davis, M. A. (2009). *Invasion Biology.* New York: Oxford University Press.

Dewine, J. M., & Cooper, D. J. (2008). Canopy shade and the successional replacement of tamarisk by native box elder. *Journal of Applied Ecology, 45,* 505–514.

Di Castri, F. (1989). History of biological invasions with special emphasis on the Old World. In J. A. Drake, H. A. Mooney, F. Di Castri, R. H. Groves, F. J. Krüger, M. Rejmanek, & M. Williamson (Eds.), *Biological Invasions: A Global Perspective* (pp. 1–30). Chichester: Wiley.

Drake, J. M., & Lodge, D. M. (2004). Global hot spots of biological invasions: evaluating options for ballast-water management. *Proceedings of the Royal Society B: Biological Sciences, 271,* 575–580.

Elliott, P., Cantab, M. A., Aldridge, D. C., & Moggrldge, P. G. D. (2005). The increasing effects of zebra mussels on water installations in England. *Water and Environment Journal, 19,* 367–375.

Elton, C. S. (1958). *The ecology of invasions by animals and plants.* Chicago: University of Chicago Press.

Ernst, W. H. O. (1998). Invasion, dispersal and ecology of the South African neophyte *Senecio inaequidens* in The Netherlands: From wool alien to railway and road alien. *Acta Botanica Neerlandica, 47,* 131–151.

Essl, F., Dullinger, S., Rabitsch, W., Hulme, P. E., Hülber, K., Jarošík, V., et al. (2011). Socioeconomic legacy yields an invasion debt. *Proceedings of the National Academy of Sciences of the United States of America, 108,* 203–207.

EU. (2016). *Natura 2000 barometer.* http://ec.europa.eu/environment/nature/natura2000/barometer/index_en.htm

Fagan, W. F., Lewis, M. A., Neubert, M. G., & Van Den Driessche, P. (2002). Invasion theory and biological control. *Ecology Letters, 5,* 148–157.

Gelbard, J. L., & Belnap, J. (2003). Roads as conduits for exotic plant invasions in a semiarid landscape. *Conservation Biology, 17,* 420–432.

Hansen, M. J., & Clevenger, A. P. (2005). The influence of disturbance and habitat on the presence of non-native plant species along transport corridors. *Biological Conservation, 125,* 249–259.

Harris, S. A. (2002). Introduction of Oxford ragwort, *Senecio squalidus* L. (Asteraceae), to the United Kingdom. *Watsonia, 24,* 31–43.

Heger, T., & Böhmer, H. J. (2005). The invasion of Central Europe by *Senecio inaequidens* DC— A complex biogeographical problem. *Erdkunde, 59,* 34–49.

Heger, T., & Böhmer, H. (2006). *NOBANIS—Invasive alien species fact sheet—Senecio inaequidens.* Online database of the European Network on invasive alien species— NOBANIS. url:www.nobanis.org

Hodkinson, D. J., & Thompson, K. (1997). Plant dispersal: The role of man. *Journal of Applied Ecology, 34,* 1484–1496.

Hofmann, J. E. (2009). Records of nine-banded armadillos, *Dasypus novemcinctus,* in Illinois. *Transactions of the Illinois State Academy of Science, 102,* 95–106.

Hulme, P. E. (2009). Trade, transport and trouble: Managing invasive species pathways in an era of globalization. *Journal of Applied Ecology, 46,* 10–18.

Hulme, P. E., Bacher, S., Kenis, M., Klotz, S., Kühn, I., Minchin, D., et al. (2008). Grasping at the routes of biological invasions: A framework for integrating pathways into policy. *Journal of Applied Ecology, 45,* 403–414.

Humair, F., Humair, L., Kuhn, F., & Kueffer, C. (2015). E-commerce trade in invasive plants. *Conservation Biology: The Journal of the Society for Conservation Biology, 0,* 1–8.

Jodoin, Y., Lavoie, C., Villeneuve, P., Theriault, M., Beaulieu, J., & Belzile, F. (2008). Highways as corridors and habitats for the invasive common reed *Phragmites australis* in Quebec, Canada. *Journal of Applied Ecology, 45,* 459–466.

Keller, R. P., Geist, J., Jeschke, J. M., & Kühn, I. (2011). Invasive species in Europe: Ecology, status, and policy. *Environmental Sciences Europe, 23,* 23.

Keller, R. P., Lodge, D. M., & Finnoff, D. C. (2007). Risk assessment for invasive species produces net bioeconomic benefits. *Proceedings of the National Academy of Sciences of the United States of America, 104,* 203–207.

Li, B., Wang, Y., & Zhang, M. (2007). Guard against invasion of *Rattus norvegicus* into Tibet along Qinghai-Tibet railway. *Research of Agricultural Modernization, 3,* 24.

Lockwood, J. L., Cassey, P., & Blackburn, T. M. (2009). The more you introduce the more you get: The role of colonization pressure and propagule pressure in invasion ecology. *Diversity and Distributions, 15,* 904–910.

Lounibos, L. (2010). Human disease vectors. In D. Simberloff & M. Rejmánek (Eds.), *Encyclopedia of Biological Invasions.* Berkeley: University of California Press.

Marbuah, G., Gren, I.-M., & McKie, B. (2014). Economics of harmful invasive species: A review. *Diversity, 6,* 500–523.

McAvoy, T. J., Snyder, A. L., Johnson, N., Salom, S. M., & Kok, L. T. (2012). Road survey of the invasive tree-of-heaven (*Ailanthus altissima*) in Virginia. *Invasive Plant Science and Management, 5,* 506–512.

Moody, M. E., & Mack, R. N. (1988). Controlling the spread of plant invasions: The importance of nascent foci. *Journal of Applied Ecology, 25,* 1009–1021.

Nentwig, W., & Kobelt, M. (2010). Spiders (Araneae). In *BioRisk 4: Alien Terrestrial Arthropods of Europe* (Vol. 4, pp. 131–147). Pensoft Publishers.

Noma, T., Colunga-Garcia, M., Brewer, M., Landis, J., Gooch, A., & Philip, M. (2010). *Carthusian snail Monacha cartusiana. Michigan State University's invasive species factsheets.* http://www.ipm.msu.edu/uploads/files/forecasting_invasion_risks/carthusiansnail.pdf

Parendes, L. A., & Jones, J. A. (2000). Role of light availability and dispersal in exotic plant invasion along roads and streams in the H.J. Andrews experimental forest, Oregon. *Conservation Biology, 14,* 64–75.

Peltanová, A., Petrusek, A., Kment, P., & Juřičková, L. (2011). A fast snail's pace: Colonization of Central Europe by Mediterranean gastropods. *Biological Invasions, 14,* 759–764.

Peltzer, D. A., Bellingham, P. J., Kurokawa, H., Walker, L. R., Wardle, D. A., & Yeates, G. W. (2009). Punching above their weight: Low-biomass non-native plant species alter soil properties during primary succession. *Oikos, 118,* 1001–1014.

Penone, C., Machon, N., Julliard, R., & Le Viol, I. (2012). Do railway edges provide functional connectivity for plant communities in an urban context? *Biological Conservation, 148,* 126–133.

Pimentel, D., Zuniga, R., & Morrison, D. (2005). Update on the environmental and economic costs associated with alien-invasive species in the United States. *Ecological Economics, 52,* 273–288.

Régnier, C., Fontaine, B., & Bouchet, P. (2009). Not knowing, not recording, not listing: Numerous unnoticed mollusk extinctions. *Conservation Biology, 23,* 1214–1221.

Ricciardi, A. (2007). Are modern biological invasions an unprecedented form of global change? *Conservation Biology, 21,* 329–336.

Scott, N. E., & Davison, A. W. (1982). De-icing salt and the invasion of road verges by maritime plants. *Watsonia, 14,* 41–52.

Seebens, H., Schwartz, N., Schupp, P. J., & Blasius, B. (2016). Predicting the spread of marine species introduced by global shipping. *Proceedings of the National Academy of Sciences, 113,* 5646–5651.

Sheley, R. L., Jacobs, J. S., & Carpinelli, M. F. (1998). Distribution, biology, and management of diffuse knapweed (*Centaurea diffusa*) and spotted knapweed (*Centaurea maculosa*). *Weed Technology, 12,* 353–362.

Sheppard, A. W., Shaw, R. H., & Sforza, R. (2006). Top 20 environmental weeds for classical biological control in Europe: A review of opportunities, regulations and other barriers to adoption. *Weed Research, 46,* 93–117.

Simberloff, D., Martin, J. L., Genovesi, P., Maris, V., Wardle, D. A., Aronson, J., et al. (2013). Impacts of biological invasions: What's what and the way forward. *Trends in Ecology & Evolution, 28,* 58–66.

Suárez-Esteban, A., Fahrig, L., Delibes, M., & Fedriani, J. M. (2016). Can anthropogenic linear gaps increase plant abundance and diversity? *Landscape Ecology, 31,* 721–729.

Suominen, J. (1970). On *Elymus arenarius* (Gramineae) and its spread in Finnish inland areas. *Annales Botanici Fennici, 7,* 143–156.

Tikka, P. M., Högmander, H., & Koski, P. S. (2001). Road and railway verges serve as dispersal corridors for grassland plants. *Landscape Ecology, 16,* 659–666.

Vanderploeg, H. A., Liebig, J. R., Carmichael, W. W., Agy, M. A., Johengen, T. H., Fahnenstiel, G. L., et al. (2001). Zebra mussel (*Dreissena polymorpha*) selective filtration promoted toxic microcystis blooms in Saginaw Bay (Lake Huron) and Lake Erie. *Canadian Journal of Fisheries and Aquatic Sciences, 58,* 1208–1221.

Vettraino, A. M., Morel, O., Perlerou, C., Robin, C., Diamandis, S., & Vannini, A. (2005). Occurrence and distribution of Phytophthora species in European chestnut stands, and their association with ink disease and crown decline. *European Journal of Plant Pathology, 111,* 169–180.

Vicente, J., Randin, C. F., Gonçalves, J., Metzger, M. J., Lomba, Â., Honrado, J., et al. (2011). Where will conflicts between alien and rare species occur after climate and land-use change? A test with a novel combined modelling approach. *Biological Invasions, 13,* 1209–1227.

Vidal-Figueroa, T., & Delibes, M. (1987). Primeros datos sobre el visón americano (*Mustela vison*) en el Suroeste de Galicia y Noroeste de Portugal. *Ecologia, 1,* 145–152.

Watkins, R. Z., Chen, J., Pickens, J., & Brosofske, K. D. (2003). Effects of forest roads on understory plants in a managed hardwood landscape. *Conservation Biology, 17,* 411–419.

White, T. C. R. (1973). The establishment, spread and host range of *Paropsis charybdis* Stal. *Pacific Insects, 15,* 59–66.

Wilcox, D. A. (1989). Migration and control of purple loosestrife (*Lythrum salicaria* L.) along highway corridors. *Environmental Management, 13,* 365–370.

Wiles, G. J., Bart, J., Beck, R. E., & Aguon, C. F. (2003). Impacts of the brown tree snake: Patterns of decline and species persistence in Guam's avifauna. *Conservation Biology, 17,* 1350–1360.

Williams, P. A., & Newfield, M. (2002). *A weed risk assessment system for new conservation weeds in New Zealand.* Wellington, New Zealand: Department of Conservation.

Wilsey, B. J., Teaschner, T. B., Daneshgar, P. P., Isbell, F. I., & Polley, H. W. (2009). Biodiversity maintenance mechanisms differ between native and novel exotic-dominated communities. *Ecology Letters, 12,* 432–442.

Wilson, J. R. U., Dormontt, E. E., Prentis, P. J., Lowe, A. J., & Richardson, D. M. (2009). Something in the way you move: Dispersal pathways affect invasion success. *Trends in Ecology & Evolution, 24,* 136–144.

With, K. A. (2002). The landscape ecology of invasive spread. *Conservation Biology, 16,* 1192–1203.

Chapter 6
Railway Disturbances on Wildlife: Types, Effects, and Mitigation Measures

Priscila Silva Lucas, Ramon Gomes de Carvalho and Clara Grilo

Abstract In this chapter, we review the level of disturbance caused by railways due to noise and vibration, air, soil and water pollution, and soil erosion. There is evidence that soil and hydrology contamination may affect vegetation and aquatic fauna while noise can affect terrestrial vertebrates. In fact, noise, light, and vibration due to railways have been observed to reduce the abundance and richness of some insects, amphibians, and birds, and to cause avoidance behaviour on predators. Interestingly, reptiles, some bird species, small mammals, and large mammals seem to ignore rail traffic and benefit from the vegetation planted in the railway verges that provide food and shelter. Some engineering structures have been implemented to reduce the effects of railway disturbance: rail fastenings, rail dampers, under-sleeper pads, and noise barriers are applied to minimize noise and vibration; washing with water and cleaning the ballast are used to mitigate soil pollution; and grass plantation, the use of gypsum and application of compost/mulch coverage, are applied to control soil erosion.

Keywords Noise and vibration · Air and soil pollution · Water pollution · Soil erosion · Terrestrial vertebrates · Species richness · Avoidance effect

Introduction

The construction and operation of railways implies changes in the surrounded landscape that alter the microclimate, soil, and hydrological dynamics, contributing to the degradation of the natural habitat for many species (Forman and Alexander 1998; Eigenbrod et al. 2009). During operation, the main disturbances caused by railways are air, soil and water pollution, as well as noise and vibration, which

P.S. Lucas (✉) · R.G. de Carvalho · C. Grilo
Centro Brasileiro de Estudos em Ecologia de Estradas,
Universidade Federal de Lavras, Lavras 37200-000, Brazil
e-mail: prilucass@gmail.com

C. Grilo
Setor Ecologia, Departamento Biologia, Universidade Federal
de Lavras, Lavras 37200-000, Brazil

© The Author(s) 2017
L. Borda-de-Água et al. (eds.), *Railway Ecology*,
DOI 10.1007/978-3-319-57496-7_6

81

may alter species richness and species abundance (e.g. Penone et al. 2012; Clauzel et al. 2013). In this chapter, we review these disturbance sources, their effects on wildlife and the main actions to avoid or minimize their effects.

Main Railway Disturbances: Noise and Vibration, and Air, Soil and Water Pollution

The two most known disturbances of railways are the noise and vibrations caused by passing trains. However, railways are also responsible for a large amount of emissions that cover a wide range of pollutants and toxic substances that affect the atmosphere, soil and water worldwide (Plakhotnik et al. 2005). Another impact resulting from the construction and establishment of the railways is soil erosion. Here, we describe the main railway disturbances that can potentially affect wildlife.

Noise and Vibration

Railway noise pollution can be either from airborne sound or from vibration-induced as a result of rail traffic (Czop and Mendrok 2011; Palacin et al. 2014). The main source of railway noise comes from freight wagons, followed by high-speed trains and inner-urban railways (Guarinoni et al. 2012). However, locomotives passing and accelerating, freight wagons braking, vibrations from rail corrugation, and out-of-round wheels or vehicle coupling in shunting yards, are other sources of noise (Clausen et al. 2010). Noise levels vary, depending on the landscape and weather; open and flat areas allow noise to travel further than forest or mountains areas. In mountainou areas, the effect of noise is greater within valleys, when their width is less than the height of their walls, reducing the attenuating effect of noise (Chiocchia et al. 2010). Frost can make the ground hard and impede sound absorption, but fog prevents noise from dissipating.

Noise above 55 dB(A), where dB(A) is a measure that attempts to correct the way the human ear perceives loudness, is considered noise pollution for humans, and the sound values in the range 65–75 dB(A) cause stress to the body, leading to arterial hypertension (high blood pressure), cardiovascular disease, and heart attacks (Berglund et al. 1999). In Canada, the sound level of a passing train reaches values up to 85 dB(A), but between trains the sound levels drop to 43–53 dB(A) (CTA 2015). Measurement campaigns on high speed trains in several European countries over 10 years revealed sound values ranging from 85.5 to 97 dB(A) when the train speed was between 250 and 350 km/h (Gautier et al. 2008). In Japan, Matsumoto et al. (2005) compared the noise as a function of distance and observed a high noise level of 64 dB at 200 m from a railway in the countryside, a value similar to that near residential areas (65.7 dB). In fact, in Japan, noise can still reach 72 dB at 50 m from the track, i.e., higher than the Japanese permissible standards of 70 dB(A) (Kanda et al. 2007).

Air Pollution and Emission

The emission of gases from traffic constitutes an important source of environmental pollution all over the world (Hofman et al. 2014). These emissions depend mainly on the type of transport and fuel. Potential sources of contaminants associated with railways include diesel exhaust, and the abrasion of brakes, wheels, and rails, as well as dust from the transport of minerals and treated railway ties (Levengood et al. 2015).

The main pollutants emitted from the diesel-powered locomotives are carbon dioxide (CO_2), methane (CH_4), carbon monoxide (CO), nitrogen oxides (NO_x), nitrous oxide (N_2O), sulphur dioxide (SO_2), non-methane volatile organic compounds (NMVOC), particulate matter (PM) and hydrocarbon (HC) (Plakhotnik et al. 2005; Cheng and Yan 2011). Some studies reported higher levels of PM_{10} (where the subscript indicates the largest diameter of the particles in microns) and $PM_{2.5}$ near railways, higher than the standard level allowed (Beychok 2011) for the USA, Europe, and Asia (Park and Ha 2008; Kamani et al. 2014).

A growing number of monitoring studies have used bioindicator plant species as surrogates of air pollution across railways (e.g., Rani et al. 2006; Hofman et al. 2014). For example, Rani et al. (2006) studied the micromorphology of leaf parts of *Croton bonplandianum*, *Cannabis sativa* and *Calotropis procera* along a gradient of distances from the railway and concluded that the number of stomata and epidermal cells were lower near railways than at 4 kilometers away from the railway. However, no statistical tests were used to evaluate the correlation between railway distance and the number of stomata and epidermal cells. To the best of our knowledge, Rybak and Olejniczak (2013) authored the only published study that measured the accumulation of polycyclic aromatic hydrocarbons (PAH) in animal species. Using *Agelenids* spider webs to collect dust suspended in the air, they concluded that spiders are efficient indicators of PAHs in roads, but not in railway viaducts due to heavier traffic in the former.

Soil Pollution

With the increase in the human population and vehicles, emissions arising from transportation have become one of the most important sources of heavy metal, PAHs and herbicides in the soil (Malawaka and Wilkomirski 2001; Böjersson et al. 2004). Fuel combustion, vehicular and track material abrasion, and leaked cargo emit particles containing metals that are deposited in the soils, where they can remain for many years due to their low biodegradability (Zhang et al. 2012).

As most products of vehicle emissions are neither biologically nor chemically degraded, they can affect the growth of plants and ecosystems (Chen et al. 2014a, b). In fact, plants and soil organisms are the first recipients of the emission pollutants (Malawaka and Wilkomirski 2001). Ongoing research indicates that plant uptake varies among species. For example, leucaena (*Leucaena leucocephala*), annual meadow grass (*Poa annua*), and indigofera (*Indigofera amblyantha*) are known to have high translocation capacity for some metals, such as lead (Pb) and cadmium (Cd), but limited translocation of heavy metals from the roots to the aboveground tissues (Chen et al. 2014a, b), while others show the opposite trend (Ge et al. 2003).

High concentration levels of heavy metals are often found in the vicinity of railways. Wiłkomirski et al. (2011) reported the same concentration of molybdenum (Mo) close to and approximately 2 km from the railways in Poland. For instance, the concentration of nickel, cadmium, cobalt, and lead in the moss (*Pleurozium schreberi*) was the same 30 m from the railway as it was at points near the railways (Mazur et al. 2013). In addition, high concentrations of PAHs were found in the aerial parts of plant species near the railway and up to distances of 30 m from the railway (Malawaka and Wilkomirski 2001; Wiłkomirski et al. 2011). On railways, in particular, the biodegradation of PAHs and herbicides is extremely low and can persist over decades (Wilkomirski et al. 2012).

Water Pollution

Infrastructures associated with railways (e.g., leakages of petroleum products from fuel storage tanks) contribute, together with pollutants, to aquatic ecosystems (Schweinsberg et al. 1999; Vo et al. 2015). Levengood et al. (2015) documented high concentrations of PAHs and heavy metals in waterways bisected or bordered by railways. They showed that the PAH concentration was higher downstream than upstream of the railway (Levengood et al. 2015). They also found that phenanthrene and dibenzo (a, h) anthracene (a PAH element) concentrations at some sites represented a risk to aquatic life, whereas the chromium (Cr) values were still below the levels of concern for aquatic life (Levengood et al. 2015).

Herbicides and pesticides are other sources of water pollution. For herbicides, Schweinsberg et al. (1999) discovered that in Germany before the 1990s, a much higher total amount of these compounds were applied on railway tracks than in agriculture. Recently, Vo et al. (2015) showed that many herbicides applied during the operation of the railway are at concentrations that are lethal to most of the aquatic fauna, particularly fish populations; they indicate that compounds such as Imazapyr or Diuron concentrations can take 6 and 48 months, respectively, to drop below 50% of their original levels.

Soil Erosion and Changes in Hydrology

The abrupt change of soil required to establish the railway embankment leads to vegetation loss, compresses the soil, and compromises water drainage (Ferrell and

Lautala 2010). Thus, soil becomes exposed and subject to an increasing runoff that promotes its erosion (Chen et al. 2015). The erosion of rail embankments can result in a washing out of sediments (Jin et al. 2008) that cause water pollution. Furthermore, Gregorich et al. (1998) noted that soil erosion and deposition alter the biological process of carbon mineralization in soil landscapes, which affects the soil quality and, hence, the vegetation.

Railway construction parallel to streams can result in hydrological disconnections (Pennington et al. 2010) that dry the soils (Blanton and Marcus 2009). Such disconnections can have a significant impact on the ecological function of riparian landscapes by negatively affecting floodplain evolution, riparian ecosystem processes, and associated biodiversity (Snyder et al. 2002). Although urban riparian areas could harbor great diversity of native canopy and shrub species, the richness of native canopy species was lower near railways (Snyder et al. 2002).

Effects of Railway Disturbance on Wildlife

There is evidence that disturbance from noise, lights, and vibrations associated with the construction and operation of the railway affect some species, and this can occur at various life cycles (van Rooyen 2009; Wiacek et al. 2015). In contrast, other studies suggest that wildlife ignores or adapts to railway disturbances (Ghosh et al. 2010; Mundahl et al. 2013). As observed for roads, the severity of railway disturbance depends on the species' bio-ecological features and on the degree of the disturbance (e.g., Rytwinski and Fahrig 2012). However, little is known about their role in species viability. We now review the main findings on wildlife behaviour responses to railway disturbance.

Insects

To the best of our knowledge, only the study by Penone et al. (2012) assessed the impact of railways on insects. They studied Tettigoniidae (bush crickets) at railway edges in an urban context (Penone et al. 2012), and found that species richness and abundance was explained by a quadratic effect of train speed (described by a hump-backed curve). This is especially true of mobile species at larger scales than on sedentary species. However, vegetated railway verges had a significant positive effect on most species, suggesting that suitable habitats with appropriate management can overcome the negative effects of disturbance on larger scales (Penone et al. 2012).

Herpetofauna

Among vertebrates, amphibians are one of the groups most affected by linear infrastructures (e.g., Fahrig and Rytwinski 2009). Their dependence on aquatic habitats and seasonal migrations make amphibians particularly vulnerable to the negative impacts of these structures (Hels and Buchwald 2001; Hamer and McDonnell 2008). However, to the best of our knowledge no studies have been carried out on the impact of railway disturbances on amphibians.

In contrast, there have been several studies that show that reptiles ignore railway disturbances. For example, in Belgium, all autochthonous reptile species have colonized the railway embankments, but their presence and abundance largely depend on the region and on the kind of railway (Graitson 2006). Railway embankments can provide important novel habitats for reptiles, mainly in highly human-altered landscapes (EN 2004). Some active lines in sunny areas of large valleys and some large switchyards, as well as unused railways not dismounted, had a particularly high richness of reptiles (Graitson 2006). In fact, those railways may have contributed to the local dispersal of at least five species of reptiles: slow worm (*Anguis fragilis*), sand lizard (*Lacerta agilis*), common wall lizards (*Podarcis muralis*), viviparous lizard (*Zootoca vivipara*), and smooth snake (*Coronella austriaca*) (Graitson 2006). Likewise, the rainbow lizard (*Cnemidophorus lemniscatus*) was found in the weeds growing along the Florida East Coast Railway line occupied by industrial buildings (Wilson and Porras 1983). In Switzerland, a relatively great abundance of reptile species can be found in the vicinity of the railways, especially sand lizards (Stoll 2013). However, the abundance of reptiles in the railway banks can be affected by two key factors: the inter-specific competition with the common wall lizard *Podarcis muralis*, and predation through domestic cats *Felis catus* (Stoll 2013).

Birds

There have been several studies that show that railway disturbance affects bird richness, abundance, and behaviour. Noise can affect acoustic communication among bird species that use calls and songs to attract and bond with mates, defend territories from rivals, maintain contact with social groups, beg for food, and warn of danger from approaching predators (Collins 2004; Marler 2004). Noise disturbance was shown to alter the behaviour of many bird species, especially for breeding birds, when territories are being defended, and during incubation (Reijnen et al. 1995). This disturbance can cause birds to accelerate hatching, abandon their occupied territories, nests and broods, and lead to hearing loss (Hanson 2007).

For example, noise emission from railway traffic has a negative effect on the density of all meadow birds in the Netherlands (Waterman et al. 2002). The threshold noise level from which densities were affected was around 42–49 dB(A)

for the black-tailed godwit (*Limosa limosa*), skylark (*Alauda arvensis*), and garganey (*Anas querquedula*) (Waterman et al. 2002). There was also evidence that
successful bald eagle (*Haliaeetus leucocephalus*) nests were farther away from
highways and railways than unsuccessful ones (Mundahl et al. 2013).

Certain bird species seem to ignore railway disturbances. For example, brants
(*Branta* asp.) seem to disregard the trains passing 50 m away (Owens 1977).
Likewise, although Waterman et al. (2002) found a negative effect on the density of
birds on railways in general, they did not find differences in bird density between
quiet and busy railways. Indeed, there is evidence that bird species tolerate the
railway disturbance due to the attractiveness of numerous railway-related features.
Species whose predators show negative responses to train disturbance may also
benefit from the railway vicinity (Rytwinski and Fahrig 2012). Railway verges may
create edge effects that can increase biodiversity. In fact, special microclimates—
thanks to different temperatures or insolation and new habitat availability along the
edges—can enhance habitat heterogeneity in homogeneous landscapes (Delgado
et al. 2007). In turn, these new habitats may improve resting and foraging opportunities for some bird species (Morelli et al. 2014). Also, railways can be a useful
source for gastroliths' digestive purposes, and a source of sand bathing locations
used by birds to clean the feathers. Ghosh et al. (2010) showed that apparently due to
the food availability, house sparrows (*Passer domesticus*) adapted to loud noise
(between 35 and 95 dB), being undisturbed by passing trains at the railway station
study site. Equally, Li et al. (2010) observed the abundance and richness of seven
ground-dwelling bird species, namely the Tibetan ground tit (*Pseudopodoces
humilis*), Tibetan lark (*Melanocorypha maxima*), horned lark (*Eremophila alpestris*), white-winged snowfinch (*Montifringillla nivalis*), plain-backed snowfinch (*M.
blandfordi*), white-rumped snowfinch (*M. taczanowskii*), and rufous-necked snowfinch (*M. ruficollis*). They found greater numbers of individuals of these species near
the Qinghai-Tibet railway (less than 300 m) than farther away (from 300 to
1,200 m), probably due to verges provide nesting sites and foraging opportunities.

The non-continuous nature of the noise near railways due to the intermittent flow of
train traffic may constitute the main reason that some birds ignore the railway. This was
hypothesized by Wiacek et al. (2015) to explain the greater abundance of the breeding
community of woodland birds near a busy railway line in Poland when they assessed the
effect of noise at three different distances from the track (30, 280, and 530 m). Species
with low frequency calls, such as the wood pigeon (*Columba palumbus*) and common
cuckoo (*Cuculus canorus*), also occurred in large numbers near railways. Other bird
species also seem to ignore the railway disturbance. For example, more than 90% of bald
eagle nests were built near human infrastructures, including railways that had more than
1,000 railcars going by each day (Mundahl et al. 2013).

Mammals

Railway disturbance impacts differ among mammal species. Besides the higher intensity of light and noise (it can achieve 120 dB), small mammals can be found near the railways if there is no grazing disturbance (Qian et al. 2009), or if the unnatural nature of rail bed and tracks is not an obstacle to their movements (van der Grift 1999). In fact, the richness of small mammal species was high at the railway verge in the Atlantic forest in southern Brazil (Cerboncini 2012), probably because the noise from the trains can force predators out of the area, which can favor small mammals' appearance (Cerboncini 2012). Similarly, railways may have a positive role in maintaining the common bat populations in highly humanized landscapes, such as intensive agriculture. In this case, railway verges seem to be used as shelters for bat species despite the traffic noise; only *Myotis* sp. foraging behaviour was negatively affected by railway verges (Vandevelde et al. 2014).

Although the response of large mammals to the effects of railway disturbance varies among species, they seem to ignore railway disturbances. However, there is evidence that the presence of railways had a subtle effect on some species' behaviour. For example, railways had little effect on the distance or direction of fox dispersal movements (Trewhella and Harris 1990); however, the railway disturbance may have influenced the fox movements within their territories (Trewhella and Harris 1990). Similarly, the Canadian Pacific Railway in Banff National Park seems to redirect wolf movements (individuals follow the railway), particularly when the snow is deep (Paquet and Callaghan 1996), while it seems to define the boundary of bears' home ranges (Kaczensky et al. 2003). As detected for other species groups, changes in the railway verges can attract large mammals. In Banff National Park, railway verges attracted black and grizzly bears (*Ursus americanus* and *U. arctos,* respectively*)* due to the berry-producing areas within the verges (Gibeau and Herrero 1998). Grain spills along the Canadian Pacific rail line also attract bears (Gibeau and Herrero 1998), while food spills seem to increase the abundance of mice, which may attract their predators, such as the coyote (Wells 1996).

By contrast, railways seem to be avoided by two large ungulates: Mongolian gazelles (*Procopra gutturosa*) and Tibetan antelopes (*Pantholops hodgsoni*). No observations of Mongolian gazelle crossings during dispersal were detected, and most of the individuals were found 300 m from the railway (Ito et al. 2005). Likewise, there were no Tibetan antelope crossings (Xia et al. 2007) because these antelopes hesitated to cross the railway, most likely because of the slope of the rail bed.

Mitigation Measures to Reduce Railway Disturbances

There are several measures to reduce the main negative effects of railway distur-bances (Schulte-Werning et al. 2008; Maeda et al. 2012; Nielsen et al. 2015). Although some of these measures are known to effectively reduce the major negative effects on biodiversity, the effectiveness has not yet been evaluated for a few other measures. Here, we describe the commonly used measures to reduce or minimize the impacts of railway disturbance for noise and vibration effects, and soil pollution and erosion that are found in the literature.

Noise and Vibration

In general, there are two approaches to minimizing rail traffic noise and vibration: (1) reduce the noise at the source, and (2) reduce its propagation (Lakušić and Ahac 2012; Tiwari et al. 2013). The reduction of noise levels can be achieved by decreasing the speed of rail vehicles and reducing the intensity of the radiated sound through regular maintenance in order to keep smooth rails and wheels (Lakušić and Ahac 2012; Schulte-Werning et al. 2012). Currently, rail fastenings, rail dampers, under-sleeper pads, and noise barriers (Fig. 6.1) are the most commonly used techniques to directly reduce the noise and vibration in railways, techniques that are also used in roads and motorways. The increased flexibility of railway elements (e.g., fastenings, sleepers, ballasts), have increased their ability to absorb vibrations generated at wheel-rail interface (Lakušić and Ahac 2012). The use of resilient rail fastening systems are designed to decrease the low-frequency ground borne or in structures with noise above 30 Hz and can be effective in reducing wayside noise

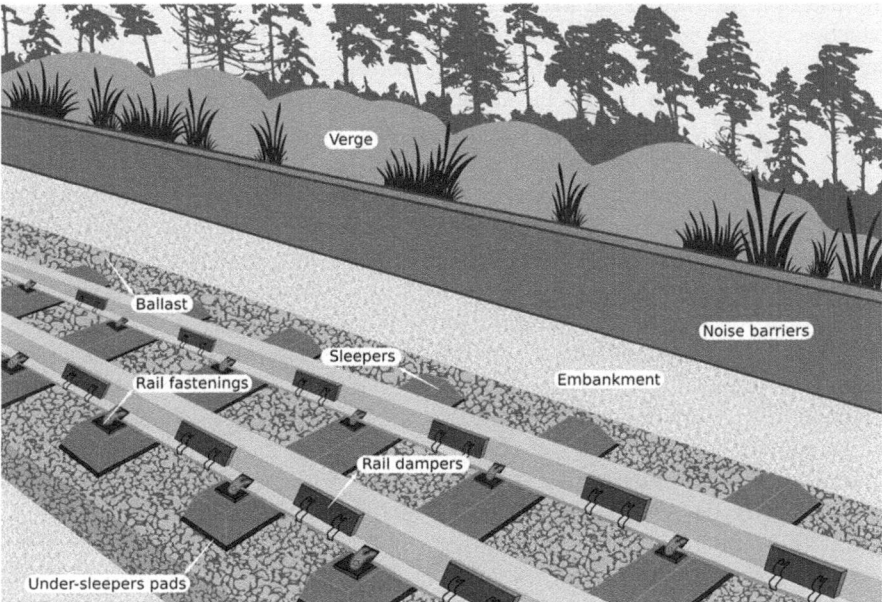

Fig. 6.1 Schematic drawing of a railway showing some measures to minimize noise and vibration: rail fastenings; rail dampers; under-sleepers pads; and noise barriers (not at scale)

radiated from steel. This structure provides noise reduction from 3 to 6 dB(A) through the elastic elements supporting the rail, which prevents direct contact between the rail foot and sleeper (Nelson 1997). However, systems that use concrete ties with spring clips may not benefit from the use of resilient fasteners, because the spring clips already eliminate any looseness between the rail and the tie (Nelson 1997).

Another technique is the use of rail dampers (Fig. 6.1). This structure consists of steel components and elastomeric material that absorbs the energy of rail (springs) vibrations (Lakušić and Ahac 2012). The damping material reduces the displacement of the vibration waves along the rail, which results in the reduction of the rail noise. Dampers are known to be an efficient way to reduce noise emission in railway networks (Lakušić and Ahac 2012). Studies performed at the rail track sections with rail dampers installed showed a reduction in noise from 4 to 6 dB(A), and vibration up to 9 dB (Benton 2006; Koller et al. 2012).

Aside from rail noise, non-audible vibrations are also generated by the trains, transmitted via the tracks, and transferred to the soil. In surface railway lines, sleepers with elastic supports (under-sleeper pads—USPs) (Fig. 6.1) are an alternative with moderate costs (compared to floating slab track systems and ballast mats) that increase track quality and achieve a significant reduction in vibrations and railway noise (Schulte-Werning et al. 2012). USPs are resilient pads attached to the bottom surface of sleepers to provide an intermediate elastic layer between the sleeper and the ballast. The USPs are normally made of polyurethane elastomer with a foam structure that includes encapsulated air voids (Johansson et al. 2008). The use of USPs causes an average reduction of vibration of 16 dB due to the reduction of the contact between the sleeper and the ballast, which increases elasticity of the track (Lakušic et al. 2010; Lakušić and Ahac 2012).

Commonly used to isolate the noise on railways (and roads), noise barriers (Fig. 6.1) can be an important tool for minimizing the negative effects of sound on wildlife, especially for species that are extremely sensitive to it. These structures can be constructed from soil, wood, concrete, or metal (FHWA 2011; Morgan and Peeling 2012), or can be just the dense vegetation along the rails, which can, in some cases, form an almost perfect noise barrier (Tiwari et al. 2013; Bashir et al. 2015). Soil verges along roads and railways can reduce, on average, 3 dB more than vertical walls of the same height. However, the construction of soil verges can require a huge area, especially if they are very extensive and elevated. The construction of artificial walls requires less space, but they are usually limited to eight meters in height for structural and aesthetic reasons (FHWA 2011). Vertical walls can be applied, together with soil verges, in order to further reduce noise.

Studies by Van Renterghem and Botteldooren (2012) show that a soil verge can reduce, on average, noise levels of 11.1 dB, and walls can reduce, on average 7.7–8.3 dB. According to the U.S. Department of Transportation, effective noise barriers can typically reduce noise levels by 5–10 dB (FHWA 2011). Noise barriers have been mainly used to reduce the effect of noise on colonies of nesting birds (Bank et al. 2002). However, when applied without planning, noise barriers can cause various negative impacts on wildlife, such as the isolation of populations

(Bank et al. 2002). Thus, it is recommended to apply noise barriers together with wildlife passages to promote crossings between railway sides in order to maintain dispersal processes and ensure the long-term persistence of the species (Bank et al. 2002; Schulte-Werning et al. 2012). However, reducing the noise disturbance may reduce the ability of wildlife to perceive incoming trains, and consequently increase the risk of collision (see Chap. 2).

Soil Pollution

Cleaning the ballast using ethylenediaminetetraacetate (EDTA) with water or thermosetting plastic resin are techniques that allow the recycling of pollutant components in the surroundings of the railway network. Cleaning it can typically involve the use of products such as solvents or surfactants, which vary in efficiency and potential environmental impacts (Anderson et al. 2002). The ballast cleaning technology with aqueous solutions of disodium EDTA can ensure metal extraction without altering the stones' mechanical and physical characteristics, allowing reuse of the ballast (Di Palma and Petrucci 2015). The use of EDTA offers a high extraction efficiency of heavy metals (Di Palma et al. 2011); however, this chemical agent should be administered with caution because it has low biodegradability in the environment (Bucheli-Witschel and Egli 2001).

Using water to extract the ballast pollutants is very common in many countries (Di Palma et al. 2012). However, this method requires a great amount of surfactant water, and the disposal of the waste produced after cleaning must be treated appropriately (Cho et al. 2008). Incineration, landfilling, or recycling of the waste are the options available, with the choice depending on the solvent system and on the nature of the waste (Anderson et al. 2002). Dry cleaning that uses thermosetting plastic resin can successfully remove ballast pollutants in a very short time (Cho et al. 2008). In areas with high levels of contamination, the support material should be renovated regularly, including changing the ballast, and wood sleepers should be replaced by concrete (Wiłkomirski et al. 2011).

Soil Erosion

There are techniques commonly used to minimize the effects of soil erosion on railways and highways: grass plantation, use of gypsum, application of compost/mulch coverage, use of concrete prefabricated panes, lattice plots, and interception and drainage. There is evidence that as the grass cover level increases, the soil erosion rate decreases (Gyasi-Agyei et al. 2001). For example, buffel grass (*Cenchrus ciliaris*) is one of the preferred species for revegetation of railway embankments in Australia (Bhattarai et al. 2008); in some of the areas (Central Australia and Western Queensland), this grass spreads readily in soils with crumbly

Fig. 6.2 Spatial responses of wildlife to railway disturbances: the length of the *white bars* are proportional to the distances up to which the railway has an effect on a given taxon, while short and *black bars* correspond to species' occurrence in the railway verges

or soft surfaces, but it needs to be handled with care because it has the potential to become a weed (Cameron 2004). This species, in combination with Japanese millet (*Echinochloa esculenta*), has also been used successfully in controlling soil erosion in the semi-arid tropics of Central Queensland. Japanese millet develop rapidly in poor soils and can minimize the invasion of other unwanted weed species (Fox et al. 2010). However, buffel grass is an invasive weed species that poses serious threats to biodiversity in many other environments, such as in the Australian regions mentioned above (Marshall et al. 2012) (see Chap. 4).

The use of gypsum on calcium-deficient soil before seeding can reduce the erosion rate by 25% (Gyasi-Agyei et al. 2001). This material has advantages, such as low cost, availability, pH neutrality, and ease of handling (Gyasi-Agyei et al. 2001). Once applied to the soil, the gypsum will increase the porosity and water infiltration ability. This phenomenon will avoid water runoff, thus preventing soil erosion (Beckett et al. 1989). The use of this mineral also promotes root penetration and the emergence of seedlings (Chen and Dick 2011).

Compost/mulch coverage is considered to be one of the best management practices in both active construction and established areas prone to soil erosion on roadsides (Bakr et al. 2012), and it can be used in railway verges as well. In Louisiana, USA, Bakr et al. (2012) found that compost/mulch covers were highly effective in reducing runoff, total suspended solids (reduction between 70 and

74%), and turbidity from soils susceptible to high-intensity storms. A study performed by Persyn et al. (2004) showed that the use of various compounds applied to the soil might also reduce erosion rates. All compost treatments (biosolids, yard waste, and bio-industrial by-products) were effective at reducing runoff and inter-rill erosion rates under the conditions simulated in the study conducted on a highway embankment after construction in central Iowa (Persyn et al. 2004).

There are also engineering measures, such as concrete prefabricated panes, geocells (three-dimensional honeycomb-like structures to reduce runoff and sediment transport), or interception and drainage that can also reduce the soil loss along highways and railways (e.g., Xu et al. 2006). These measures were effective in the short term by decreasing between 40 and 60% of the runoff that had effectively reduced soil loss from sideslopes of highways in the Tibetan Plateau (Xu et al. 2006). However, this measure needs to be applied with some caution, since weak construction may lead to water leaking from improperly sealed cracks or holes, loose contact parts, poorly rammed structures, or runoff converging along a flow line, causing adverse effects on the road embankment and in the surrounding environment (Xu et al. 2006).

Conclusions

There have been several studies that quantified the level of various disturbances and their impacts on wildlife populations. Plants are, in general, the first to be described in relation to air and soil contamination, perhaps due to the facility of realizing experimental analysis, or of using plants as bio-indicators. Nevertheless, there is still little scant knowledge about the mechanisms and processes underlying the behaviour of vertebrates, because most studies concerned only a small number of species (van der Grift 1999). There is strong evidence that noise, light, and vibrations that can reach from 85.5 to 97 dB(A), can affect insects, amphibians and birds (Fig. 6.2). In contrast, the availability of food and vegetation in the railway verges seems to overcome the noise pollution and seems to attract reptiles, some bird species, and several mammals (Fig. 6.2). There is a wide diversity of measures to minimize railway disturbances, but further studies are needed to understand the effectiveness of these measures in reducing railway disturbance of wildlife.

Acknowledgments This research for this chapter was conducted within the framework of the project "Road macroecology: analysis tools to assess impacts on biodiversity and landscape structure," that was funded by CNPq in Brazil (N° 401171/2014-0). CG was supported by the Grant (300021/2015-1).

References

Anderson, P., Cunningham, C. J., & Barry, D. A. (2002). Efficiency and potential environmental impacts of different cleaning agents used on contaminated railway ballast. *Land Contamination & Reclamation, 10,* 71–78.

Bakr, N., Weindorf, D. C., Zhu, Y., Arceneaux, A. E., & Selim, H. M. (2012). Evaluation of compost/mulch as highway embankment erosion control in Louisiana at the plot-scale. *Journal of Hydrology, 468,* 257–267.

Bank, F. G., Irwin, C. L., Evink, G. L., Gray, M. E., Hagood, S., Kinar, J. R., et al. (2002). *Wildlife habitat connectivity across European highways* (No. FHWA-PL-02-011).

Bashir, I., Taherzadeh, S., Shin, H. C., & Attenborough, K. (2015). Sound propagation over soft ground without and with crops and potential for surface transport noise attenuation. *The Journal of the Acoustical Society of America, 137,* 154–164.

Beckett, P. H. T., Bouma, J., Farina, M. P. W., Fey, M. V., Miller, W. P., Pavan, M. A., et al. (1989). *Advances in soil science* (p. 9). New York: Springer.

Benton, D. (2006). Engineering aspects of rail damper design and installation. Rail Noise 2006: Mitigazione del Rumore Ferroviario, GAA, Arpat news, Pisa, 179.

Berglund, B., Lindvall, T., & Schwela, D. H. (1999). *Guidelines for community noise.* World Health Organization. Retrieved February 3, 2016, from http://www.who.int/docstore/peh/noise/guidelines2.html

Beychok, M. (2011). *U. S. National ambient air quality standards.* Retrieved February 3, 2016, from http://www.eoearth.org/view/article/170853/

Bhattarai, S. P., Fox, J., & Gyasi-Agyei, Y. (2008). Enhancing buffel grass seed germination by acid treatment for rapid vegetation establishment on railway batters. *Journal of Arid Environments, 72,* 255–262.

Blanton, P., & Marcus, W. A. (2009). Railroads, roads and lateral disconnection in the river landscapes of the continental United States. *Geomorphology, 112,* 212–227.

Böjersson, E., Lennart, T., & Stenström, J. (2004). The fate of imazapyr in a Swedish railway embankment. *Pest Management Science, 60,* 544–549.

Bucheli-Witschel, M., & Egli, T. (2001). Environmental fate and microbial degradation of aminopolycarboxylic acids. *FEMS Microbiology Reviews, 25,* 69–106.

Cameron, A. G. (2004). *Buffel grass (A pasture grass for sandy soils)* (online). Northern Territory Government, Australia. Retrieved February 9, 2016, from http://www.nt.gov.au/d/Content/File/p/Pasture/283.pdf

Cerboncini, R. (2012). Respostas de pequenos mamíferos ao efeito de borda da ferrovia Paranaguá-Curitiba no parque estadual pico do marumbi, morretes. Dissertation. University of Paraná.

Chen, L., & Dick, W. A. (2011). *Gypsum as an agricultural amendment: General use guidelines.* Ohio State University Extension.

Chen, Z., Ai, Y., Fang, C., Wang, K., Li, W., Liu, S., et al. (2014b). Distribution and phytoavailability of heavy metal chemical fractions in artificial soil on rock cut slopes alongside railways. *Journal of Hazardous Materials, 273,* 165–173.

Chen, Z., Luo, R., Huang, Z., Tu, W., Chen, J., Li, W., et al. (2015). Effects of different backfill soils on artificial soil quality for cut slope revegetation: Soil structure, soil erosion, moisture retention and soil C stock. *Ecological Engineering, 83,* 5–12.

Chen, Z., Wang, K., Ai, Y., Li, W., Gao, H., & Fang, C. (2014a). The effects of railway transportation on the enrichment of heavy metals in the artificial soil on railway cut slopes. *Environmental Monitoring and Assessment, 186,* 1039–1049.

Cheng, Y. H., & Yan, J. W. (2011). Comparisons of particulate matter, CO, and CO_2 levels in underground and ground-level stations in the Taipei mass rapid transit system. *Atmospheric Environment, 45,* 4882–4891.

Chiocchia, G., Clerico, M., Salizzoni, P., & Marro, M. (2010). Impact assessment of a railway noise in an alpine valley. In *10ème Congrès Français d'Acoustique*.

Cho, Y., Park, D. S., Lee, J. Y., Yung, W. S., Kim, H. M., & Lim, J. I. (2008). Dry cleaning of railroad ballast gravels by blasting with thermosetting plastic resin powder. In *Proceedings of 8th World Congress on Railway Research, Coex, Seoul, Korea* (pp. 18–22).

Clausen, U., Doll, C., Franklin, F. J., Heinrichmeyer, H., Kochsiek, J., Rothengatter, W., et al. (2010). *Reducing railway noise pollution.* Directorate General for Internal Policies, Policy Department B: Structural and cohesion policies. Transport and Tourism, European Parliament.

Clauzel, C., Girardet, X., & Foltête, J. C. (2013). Impact assessment of a high-speed railway line on species distribution: Application to the European tree frog (*Hyla arborea*) in Franche-Comté. *Journal of Environmental Management, 127,* 125–134.

Collins, S. (2004). Vocal fighting and flirting: The functions of birdsong. In P. Marler & H. Slabbekoorn (Eds.), *Nature's music: The science of birdsong* (pp. 39–79). San Diego, California: Academic/Elsevier.

CTA. (2015). *Railway noise measurement and reporting methodology.* Retrieved February 3, 2016, from https://www.otc-cta.gc.ca/eng/railway_noise_measurement

Czop, P., & Mendrok, K. (2011). A new method for operational monitoring of railway tracks to reduce environmental noise. *Polish Journal of Environmental Studies, 20,* 311–316.

Delgado, J. D., Arroyo, N. L., Arevalo, J. R., & Fernandez-Palacios, J. (2007). Edge effect of roads on temperature, light, canopy cover, and canopy height in laurel and pine forests. *Landscape and Urban Planning, 81,* 328–340.

Di Palma, L., Gonzini, O., & Mecozzi, R. (2011). Use of different chelating agents for heavy metal extraction from contaminated harbour sediment. *Chemistry and Ecology, 27,* 97–106.

Di Palma, L., Mancini, D., & Petrucci, E. (2012). Experimental assessment of chromium mobilization from polluted soil by washing. *Chemical Engineering Transactions, 28,* 145–150.

Di Palma, L., & Petrucci, E. (2015). Treatment and recovery of contaminated railway ballast. *Turkish Journal of Engineering and Environmental Sciences, 38,* 248–255.

Eigenbrod, F., Hecnar, S. J., & Fahrig, L. (2009). Quantifying the road effect zone: Threshold effects of a motorway on anuran populations in Ontario, Canada. *Ecology and Society, 14,* 18–24.

EN. (2004). *Reptiles: Guidelines for developers.* English Nature. Northampton: Belmont Press.

Fahrig, L., & Rytwinski, T. (2009). Effects of roads on animal abundance: An empirical review and synthesis. *Ecology and Society, 14,* 21.

Ferrell, S. M., & Lautala, P. T. (2010). *Rail embankment stabilization on permafrost.* London: Global Experiences.

FHWA, Federal Highway Administration. (2011). *Highway traffic noise barriers at a glance.* Washington, DC: U. S. Department of Transportation.

Forman, R. T., & Alexander, L. E. (1998). Roads and their major ecological effects. *Annual Review of Ecology and Systematics,* 207-C2.

Fox, J. L., Bhattarai, S. P., & Gyasi-Agyei, Y. (2010). Evaluation of different seed mixtures for grass establishment to mitigate soil erosion on steep slopes of railway batters. *Journal of Irrigation and Drainage Engineering, 137,* 624–631.

Gautier, P. E., Poisson, F., & Letourneaux, F. (2008). High speed trains external noise: A review of measurements and source models for the TGV case up to 360 km/h. In *Proceedings of the 8th world congress on railway research* (pp. 1-1).

Ge, Y., Murray, P., Sauvé, S., & Hendershot, W. (2003). Low metal bioavailability in a contaminated urban site. *Environmental Toxicology and Chemistry, 21,* 954–961.

Gibeau, M. L., & Herrero, S. (1998). Roads, rails and grizzly bears in the Bow River Valley, Alberta. In G. L. Evink (Ed.), *Proceedings of the international conference on ecology and transportation* (pp. 104–108), Florida Dept. of Transportation, Tallahassee, Florida, USA.

Graitson, E. (2006). Répartition et écologie des reptiles sur le réseau ferroviaire en Wallonie. *Bulletin de la Societe Herpetologique, 120,* 15–32.

Gregorich, E. G., Greer, K. L., Anderson, D. W., & Liang, B. C. (1998). Carbon distribution and losses: Erosion and deposition effects. *Soil and Tillage Research, 47,* 291–302.

Ghosh, S., Kim, K., & Bhattacharya, R. (2010). A survey on house sparrow population decline at Bandel, West Bengal, India. *Journal Korean Earth Science Society, 31,* 448–453.

Guarinoni, M., Ganzleben, C., Murphy, E., & Jurkiewicz, K. (2012). *Towards a comprehensive noise strategy*. Brussels: European Union.

Gyasi-Agyei, Y., Sibley, J., & Ashwath, N. (2001). Quantitative evaluation of strategies for erosion control on a railway embankment batter. *Hydrological Processes, 15*, 3249–3268.

Hamer, A. J., & McDonnell, M. J. (2008). Amphibian ecology and conservation in the urbanising world: A review. *Biological Conservation, 141*, 2432–2449.

Hanson, C. E. (2007). High speed train noise effects on wildlife and domestic livestock. In B. S. Werning, D. Thompson, P. E. Gautier, C. Hanson, B. Hemsworth, J. Nelson, T. Maeda, & P. Vos (Eds.), *Noise and vibration mitigation for rail transportation systems* (pp. 26–32). New York: Springer.

Hels, T., & Buchwald, E. (2001). The effect of road kills on amphibian populations. *Biological Conservation, 99*, 331–340.

Hofman, J., Lefebvre, W., Janssen, S., Nackaerts, R., Nuyts, S., Mattheyses, L., et al. (2014). Increasing the spatial resolution of air quality assessments in urban areas: A comparison of biomagnetic monitoring and urban scale modelling. *Atmospheric Environment, 92*, 130–140.

Ito, T., Miura, N., Lhagvasuren, B., Enkhbileg, D., Takatsuki, S., Tsunekawa, A., et al. (2005). Preliminary evidence of a barrier effect of a railroad on the migration of Mongolian gazelles. *Conservation Biology, 19*, 945–948.

Jin, H. J., Yu, Q. H., Wang, S. L., & Lü, L. Z. (2008). Changes in permafrost environments along the Qinghai-Tibet engineering corridor induced by anthropogenic activities and climate warming. *Cold Regions Science and Technology, 53*, 317–333.

Johansson, A., Nielsen, J. C., Bolmsvik, R., Karlström, A., & Lundén, R. (2008). Under sleeper pads—Influence on dynamic train–track interaction. *Wear, 265*, 1479–1487.

Kaczensky, P., Knauer, F., Krze, B., Jonozovic, M., Adamic, M., & Gossow, H. (2003). The impact of high speed, high volume traffic axes on brown bears in Slovenia. *Biological Conservation, 111*, 191–204.

Kamani, H., Hoseini, M., Seyedsalehi, M., Mahdavi, Y., Jaafari, J., & Safari, G. H. (2014). Concentration and characterization of airbone particles in Tehran's subways system. *Environmental Science and Pollution Research, 21*, 7319–7328.

Kanda, H., Tsuda, H., Ichikawa, K., & Yoshida, S. (2007). Environmental noise reduction of Tokaido Shinkansen and future prospect. In B. S. Werning, D. Thompson, P. E. Gautier, C. Hanson, B. Hemsworth, J. Nelson, T. Maeda, & P. Vos (Eds.), *Noise and vibration mitigation for rail transportation systems* (pp. 1–8). Berlin, Heidelberg: Springer.

Koller, G., Kalivoda, M. T., Jaksch, M., Muncke, M., Oguchi, T., & Matsuda, Y. (2012). Railway noise reduction technology using a damping material. In T. Maeda, P. E. Gautier, C. Hanson, B. Hemsworth, J. T. Nelson, B. Schulte-Werning, D. Thompson, & P. Vos (Eds.), *Noise and vibration mitigation for rail transportation systems* (pp. 159–166). Tokyo: Springer.

Lakušić, S., & Ahac, M. (2012). Rail traffic noise and vibration mitigation measures in urban areas. *Technical Gazette, 19*, 427–435.

Lakušić, S., Ahac, M., & Haladin, I. (2010). Experimental investigation of railway track with under sleeper pad. In: *10th Slovenian road and transportation congress* (pp. 386–393). Vilhar Matija (ur.). Ljubljana.

Levengood, J. M., Heske, E. J., Wilkins, P. M., & Scott, J. W. (2015). Polyaromatic hydrocarbons and elements in sediments associated with a suburban railway. *Environmental Monitoring and Assessment, 187*, 1–12.

Li, Z., Ge, C., Li, J., Li, Y., Xu, A., Zhou, K., et al. (2010). Ground-dwelling birds near the Qinghai–Tibet highway and railway. *Transportation Research Part D, 15*, 525–528.

Maeda, T., Gautier, P. E., Hanson, C., Hemsworth, B., Nelson, J. T., Schulte-Werning, B., et al. (2012). Noise and vibration mitigation for rail Transportation systems. In *Notes on numerical fluid mechanics and multidisciplinary design* (Vol. 118).

Malawaka, M., & Wilkomirski, B. (2001). An analysis of soil and plant (*Taraxacum officinale*) contamination with heavy metals and polycyclic aromatic hydrocarbons (PAHs) in the area of the railway junction IŁAWA GŁÓWNA, Poland. *Water, Air, and Soil Pollution, 127*, 339–349.

Marler, P. (2004). Bird calls: A cornucopia for communication. In P. Marler & H. Slabbekoorn (Eds.), *Nature's music: The science of birdsong* (pp. 132–177). San Diego, California: Academic/Elsevier.

Marshall, V. M., Lewis, M. M., & Ostendorf, B. (2012). Buffel grass (*Cenchrus ciliaris*) as an invader and threat to biodiversity in arid environments: A review. *Journal of Arid Environments, 78,* 1–12.

Matsumoto, S., Shao, X., & Shino, K. (2005). Environmental noise in Middle East region of Kochi Prefecture [Japan]. *Transactions of the Japanese Society of Irrigation, Drainage and Reclamation Engineering* (Japan).

Mazur, Z., Radziemska, M., Maczuga, O., & Makuch, A. (2013). Heavy metal concentrations in soil and moss (*Pleurozium schreberi*) near railroad lines in Olsztyn (Poland). *Fresenius Environmental Bulletin, 22,* 955–961.

Morelli, F., Beim, M., Jerzak, L., Jones, D., & Tryjanowski, P. (2014). Can roads, railways and related structures have positive effects on birds? A review. *Transportation Research Part D, 30,* 21–31.

Morgan, P. A., & Peeling, J. (2012). *Railway noise mitigation factsheet 01: Overview of railway noise*. HS2 Rail Project Group.

Mundahl, N. D., Bilyeu, A. G., & Maas, L. (2013). Bald Eagle nesting habitats in the Upper Mississippi River National Wildlife and Fish Refuge. *Journal of Fish and Wildlife Management, 4,* 362–376.

Nelson, J. T. (1997). *Wheel/rail noise, control manual* (No. Project C-3 FY'93), Washington, DC.

Nielsen, J., Anderson, D. F., Gautier, P.-E., Iida, M., Nelson, J. T., Thompson, D., et al. (2015). Noise and vibration mitigation for rail transportation systems. In *Notes on numerical fluid mechanics and multidisciplinary design* (Vol. 126).

Owens, N. W. (1977). Responses of wintering brent geese to human disturbance. *Wildfowl, 28,* 5–14.

Palacin, R., Correia, J., Zdziech, M., Cassese, T., & Chitakova, T. (2014). Rail environmental impact: Energy consumption and noise pollution assessment of different transport modes connecting Big Ben (London, UK) and Eiffel Tower (Paris, Fr). *Transport Problems: An International Scientific Journal, 9.*

Paquet, P., & Callaghan, C. (1996). Effects of linear developments on winter movements of gray wolves in the Bow river Valley of Banff National Park, Alberta. In G. L. Evink, P. Garrett, D. Zeigler, & J. Berry (Eds.), *Trends in addressing transportation related wildlife mortality* (pp. 21). Proceedings of the Transportation Related Wildlife Mortality Seminar, Orlando, Florida.

Park, D. U., & Ha, K. C. (2008). Characteristics of PM_{10}, $PM_{2.5}$, CO_2 and CO monitored in interiors and platforms of subway train in Seoul, Korea. *Environment International, 34,* 629–634.

Pennington, D. N., Hansel, J. R., & Gorchov, D. L. (2010). Urbanization and riparian forest woody communities: Diversity, composition, and structure within a metropolitan landscape. *Biological Conservation, 143,* 182–194.

Penone, C., Kerbiriou, C., Julien, J. F., Julliard, R., Machon, N., & Viol, I. (2012). Urbanisation effect on orthoptera: Which scale matters? *Insect Conservation and Diversity, 6,* 319–327.

Persyn, R. A., Glanville, T. D., Richard, T. L., Laflen, J. M., & Dixon, P. M. (2004). Environmental effects of applying composted organics to new highway embankments: Part 1. interrill runoff and erosion. *Transactions of the American Society of Agricultural Engineers, 47,* 463–469.

Plakhotnik, V. N., Onyshchenko, J. V., & Yaryshkina, L. A. (2005). The environmental impacts of railway transportation in the Ukraine. *Transportation Research D, 10,* 263–268.

Qian, Z., Lin, X., Jun, M., Pan-Wen, W., & Qi-Sen, Y. (2009). Effects of the Qinghai–Tibet Railway on the community structure of rodents in Qaidam desert region. *Acta Ecologica Sinica, 29,* 267–271.

Rani, M., Pal, N., & Sharma, R. K. (2006). Effect of railway engines emission on the micromorphology of some field plants. *Journal of Environmental Biology, 27,* 373–376.

Reijnen, R., Foppen, R., Braak, C. T., & Thissen, J. (1995). The effects of car traffic on breeding bird populations in woodland. III. Reduction of density in relation to the proximity of main roads. *Journal of Applied Ecology, 32*, 187–202.

Rybak, J., & Olejniczak, T. (2013). Accumulation of polycyclic aromatic hydrocarbons (PAHs) on the spider webs in the vicinity of road traffic emissions. *Environmental Science and Pollution Research, 21*, 2313–2324.

Rytwinski, T., & Fahrig, L. (2012). Do species life history traits explain population responses to roads? A meta-analysis. *Biological Conservation, 147*, 87–98.

Schulte-Werning, B., Asmussen, B., Befr, W., Degen, K. G., & Garburg, R. (2012). Advancements in noise and vibration abatement to support the noise reduction strategy of Deutsche Bahn. In T. Maeda, P. E. Gautier, C. Hanson, B. Hemsworth, J. T. Nelson, B. Schulte-Werning, D. Thompson, & P. Vos (Eds.), *Noise and vibration mitigation for rail transportation systems* (pp. 9–16). New York: Springer.

Schulte-Werning, B., Thompson, D., Gautier, P. E., Hanson, C., Hemsworth, B., Nelson, J., et al. (2008). Noise and vibration mitigation for rail transportation systems. In *Notes on numerical fluid mechanics and multidisciplinary design* (Vol. 99).

Schweinsberg, F., Abke, W., Rieth, K., Rohmann, U., & Zullei-Seibert, N. (1999). Herbicide use on railway tracks for safety reasons in Germany? *Toxicology Letters, 107*, 201–205.

Snyder, E. B., Arango, C. P., Eitemiller, D. J., Stanford, J. A., & Uebelacker, M. L. (2002). Floodplain hydrologic connectivity and fisheries restoration in the Yakima River, USA. *Internationale Vereinigung für Limnologie, 28*, 1653–1657.

Stoll, S. (2013). How site characteristics, competition and predation influence site specific abundance of sand lizard on railway banks. Dissertation. Universität Bern.

Tiwari, A., Kadu, P., & Mishra, A. (2013). Study of noise pollution due to railways and vehicular traffic at level crossing and its remedial measures. *American Journal of Engineering Research, 2*, 16–19.

Trewhella, W. J., & Harris, S. (1990). The effect of railway lines on urban fox (*Vulpes vulpes*) numbers and dispersal movements. *Journal of Zoology, 221*, 321–326.

van der Grift, E. A. (1999). Mammals and railroads: Impacts and management implications. *Lutra, 42*, 77–79.

Vandevelde, J. C., Bouhoursc, A., Julien, J. F., Couvet, D., & Kerbiriou, C. (2014). Activity of European common bats along railway verges. *Ecological Engineering, 64*, 49–56.

Van Renterghem, T., & Botteldooren, D. (2012). On the choice between walls and berms for road traffic noise shielding including wind effects. *Landscape and Urban Planning, 105*, 199–210.

van Rooyen, C. (2009). *Bird impact assessment study Kusile railway line and associated infrastructure*. Report, Robindale.

Vo, P. T., Ngo, H. H., Guo, W., Zhou, J. L., Listowski, A., Du, B., et al. (2015). Stormwater quality management in rail transportation—Past, present and future. *Science of the Total Environmnet, 512*, 353–363.

Waterman, E., Tulp, I., Reijnen, R., Krijgsveld, K., & Braak, C. (2002). Disturbance of meadow birds by railway noise in The Netherlands. *Geluid, 1*, 2–3.

Wells, P. (1996). Wildlife mortality on the Canadian Pacific Railway between Field and Revelstoke, British Columbia. Unpublished report.

Wiacek, J., Polak, M., Filipiuk, M., Kucharczyk, M., & Bohatkiewicz, J. (2015). Do birds avoid railroads as has been found for roads? *Environmental Management, 56*, 643–652.

Wiłkomirski, B., Sudnik-Wójcikowska, B., Galera, H., Wierzbicka, M., & Malawska, M. (2011). Railway transportation as a serious source of organic and inorganic pollution. *Water, Air, and Soil pollution, 218*, 333–345.

Wilkomirski, B., Galera, H., Sudnik-Wójcikowska, B., Staszewski, T., & Malawska, M. (2012). Railway tracks-habitat conditions, contamination, floristic settlement-a review. *Environment and Natural Resources Research, 2*, 86.

Wilson, L. D., & Porras, L. (1983). The ecological impact of man on the south Florida herpetofauna. *University of Kansas Museum of Natural History, Special Publication, 9*, 89.

Xia, L., Yang Q., Li Z., Wu, Y., & Feng, Z. (2007). The effect of the Qinghai-Tibet railway on the migration of Tibetan antelope Pantholops hodgsonii in Hoh-xil National Nature Reserve, China. *Oryx, 41,* 3.

Xu, X., Zhang, K., Kong, Y., Chen, J., & Yu, B. (2006). Effectiveness of erosion control measures along the Qinghai–Tibet highway, Tibetan plateau, china. *Transportation Research Part D: Transport and Environment, 11,* 302–309.

Zhang, H., Wang, Z. F., Zhang, Y. L., & Hu, Z. J. (2012). The effects of the Qinghai–Tibet railway on heavy metals enrichment in soils. *Science of the Total Environment, 439,* 240–248.

Part II
Case Studies

Chapter 7
Bird Collisions in a Railway Crossing a Wetland of International Importance (Sado Estuary, Portugal)

Carlos Godinho, João T. Marques, Pedro Salgueiro, Luísa Catarino, Cândida Osório de Castro, António Mira and Pedro Beja

Abstract Many studies have evaluated bird mortality in relation to roads and other human structures, but little is known about the potential impacts of railways. In particular, it is uncertain whether railways are an important mortality source when crossing wetlands heavily used by aquatic birds. Here we analyze bird collisions in a railway that crosses the Nature Reserve of the Sado Estuary (Portugal) over an annual cycle, documenting bird mortality and the flight behaviour of aquatic birds in relation to a bowstring bridge. During monthly surveys conducted on 16.3 km of railway, we found 5.8 dead birds/km/10 survey days in the section crossing wetland habitats (6.3 km), while <0.5 dead birds/km/10 survey days were found in two sections crossing only forested habitats. Most birds recorded were small songbirds (Passeriformes), while there was only a small number of aquatic birds (common moorhen, mallard, flamingo, great cormorant, gulls) and other non-passerines associated with wetlands (white stork). During nearly 400 h of observations, we

C. Godinho (✉) · L. Catarino
LabOr—Laboratório de Ornitologia, ICAAM—Instituto de Ciências Agrárias e Ambientais Mediterrânicas, Universidade de Évora, Mitra, 7002-554 Évora, Portugal
e-mail: capg@uevora.pt

J.T. Marques · P. Salgueiro · A. Mira
Unidade de Biologia da Conservação, Departamento de Biologia, Universidade de Évora, Mitra, 7002-554 Évora, Portugal

J.T. Marques · P. Salgueiro · A. Mira
CIBIO/InBIO, Centro de Investigação em Biodiversidade e Recursos Genéticos, Universidade de Évora, Pólo de Évora, Casa do Cordovil 2º Andar, 7000-890 Évora, Portugal

C.O. de Castro
Direção de Engenharia e Ambiente. Infraestruturas de Portugal, Campus do Pragal—Ed. 1, Praça da Portagem, 2801-013 Almada, Portugal

P. Beja
CIBIO/InBIO, Centro de Investigação em Biodiversidade e Recursos Genéticos, Universidade do Porto, Campus Agrário de Vairão, Rua Padre Armando Quintas, 4485-661 Vairão, Portugal

P. Beja
CEABN/InBIO, Centro de Ecologia Aplicada "Professor Baeta Neves", Instituto Superior de Agronomia, Universidade de Lisboa, Tapada da Ajuda, 1349-017 Lisboa, Portugal

© The Author(s) 2017
L. Borda-de-Água et al. (eds.), *Railway Ecology*,
DOI 10.1007/978-3-319-57496-7_7

recorded 27,000 movements of aquatic birds across the Sado bridge, particularly in autumn and winter. However, only <1% of movements were within the area of collision risk with trains, while about 91% were above the collision risk area, and 8% were below the bridge. Overall, our case study suggests that bird collisions may be far more numerous in railways crossing wetland habitats than elsewhere, although the risk to aquatic birds may be relatively low. Information from additional study systems would be required to evaluate whether our conclusions apply to other wetlands and railway lines.

Keywords Anthropogenic mortality · Aquatic birds · Collision risk · Environmental impact · Wetlands · Wildlife mortality

Introduction

Collision with human structures and vehicles is an important source of wild bird mortality, killing hundreds of millions of birds each year (Loss et al. 2015). Although the population-level consequences of such mortality are poorly known (but see, e.g., Carrete et al. 2009; Borda-de-Água et al. 2014), it is generally recognized under the precautionary principle that efforts should be made to reduce the number of birds killed each year as much as possible (Loss et al. 2015). Information is thus needed on the bird species most vulnerable to collisions, and on the environmental and species-specific factors affecting such vulnerability (Barrios and Rodríguez 2004; Santos et al. 2016), which are essential for developing management guidelines aimed at reducing collision risk (Barrientos et al. 2011; May et al. 2015).

To collect baseline information for mitigating impacts, bird collision risk has been the subject of intensive research over the past two decades (Loss et al. 2015), with many studies documenting bird mortality and its correlates in relation to roads (Santos et al. 2016), wind farms (Barrios and Rodríguez 2004; Drewitt and Langston 2006), power lines (Barrientos et al. 2011, 2012), and buildings (Loss et al. 2015). Surprisingly, however, very few studies have analyzed bird collision in relation to railways, although these linear infrastructures extend over tens of thousands of kilometers across the world (see Chap. 2). Railways present a number of risks to birds from potential collisions with circulating trains, but there are also risks regarding collision with catenary wires, electrocution, and barotrauma by the train movement (Dorsey et al. 2015, and see Chap. 2). The few studies addressing these problems reported that birds can account for over 50% of the vertebrates killed in railways (SCV 1996; van der Grift and Kuijsters 1998), and that collisions may often involve species of conservation concern such as owls and birds of prey (Peña and Llama 1997; SCV 1996; Schaub et al. 2010). For some of these species, railway-related mortality may represent a considerable proportion of the overall mortality (van der Grift and Kuijsters 1998; Schaub et al. 2010), and thus may be a risk worth considering when designing or managing railways. However, the few studies conducted so far have covered just a very limited range of species and environmental conditions, making it difficult to draw generalizations (see Chap. 2).

Here we provide a case study on bird collisions in a railway that crosses a wetland of international importance, the Sado Estuary (Portugal), which is a RAMSAR site and a Special Protection Area for birds classified under the European Union Directive 79/409/EEC. Especially from autumn to mid-spring, this wetland is home to thousands of waders, wildfowl, flamingos *Phoenicopterus roseus*, and other aquatic birds, which use a diversity of wet habitats including open water, mud flats, rice fields, and salt pans (Lourenço et al. 2009; Alves et al. 2011). These birds were expected to cross the railway area on a daily basis, particularly the bridge crossing the Sado River, due to movements from roosts to feeding areas, and among feeding areas (e.g., Dias et al. 2006). Therefore, it was feared that they were exposed to a high risk of collision, with the possibility of mass mortality events occurring due to the collision of large flocks with bridge structures and circulating trains. Our study was designed to clarify this issue, aiming at: (1) quantifying bird mortality patterns due to collisions associated with the presence of the railway; (2) characterizing the movements of wetland birds crossing the railway bridge along the circadian and the annual cycles; and (3) estimating the seriousness of the risk of wetland bird mortality associated with this railway, as well as which bird species are most at risk. To the best of our knowledge, this is the first published study documenting the impacts of railways on wetland bird mortality.

Methods

The study was carried out in Portugal, focusing on the "Variante de Alcácer" railway, whose construction started in February 2007, with train circulation beginning in December 2010 (Figs. 7.1 and 7.2). This is a single-track electrified railway extending over 29 km, that is part of the network connecting Lisbon to the south of the country. It crosses the Natural Reserve of the Sado Estuary (Portugal) in a section of about 2.6 km (38° 24′N, 8° 36′W), most of which is occupied by wetland habitats including the Sado River, rice fields and salt pans (1.6 km), while the rest is dominated by cork oak *Quercus suber* and stone pine *Pinus pinea* woodlands. The crossing is made through a bowstring bridge built between 2007 and 2010, with a length of 2735 m and three 160 m high arcs (Fig. 7.2). Another viaduct of this railway (about 0.8 km) crosses a small wetland outside the Natural Reserve, which corresponds to the São Martinho stream and the adjacent rice fields (Fig. 7.2). During the period of observations conducted over one year (see below), the bridge was crossed by 26 ± 4 [SD] trains/day. The train speed ranged from 44 km/h (charcoal trains) to 170–200 km/h (passenger trains).

Bird mortality due to collisions with trains and railway structures was estimated at three sections of this railway, with a total length of 16.3 km. One of these sections (Section 1: 6.3 km) included the sectors crossing the Sado (2.6 km) and São Martinho (0.8 km) wetlands, while the rest (2.9 km) crossed forested habitats. The other two sections (Section 2: 5.8 km; Section 3: 4.2 km) crossed only forested habitats, and they were used as controls to estimate how the presence of wetlands affected bird collision rates. The railway tracks and the surrounding areas were thoroughly surveyed for vertebrate corpses in the three sections, between

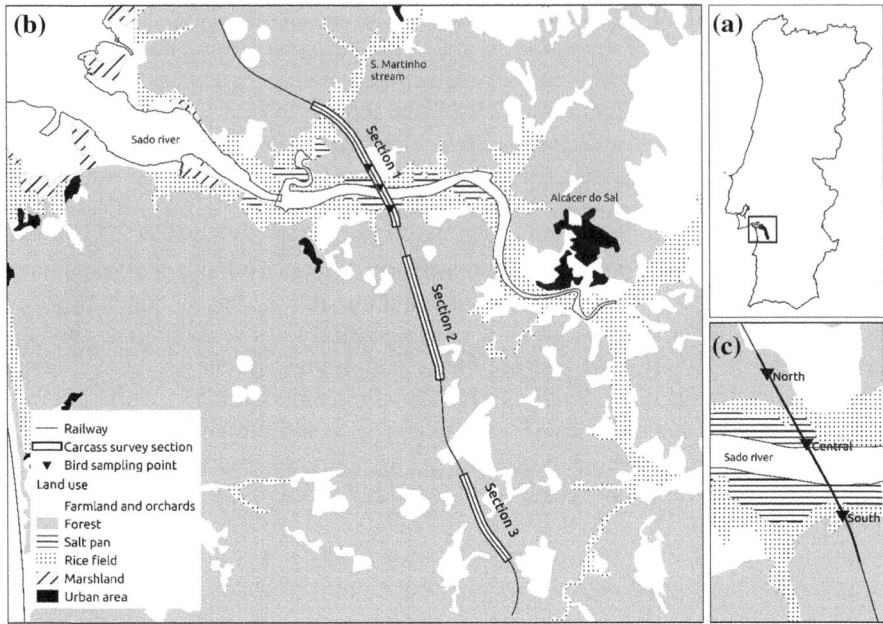

Fig. 7.1 **a** Location of the Sado Estuary in Portugal, **b** map of the study area showing the location of the three sections surveyed for bird mortality in the Variante de Alcácer" railway line, and **c** detail of the section 1 crossing the Sado River, with the location of the three bird counting points

November 2012 and October 2013. In section 1, surveys were conducted once a each month, and on two consecutive days per month. In sections 2 and 3, surveys were conducted in the same months as Section 1, but they were carried out in a single day each month. For safety reasons, surveys were conducted only when accompanied by staff of the railway company and following all safety procedures. Each bird corpse located during surveys was identified to the lowest possible taxonomic level and its position was recorded with a GPS receiver.

To evaluate bird movements across the Sado bridge, observations were made within 500-m buffers at three sampling points set 1 km apart along the bridge, and covering all deck (Figs. 7.1 and 7.2). The points were set to cover the main habitats available in the area: North—forested areas; Central—river and wetland habitats; South—forest and wetland habitats (Fig. 7.2). Twenty-two survey sessions took place between November 2012 and September 2013 by three observers simultaneously. The number of birds crossing the railway were counted during the day in two 6-hours sessions per month (i.e., ≈400 h of observation homogeneously distributed during the study period), to cover the tidal cycle. Observers rotated between observation points every 3 h to avoid hour-related biases. Species, number of individuals, hour, flight direction and height (based on structural elements of the bridge: (deck, electric cables and bridge arcs) were recorded for each crossing. Particular care was taken to estimate the number of birds crossing at the train collision risk zone, corresponding to the area between the deck and the catenary lines.

Fig. 7.2 Photographs illustrating aspects of the study area. *Upper panel* railway crossing rice fields (São Martinho stream); *middle panel* bowstring railway bridge crossing the Sado River; *lower panel* flock of flamingos flying close to the Sado railway bridge

Results

Bird Mortality

Overall, we found 124 vertebrate corpses during mortality surveys, most of which (75.8%) were birds, although there were also a few amphibians (8.9%), reptiles

(3.2%), mammals (8.9%), and unidentified vertebrates (3.2%) (Table 7.1). Most birds were passerines (Passeriformes), which accounted for 15 of the 25 species identified, and 67.9% of 53 individuals identified to species level. White wagtails (*Motacilla alba*) accounted for about 35% of the individuals identified, while for 18 species we found only a single individual. Only eight individuals of strictly aquatic species were identified, including two common moorhens (*Gallinula chloropus*), and one each of mallard (*Anas plathyrhynchos*), flamingo, great cormorant (*Phalacrocorax carbo*) and lesser black-backed gull (*Larus fuscus*). To these should be added five white storks (*Ciconia ciconia*), which are frequently found in flooded rice fields in the area.

Table 7.1 Summary results of vertebrate mortality detected in the "Variante de Alcácer" railway (southern Portugal), between November 2012 and October 2013

Species	Section 1		Section 2		Section 3	
	Wetland/Forest (6.3 km)		Forest (5.8 km)		Forest (4.2 km)	
	N	N/km/10 days	N	N/km/10 days	N	N/km/10 days
Amphibians	**6**	**0.40**	**2**	**0.29**	**3**	**0.60**
Common toad *Bufo bufo*	–	–	1	0.14	–	–
Natterjack toad *Bufo calamita*	–	–	–	–	2	0.40
Common/Natterjack toad *Bufo bufo/calamita*	–	–	–	–	1	0.20
Treefrog *Hyla* sp.	1	0.07	1	0.14	–	–
Perez's frog *Rana perezi*	1	0.07	–	–	–	–
Fire salamander *Salamandra salamandra*	1	0.07	–	–	–	–
Crested newt *Triturus marmoratus*	1	0.07	–	–	–	–
Unidentified amphibian	2	0.13	–	–	–	–
Reptiles	**2**	**0.13**	**–**	**–**	**2**	**0.40**
Large psammodromus *Psammodromus algirus*	–	–	–	–	2	0.40
Montpellier snake *Malpolon monspessulanus*	2	0.13	–	–	–	–
Birds	**89**	**5.89**	**3**	**0.43**	**2**	**0.40**
Mallard *Anas plathyrhychos*	1	0.07	–	–	–	–
Greater flamingo *Phoenicopterus roseus*	1	0.07	–	–	–	–
White stork *Ciconia ciconia*	5	0.33	–	–	–	–
Great cormorant *Phalacrocorax carbo*	1	0.07	–	–	–	–
Cattle egret *Bubulcus ibis*	1	0.07	–	–	–	–
Moorhen *Gallinula chloropus*	2	0.13	–	–	–	–
Lesser black-backed gull *Larus fuscus*	1	0.07	–	–	–	–
Rock dove (domestic) *Columba livia*	3	0.20	–	–	–	–

(continued)

Table 7.1 (continued)

Species	Section 1		Section 2		Section 3	
	Wetland/Forest (6.3 km)		Forest (5.8 km)		Forest (4.2 km)	
	N	N/km/10 days	N	N/km/10 days	N	N/km/10 days
Tawny owl *Strix aluco*	1	0.07	–	–	–	–
Little owl *Athene noctua*	–	–	1	0.14	–	–
Carrion crow *Corvus corone*	1	0.07	–	–	–	–
Wood lark *Lululla arborea*	–	–	1	0.14	–	–
Barn swallow *Hirundo rustica*	2	0.13	–	–	–	–
Blue tit *Cyanistes caeruleus*	1	0.07	–	–	–	–
Common chiffchaff *Phylloscopus collybita*	1	0.07	–	–	–	–
Reed warbler *Acrocephalus scirpaceus*	2	0.13	–	–	–	–
Blackcap *Sylvia atricapilla*	–	–	–	–	1	0.20
Common redstart *Phoenicurus ochruros*	1	0.07	–	–	–	–
European stonechat *Saxicola rubicola*	1	0.07	–	–	–	–
Song thrush *Turdus philomelos*	–	–	1	0.14	–	–
Mistle thrush *Turdus viscivorus*	–	–	–	–	1	0.20
White wagtail *Motacilla alba*	19	1.26	–	–	–	–
Chaffinch *Fringilla coelebs*	1	0.07	–	–	–	–
Greenfinch/goldfinch *Chloris chloris/Carduelis carduelis*	1	0.07	–	–	–	–
House sparrow *Passer domesticus*	2	0.13	–	–	–	–
Unidentified Ardeidae	1	0.07	–	–	–	–
Unidentified Columbidae	3	0.20	–	–	–	–
Unidentified aquatic bird	1	0.07	–	–	–	–
Unidentified songbird (Passeriforme)	6	0.40	–	–	–	–
Unidentified bird	30	1.98	–	–	–	–
Mammals	**8**	**0.53**	**2**	**0.29**	**1**	**0.20**
Kuhl's pipistrelle *Pipistrellus kuhlii*	–	–	–	–	1	0.20
Wood mouse *Apodemus sylvaticus*	–	–	1	0.14	–	–
Red fox *Vulpes vulpes*	1	0.07	–	–	–	–
Stone marten *Martes foina*	1	0.07	–	–	–	–
Domestic cat *Felis catus*	1	0.07	–	–	–	–
Domestic sheep *Ovis aries*	4	0.26	–	–	–	–
Unidentified carnivore	1	0.07	–	–	–	–
Unidentified mammal	–	–	1	0.14	–	–
Unidentified vertebrate	4	0.26	–	–	–	–
Total	109	7.21	7	1.01	8	1.59

Results are reported separately for three sections of the railway with different lengths and crossing different habitats. For each section and species/taxa, we provide the total number of individuals detected (N) and the number of individuals detected per km and per 10 survey days (N/km/10 days)

Most of the birds (94.6%) were found in Section 1, where we recorded 5.89 birds/km/10 survey days, while in Sections 2 and 3 we recorded 0.43 birds/km/10 days and 0.40 birds/km/10 days, respectively (Table 7.1). Within Section 1, we found 76.4% of the birds in the two sectors crossing the Sado River (n = 59) and the São Martinho stream (n = 9) (Table 7.1). These two sectors were responsible for all mortality of strictly aquatic species, and for 80% of the white storks killed. Most bird mortality was recorded in the autumn (37.2%) and winter (33.0%), with much lower values in the spring (16.0%) and summer (13.8%). The group of aquatic birds and white stork were recorded in each season except winter.

Bridge Crossing by Birds

We observed a total of about 27,000 birds crossing the bridge during the observation periods, most of which (82.7%) were waterfowl and other aquatic birds, and the others were mainly passerines. Overall, most crossings (76.6%) were detected in the central section of the bridge, followed by the south (18.7%) and north sections (4.7%). More than half the crossings were detected in winter (December–February), corresponding to 4217 crossings/month (Fig. 7.3). Spring (March–June) was the season with fewer crossings observed, with 571 crossings per month (Fig. 7.3). Less than 1% of the crossings were made between the deck and the

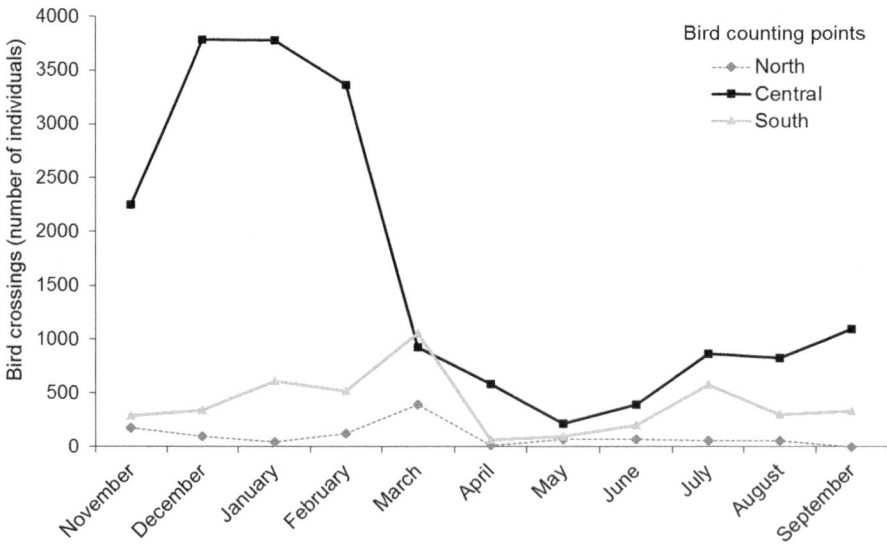

Fig. 7.3 Monthly number of aquatic birds observed crossing the railway bridge over the Sado River, between November 2012 and September 2013. Numbers are presented separately for the three sampling points covering the main habitats available in the area: north (forested areas); central (river and wetland habitats); and south (forest and wetland habitats)

Table 7.2 Aquatic bird species observed most frequently crossing the railway bridge over the Sado River, indicating the total number of birds crossing and the number of crosses within the train collision risk zone (between the deck and the catenary lines)

Species		Total crosses	Collision risk zone
Lesser black-backed gull	*Larus fuscus*	8432	20 (<1%)
Black-headed gull	*Chroirocephalus ridibundus*	7036	91 (1%)
Great cormorant	*Phalacrocorax carbo*	1982	10 (1%)
Glossy ibis	*Plegadis falcinellus*	1596	10 (<1%)
White stork	*Ciconia ciconia*	1380	25 (2%)
Little egret	*Egretta garzetta*	1074	17 (2%)
Western cattle egret	*Bubulcus ibis*	693	19 (3%)
Mallard	*Anas platyrhynchos*	205	1 (<1%)
Dunlin	*Calidris alpina*	99	15 (15%)
Black stork	*Ciconia nigra*	16	2 (13%)

catenary lines, where there is a risk of collision with trains (Fig. 7.3). Although most birds crossed the bridge above the train collision risk zone (91.3%), 39.1% of the crossings were recorded below the top of the bridge arcs (Fig. 7.3).

Most aquatic birds observed crossing the bridge were gulls (lesser black-backed gull and black-headed gull (*Chroicocephalus ridibundus*), greater flamingos and glossy ibis (*Plegadis falcinellus*) (Table 7.2). The proportion of crossings within the train collision risk height were low (≤3%) for most species, with relatively high values (>10%) recorded only for dunlin (*Calidris alpine*) and black stork (*C. nigra*) (Fig. 7.4; Table 7.2). In the case of black storks, however, this value is based on just two individuals observed flying at a low height.

Discussion

Our results suggest that bird mortality in the "Variante de Alcácer" railway was relatively low during the study period, although there were sharp peaks in the two sectors where it crossed wetland habitats. Nevertheless, there were very few aquatic birds found dead, suggesting that they were not particularly susceptible to collisions with circulating trains or the railway structures. This was in line with the observations of aquatic birds flying across the railway in the area of the bridge over the Sado River, with only a very small proportion of individuals using heights that expose them to collision risk. Overall, we suggest that this railway is unlikely to cause significant mortality to aquatic bird species, although it remains uncertain whether this result can be extrapolated to other wetlands and railway lines.

This study had some limitations and potential shortcomings, but it is unlikely that they have significantly affected our key findings. A potentially important

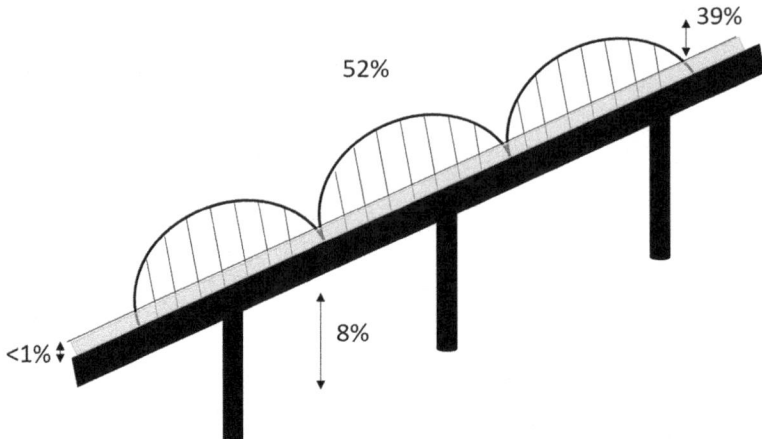

Fig. 7.4 Schematic drawing of the bowstring railway bridge across the Sado River, showing the proportion of bird crossings at various heights during one annual cycle (November 2012–September 2013). The *grey section* indicates the birds flying at the heights of collision risk with circulating trains

problem is that mortality surveys were carried out at monthly intervals and only on one or two days per month, which may influence mortality estimates (Santos et al. 2011, 2015). For instance, an eventual mass mortality event due to a flock colliding with a train might go unnoticed, if it occurs after a monthly survey and corpses decay or are removed until the next survey (Santos et al. 2011). The problem may be particularly serious for small passerines, because their corpses are likely to persist for very short periods in the railway, thereby causing underestimation of the mortality rate (Santos et al. 2011) and making it difficult to precisely estimate the location of mortality hotspots (Santos et al. 2015). However, the main focus of this study was on aquatic birds, which tend to be relatively large and thus less affected by very short persistence times. Furthermore, a collision with a flock of large birds would likely be detected by the train driver and then reported to the safety department. Also, the observation that mortality hotspots were coincident with wetland crossings is unlikely to be due to survey bias, given that the number of bird corpses per km and survey day detected therein was more than 10 times higher than anywhere else along this railway. This number might be even greater, because some birds colliding with trains on the wetland bridges could be thrown to the surrounding water or other wet habitats, where they would likely decay rapidly or otherwise go unnoticed by observers. Another potential problem is that bird movements across the bridge over the Sado River were also observed for just a few days, and we did not include days with poor visibility conditions (e.g., fog) or night time periods when there may be considerable activity by waders (Lourenço et al. 2008). Therefore, our conclusion that birds were mainly flying outside the collision risk areas should be accepted with care, because it cannot be completely ruled out that more dangerous flight patterns were taken at night or under particular weather conditions.

Despite these potential problems, our study suggests that the mortality risk for birds was higher in railway sections crossing wetlands than elsewhere, as underlined by the concentration of bird kills in the Sado and, to a lesser extent, the São Martinho wetlands. However, in contrast to expectations, most mortality involved small passerines, particularly white wagtails, while there were only a few aquatic birds killed. This may be a consequence of the high productivity of wetland habitats, which attract large numbers of birds that are then exposed to collision with circulating trains and railway structures during their daily movements. In the Mediterranean region, these wetlands are particularly important in autumn and winter, when they are used by large numbers of passerines from northern latitudes (Finlayson et al. 1992), which may be associated with the higher mortality observed during these seasons. Aquatic birds may be less exposed to collisions because they fly higher than small passerines (see below), which may explain the small number of kills recorded in our study. However, this may vary in relation to ecological conditions and railway characteristics, because apparently larger mortalities of gulls and waterfowl have been found in other study areas (van der Grift and Kuijsters 1998; Heske 2015).

As expected, we found a large number of aquatic birds crossing the bridge over the Sado River, particularly from October to February, and following a path along the river. The birds observed most frequently included a number of waterfowl, waders and other aquatic species, which are abundant both locally and across the estuary (Alves et al. 2011). The flight patterns taken by these birds suggest that they were very rarely at risk of collision with circulating trains or bridge structures. We found that only a very small proportion of crossings ($\approx$1%) were within the height of collision risk with trains, while most birds ($\approx$52%) crossed above the bridges' arcs. Actually, the presence of the arcs may have helped deflect birds away from the collision risk area, as we often observed birds avoiding these by changing flight paths. This is in line with observations showing that individual birds and bird flocks change flight paths in response to the presence of anthropogenic structures such as pole barriers (Zuberogoitia et al. 2015).

Overall, our study contributed to clarifying the bird mortality risk due to collisions in railways, adding to just a few previous studies addressing this issue (SCV 1996; Peña and Llama 1997; van der Grift and Kuijsters 1998; Schaub et al. 2010, see also Chap. 2). Our results, together with those previous studies, suggest that the risk associated with railways may be relatively low compared to other anthropogenic sources of mortality (Loss et al. 2015), although differences are difficult to judge precisely, because of the variation in methodological approaches and ecological contexts. Nevertheless, it is worth noting that there were many more collisions where railways crossed wetland habitats, suggesting that the planning of new railways should strive to avoid these habitats as much as possible. Where this is unfeasible or undesirable, monitoring programs should be implemented to evaluate the risks and provide information to design mitigation measures if necessary. Despite the relatively low numbers of dead aquatic birds recorded in our study, we believe that they should not be disregarded in monitoring programs because mortality patterns may depend on local ecological and railway characteristics. Overall, we suggest that further empirical studies are necessary on other railways and covering a wider range of ecological conditions, to help draw generalizations that can be useful for landscape planners and environmental managers.

Acknowledgements This study was supported by Infraestruturas de Portugal with the contribution of the Portuguese Science Foundation through the doctoral Grants SFRH/BD/81602/2011 (Carlos Godinho) and SFRH/BD/87177/2012 (Pedro Salgueiro). Carlos Oliveira and José Rebocho (REFER Manutenção) provided invaluable help during the mortality surveys.

References

Alves, J. A., Dias, M., Rocha, A., Barreto, B., Catry, T., Costa, H., et al. (2011). Monitorização das populações de aves aquáticas dos estuários do Tejo, Sado e Guadiana. Relatório do ano de 2010. *Anuário Ornitológico (SPEA), 8,* 118–133.

Barrientos, R., Alonso, J. C., Ponce, C., & Palacin, C. (2011). Meta-analysis of the effectiveness of marked wire in reducing avian collisions with power lines. *Conservation Biology, 25,* 893–903.

Barrientos, R., Ponce, C., Palacin, C., Martín, C. A., Martín, B., & Alonso, J. C. (2012). Wire marking results in a small but significant reduction in avian mortality at power lines: A BACI designed study. *PLoS ONE, 7,* e32569.

Barrios, L., & Rodríguez, A. (2004). Behavioral and environmental correlates of soaring-bird mortality at on-shore wind turbines. *Journal of Applied Ecology, 41,* 72–81.

Borda-de-Água, L., Grilo, C., & Pereira, H. M. (2014). Modeling the impact of road mortality on barn owl (*Tyto alba*) populations using age-structured models. *Ecological Modelling, 276,* 29–37.

Carrete, M., Sánchez-Zapata, J. A., Benítez, J. R., Lobón, M., & Donázar, J. A. (2009). Large scale risk assessment of wind-farms on population viability of a globally endangered long-lived raptor. *Biological Conservation, 142,* 2954–2961.

Dias, M. P., Granadeiro, J. P., Lecoq, M., Santos, C. D., & Palmeirim, J. M. (2006). Distance to high-tide roosts constrains the use of foraging areas by dunlins: Implications for the management of estuarine wetlands. *Biological Conservation, 131,* 446–452.

Dorsey, B., Olsson, M., & Rew, L. J. (2015). Ecological effects of railways on wildlife. In R. van der Ree, D. J. Smith, & C. Grilo (Eds.), *Handbook of Road Ecology* (pp. 219–227). West Sussex: Wiley.

Drewitt, A. L., & Langston, R. H. (2006). Assessing the impacts of wind farms on birds. *Ibis, 148,* 29–42.

Finlayson, C. M., Hollis, G. E., & Davis, T. J. (1992). *Managing Mediterranean Wetlands and Their Birds.* IWRB Special Publication 20. Slimbridge: International Wetland Research Bureau.

Heske, E. J. (2015). Blood on the tracks: Track mortality and scavenging rate in urban nature preserves. *Urban Naturalist, 4,* 1–13.

Loss, S. R., Will, T., & Marra, P. P. (2015). Direct mortality of birds from anthropogenic causes. *Annual Review of Ecology Evolution and Systematics, 46,* 99–120.

Lourenço, P. M., Groen, N., Hooijmeijer, J. C. E. W., & Piersma, T. (2009). The rice fields around the estuaries of the Tejo and Sado are a critical stopover area for the globally near-threatened Black-tailed Godwit *Limosa l. limosa*: Site description, international importance and conservation proposals. *Airo, 19,* 19–26.

Lourenço, P. M., Silva, A., Santos, C. D., Miranda, A. C., Granadeiro, J. P., & Palmeirim, J. M. (2008). The energetic importance of night foraging for waders wintering in a temperate estuary. *Acta Oecologica, 34,* 122–129.

May, R., Reitan, O., Bevanger, K., Lorentsen, S. H., & Nygård, T. (2015). Mitigating wind-turbine induced avian mortality: Sensory, aerodynamic and cognitive constraints and options. *Renewable and Sustainable Energy Reviews, 42,* 170–181.

Peña, O. L., & Llama, O. P. (1997). Mortalidad de aves en un tramo de linea de ferrocarril. Grupo Local SEO-Sierra de Guadarrama, Spain. Unpublished report. SEO/BirdLife, Grupo local SEO-Sierra de Guadarrama. Retrieved July 19, 2016, from http://www.actiweb.es/seosierradeguadarrama/archivo1.pdf

Santos, S. M., Carvalho, F., & Mira, A. (2011). How long do the dead survive on the road? Carcass persistence probability and implications for road-kill monitoring surveys. *PLoS ONE, 6,* e25383.

Santos, S. M., Marques, J. T., Lourenço, A., Medinas, D., Barbosa, A. M., Beja, P., et al. (2015). Sampling effects on the identification of roadkill hotspots: Implications for survey design. *Journal of Environmental Management, 162,* 87–95.

Santos, S. M., Mira, A., Salgueiro, P. A., Costa, P., Medinas, D., & Beja, P. (2016). Avian trait-mediated vulnerability to road traffic collisions. *Biological Conservation, 200,* 122–130.

Schaub, M., Aebischer, A., Gimenez, O., Berger, S., & Arlettaz, R. (2010). Massive immigration balances high anthropogenic mortality in a stable eagle owl population: Lessons for conservation. *Biological Conservation, 143,* 1911–1918.

S. C. V. (1996). Mortalidad de vertebrados en líneas de ferrocarril. Documentos Técnicos de Conservación SCV 1. Madrid: Sociedad Conservación de Vertebrados.

van der Grift, E. A., & Kuijsters, H. M. J. (1998). Mitigation measures to reduce habitat fragmentation by railway lines in the Netherlands. In G. L. Evink, P. Garrett, D. Zeigler, & J. Berry (Eds.), *Proceedings of the international conference on wildlife ecology and transportation* (pp. 166–170). Tallahassee: Florida Department of Transportation.

Zuberogoitia, I., del Real, J., Torres, J. J., Rodríguez, L., Alonso, M., de Alba, V., et al. (2015). Testing pole barriers as feasible mitigation measure to avoid bird vehicle collisions (BVC). *Ecological Engineering, 83,* 144–151.

Chapter 8
Cross-scale Changes in Bird Behavior Around a High Speed Railway: From Landscape Occupation to Infrastructure Use and Collision Risk

Juan E. Malo, Eladio L. García de la Morena, Israel Hervás, Cristina Mata and Jesús Herranz

Abstract Large-scale transportation infrastructures, such as high-speed railway (HSR) systems, cause changes in surrounding ecosystems, thus generating direct and indirect impacts on bird communities. Such impacts are rooted in the individual responses of birds to infrastructure components, such as habitat occupancy of railway proximities, the use of structural elements (e.g., perching or nesting sites), flights over the railway, and behavior towards approaching trains. In this chapter, we present the most important results of several studies that were carried out on bird communities, between 2011 and 2015 on a 22-km stretch of HSR built on an agrarian landscape in central Spain. Available data describe the abundance and spatial distribution of birds up to 1000 m from the railway, bird infrastructure use (e.g., embankments, catenaries), cross-flights of the railway obtained through focal sampling, and animal responses to approaching trains recorded from train cockpits. These data depict how bird species respond at various scales to the presence of the HSR, and show how the infrastructure impacts bird communities, due to both habitat changes and increases in mortality risk.

Keywords Agrarian steppe · Avian mortality · Community homogenization · Disturbance · Environmental impact assessment · Flight initiation distance · Habitat degradation · Road-kill

J.E. Malo (✉) · I. Hervás · C. Mata · J. Herranz
Terrestrial Ecology Group, Departamento de Ecología,
Universidad Autónoma de Madrid, C. Darwin 2, 28049 Madrid, Spain
e-mail: je.malo@uam.es

E.L. García de la Morena
SECIM, Servicios Especializados de Consultoría e Investigación Medioambiental,
C. Segura 2, 28410 Manzanares el Real, Madrid, Spain

© The Author(s) 2017
L. Borda-de-Água et al. (eds.), *Railway Ecology*,
DOI 10.1007/978-3-319-57496-7_8

Introduction

The development of large-scale transport infrastructure impacts surrounding ecosystems. Environmental impact assessment is the formal procedure set up to predict, evaluate, and plan for the mitigation or compensation of such impacts in advance of project development (Petts 1999). Large projects impact various kinds of species; those that affect animal populations, especially vertebrates, are prioritized due to societal biases in conservation (Wilson et al. 2007). Nonetheless, the existing knowledge about bird responses to the presence and operation of railways is extremely scarce, and the data are even more scarce with respect to high speed railways (HSRs), due to their limited and relatively recent presence (Dorsey et al. 2015). Given the expected expansion of these railways in the near future (Campos and de Rus 2009; Todorovich et al. 2011; Fu et al. 2012), it is urgent that the environmental effects of this new technology be assessed to enable the development of mitigation strategies.

There is now reliable information on the impacts of roads on wildlife and on birds in particular (Benítez-López et al. 2010). We know that the effects are numerous, beginning with habitat loss generated by the construction of the infrastructure. However, the most serious impacts occur during the operating phase of roads, such as direct mortality of individuals and indirect impacts derived from the degradation of habitats close to roads. Similarly, railways alter bird communities both by destruction and degradation of habitats, as well as by lethal collision (Spencer 1965; Mammen et al. 2002). Due to the structural characteristics of HSRs (larger embankments, catenaries, etc.), their structural effects on habitats (and thus on species) are expected to exceed those of roads, while the degradation of habitats due to light, noise, and pollution, as well as mortality, are expected to be smaller due to lower traffic intensity.

In any case, the effects of HSRs on birds will depend upon the individual responses of species to the new landscape conditions. On the one hand, species will respond to the presence of structures that alter the natural physiognomy of the area, including the development of new habitats (e.g., revegetated embankments), and structural elements (e.g., catenaries, poles, bridges) that provide new potential perching and nesting sites (Tryjanowski et al. 2013; Morelli et al. 2014). On the other hand, the risk of mortality is restricted to birds that enter the narrow corridor where trains circulate, where the habitat has undergone extreme modification, from the gravel ballast where the rails lie to the catenary wires. To explain the impacts of future projects, it is crucial to know how birds respond to the presence of the railway.

This chapter summarizes the results gathered during the monitoring of birds in an agricultural area of the Iberian Peninsula that is crossed by a high-speed railway. We present a sequential approach that addresses the presence, abundance, and activity patterns of birds in the surroundings and on an HSR, complemented with data addressing the response of birds facing approaching trains obtained through cockpit video recordings.

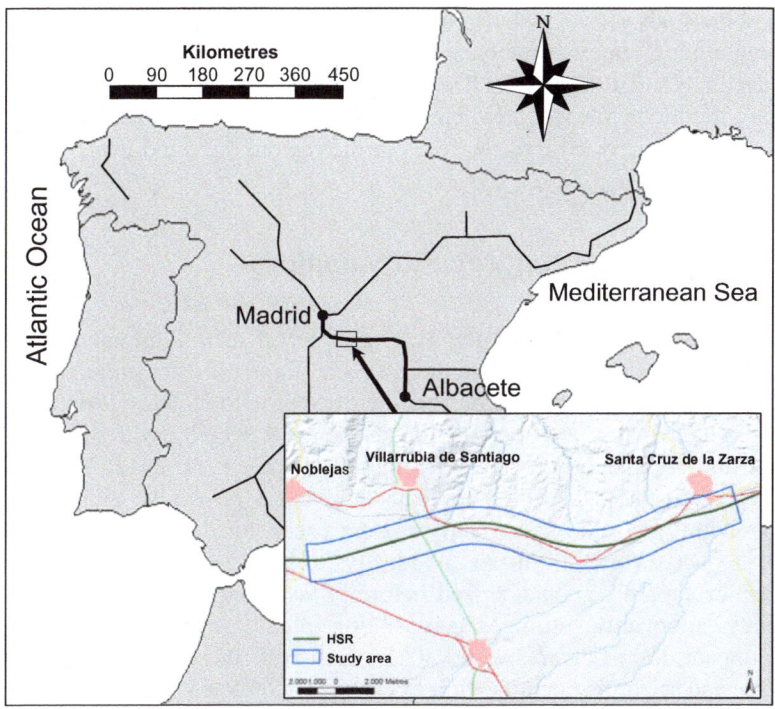

Fig. 8.1 Location of study site along the Madrid-Albacete HSR. *Inset image* shows a magnified view of the study area (*boxed-in main image*)

Study Area

The study area is the HSR traversing Madrid-Castilla La Mancha-Valencia-Murcia, specifically on the high plateau (Mesa) of Ocaña. The selected section corresponds to km mileposts 80–102 of the HSR that runs through the municipalities of Noblejas, Villarrubia de Santiago, and Santa Cruz de la Zarza, in the northeast area of the Toledo province in central Spain (Fig. 8.1).

The Mesa de Ocaña is a large flatland approximately 100 m above the Tagus River valley, with an average altitude between 750 m above sea level (a.s.l.) at the western end of the section, and 790 m a.s.l. on the eastern side. The climate is Mediterranean-continental, with an annual average temperature of 13.9 °C (average monthly minimum 7.4 °C, and maximum 20.5 °C), and an average annual rainfall of 437.8 mm (Ninyerola et al. 2005).

Of plain or slightly wavy relief, the area crossed by the railway is characterized by its open physiognomy, with a predominance of dry crops (mainly cereals, although legume crops like chickpeas, are frequently found), interspersed with olive crops and vineyards. In some areas north of the railway, the topography is somewhat more irregular, with ravines, hills, and hillocks where some patches of natural vegetation remain (oaks, pine trees, bushes, and *Stipa tenacissima* formations).

Part of the study area is included in or adjacent to natural protected areas under a conservation legal status. Notably, south of the railway, there are 10 polygons that constitute the NATURA2000 site "Áreas esteparias de la Mancha Norte," which are especially important for the great bustard (*Otis tarda*) in the Castilla-La Mancha region. Approximately 60% of the birds in the region are found here.

Bird Occupation of Railway Surroundings

Spatial patterns of birds around the HSR give broad-scale information of infrastructure impacts on the animal community because the presence and abundance of bird species along the landscape mirrors their response to it. Thus, it is foreseeable that habitat destruction and degradation in the HSR vicinity will shape the way birds use the area (Benítez-López et al. 2010; Torres et al. 2011; Wiącek et al. 2015). Comparisons of species abundance in close proximity to infrastructure locations with that in the wider landscape are useful for the assessment of infrastructure impacts (Torres et al. 2016). Ideally, to analyze human impacts, animal communities should be characterized before infrastructure construction as well as recurrently afterwards, during the operation phase, in a "before and after" control-impact design (Torres et al. 2011). However, this approach is frequently unfeasible and scientific quantification of changes in bird densities around roads has shown that most effects on birds extend up to 500 m from the infrastructure, with outlying areas largely free of impacts (Benítez-López et al. 2010).

In this case study, the characterization of bird populations potentially affected by the HSR line started before the operation phase, but after the line was already built. Thus, there is no information related to impacts associated with the construction phase. Nonetheless, control areas 500–1,000 m from the railway give a detailed picture of bird communities presumed to be negligibly affected by the presence of the project, and they are informative about local changes in bird communities by comparison with data obtained in close proximity to the HSR (Wiącek et al. 2015). The first censuses were conducted in May of 2010, coinciding with the breeding season. Afterwards, the monitoring of bird populations took place during the first three years of operation (2011–2013), with censuses in both winter and spring. A final year-round monitoring with four seasonal samplings was conducted in 2014–2015. All of these censuses involved two complementary methodologies: (1) linear transects on foot, and (2) direct counts from a car along dirt paths.

The census of passerine birds and other small-sized species was performed by walking 1 km line-transects without a defined count bandwidth. Each transect was walked by a single experienced observer, and all birds identified–visually or acoustically–were recorded. This method is frequently used in extensive surveys to estimate abundance, distribution, and patterns of habitat selection by birds (Bibby et al. 2000). The transects were regularly distributed on both sides of the railway in two lines parallel to the HSR, one next to the fence to sample the bird community affected by the railway, and the second (control) at a distance of 500 m from it.

Direct counts of mid- and large-sized birds were conducted by car, as is common (Alonso et al. 2005; Morales et al. 2008). For these counts, the car was driven slowly along dirt tracks up to 1000 m from the railway, with a fixed detection band of 1000 m to each side of the vehicle. Regular stops were made to prospect the area thoroughly, including the surroundings of the railway and control sites on both sides. All morning, censuses were carried out within four hours of sunrise, and afternoon censuses were done in the 2 h before sunset. Censuses were done on days with good visibility and no rain, and hunting days were avoided.

In the landscape traversed by the HSR, the avian community was typical of that in the (Traba et al. 2013). Outstanding species typical of steppes include the great bustard, little bustard (*Tetrax tetrax*), and calandra lark (*Melanocorypha calandra*), together with others common in agricultural areas, including the spotless starling (*Sturnus unicolor*), rock dove (*Columba livia*), common magpie (*Pica pica*), sparrows (*Passer* spp., *Petronia petronia*), and several Fringilidae and Embericidae species.

In all, 80 bird species have been detected in the study area, reflecting the presence of a rich bird community, though 40–50% of the species that have been observed in the area were detected in fewer than 10 observations within each sampling period. Eleven species of high conservation interest were present in the area, including the black kite (*Milvus milvus*), which is nationally classified as "endangered" (Madroño et al. 2004). Species classified as "vulnerable" that were observed in the area include the great bustard, pin-tailed sandgrouse (*Pterocles alchata*), black-bellied sandgrouse (*P. orientalis*) and little bustard as well as Montagu's harrier (*Circus pygargus*).

Regarding changes potentially introduced by the railway, it is noteworthy that total bird densities were reduced from approximately 550 birds/km^2 in 2010–2011 to values in the range 360–390 birds/km^2 range in the following years. This change occurred shortly after the trains began running, but there is no information from control sites remote enough to affirm that the decrease was directly associated with the railway rather than other regional-scale phenomena. Among the local bird species, some showed a constant growing trend in their densities over the 5-year study period, including the corn bunting (*Emberiza calandra*), house sparrow (*Passer domesticus*), rock dove, common magpie, and Eurasian skylark (*Alauda arvensis*) (Table 8.1). Some species, such as the crested lark (*Galerida cristata*), showed consistently high average densities. Reductions in population density estimates over time were observed for the common quail (*Coturnix coturnix*), European serin (*Serinus serinus*), common linnet (*Carduelis cannabina*), meadow pipit (*Anthus pratensis*), calandra lark, Spanish sparrow (*Passer hispaniolensis*), and red-legged partridge (*Alectoris rufa*), among others.

Analysis of the spatial distribution of birds around the railway showed a differential use across the species (Table 8.1), though strong seasonal variability makes it difficult to detect the effects of HSR proximity. This relationship was negative for a number of species, including the calandra lark, crested lark, rock dove, and spotted starling. However, it was strongly positive for several species of finches (common linnet, European serin) that take advantage of the grassland abundance on HSR embankments. Others, such as the house sparrow, benefitted from their use of railway structures to nest. In parallel, a specific analysis of large- and mid-sized bird locations showed that both the great bustard and the little

Table 8.1 Trends in bird densities in the area traversed by the HSR, 2011–2015

Species		Average density	Density near the fence
Calandra lark	*Melanocorypha calandra*	–	–
Common chaffinch	*Fringilla coelebs*	–	–
Common linnet	*Carduelis cannabina*	–	+
Common magpie	*Pica pica*	+	
Common quail	*Coturnix coturnix*	–	
Common wood pigeon	*Columba palumbus*	–	
Corn bunting	*Emberiza calandra*	+	
Eurasian skylark	*Alauda arvensis*	+	–
Eurasian stone-curlew	*Burhinus oedicnemus*	–	
European serin	*Serinus serinus*	–	+
House sparrow	*Passer domesticus*	+	+
Little bustard	*Tetrax tetrax*	–	
Meadow pipit	*Anthus pratensis*	–	
Northern wheater	*Oenanthe oenanthe*	–	
Red-legged partridge	*Alectoris rufa*	–	
Rock dove	*Columba livia*	+	–
Sardinian warbler	*Sylvia melanocephala*	–	
Spanish sparrow	*Passer hispaniolensis*	–	
Spotless starling	*Sturnus unicolor*	+	–

Average density reflects average for whole study area. Density near the fence reflects data obtained <500 m from the fence. Only species that had a clear, stable trend, either (+) positive or negative (−), over several monitoring seasons are included

bustard significantly increased their distance (by 60–150 m) from the railway once trains started running. However, for most of the species, potential trends were obscured by seasonal and inter-annual variability.

In conclusion, the construction of the HSR resulted in a general decrease in bird density in the area, and modified the small-scale spatial patterns of the avian community. Thus, certain birds typical of these pseudo-steppe open landscapes, such as the little bustard, common quail, calandra lark, and red-legged partridge appeared to move away from the HSR location (Traba et al. 2013). On the other hand, the infrastructure has led to the presence and/or increase in the numbers of certain species, mostly anthropohiles, that use it as a substrate for nesting, resting, or feeding, such as, most notably, sparrows, starlings, rock doves, and magpies (Clavel et al. 2011). The many perching possibilities enabled by the HSR led to a wide use by corn buntings as well as by some raptors and corvids. On a narrower scale, some species were found to concentrate around the railway (common linnet, European serin, house sparrow), while others tended to avoid it (calandra lark, common chaffinch). It is noteworthy that there were some incongruous observations, such as species attracted to the railway or the embankments having decreased in numbers in the area (European serin, common linnet), while others that avoid the railway increased their abundance (Eurasian skylark, spotless starling, rock dove).

Bird Use of Railway Structures

Bird species behavior is an important factor in determining whether the HSR has a positive or negative impact on proximal population density. The mortality of birds depends on the extent to which they are exposed to the risk of being over-run while flying, or being electrocuted by the HSR poles or catenary. In contrast, the provision of new elements in the landscape (e.g., poles, bridges, embankments) may facilitate the presence and/or breeding of some species (Tryjanowski et al. 2013; Morelli et al. 2014). It is therefore interesting to elucidate how birds use track environs, because that use may be beneficial, such as for nesting, or detrimental, such as by producing a mortality risk. In a sense, the above-mentioned changes in the bird community may correspond to land-scape scale impacts, al-though some impacts of the railway are associated with much more specific processes that take place on the infrastructure itself (Mainwaring 2015).

Data in this section were obtained from fixed observation stations and made seasonally between the autumn of 2014 and the summer of 2015. The sampling unit in this case was 10-min observations of all birds that make any use of a 120 m section of the HSR from the position of the observer. This distance corresponds to two spans between catenary poles, elements that aide in systematic data collection and define a distance that allows the identification of species of almost all observed individual birds. For each observed bird, the infrastructure elements used, and the time spent in them, were noted. Recorded elements were powerlines (catenary, power line and all other supporting wires), built structures (passages and bridges), ballast (stone and gravel track bed), catenary poles, fences, and embankments.

Sampling station locations were selected with the premise of concentrating on high embankments where birds have access to all the typical HSR elements and directly face the need to determine at what height to cross the railway (see below). In the Villarrubia de Santiago area, 26–28 sampling stations were distributed over approximately 3360 m. In the Santa Cruz de la Zarza area, 20 stations were distributed over 2400 m. Data from both sections were clumped together for the presentation of results. The infrastructure use by birds (nesting, rest, power, etc.) is expressed as: (1) the number of birds and (2) the time of use per kilometer of HSR during 1 h of observation (birds $km^{-1} h^{-1}$).

In total, 936 birds, belonging to 30 species, were observed with an average of 81.8 birds and 404 min of use per kilometer of HSR observed for 1 h (Table 8.2). The most common species observed in the vicinity of the HSR were the spotless starling, common linnet, rock dove, house sparrow, rock sparrow, crested lark, and common magpie (all >5 birds $km^{-1} h^{-1}$). No protected species was found using the sampled elements of the infrastructure.

The structures used most were embankments (24 species in total; 26.4 birds $km^{-1} h^{-1}$) and powerlines (12 species; 25.2 birds $km^{-1} h^{-1}$), although the numerical importance of the latter is due to frequent perching by the most abundant species—the spotless starling. In this species, 94.3% of the observations were on these elements. In order of importance, following embankments and powerlines, are

Table 8.2 Species found making use of railway elements categorized by registered intensity of use

Use categories	Species
Scarce	*Buteo buteo* (1), Clamator glandarius (1), *Streptopelia decaocto* (1), *Alauda arvensis* (2), *Sylvia undata* (2), *Lanius meridionalis* (3), *Parus major* (3), *Oenanthe hispanica* (3), *Merops apiaster* (4), *Falco tinnunculus* (4), *Corvus corone* (4), *Sylvia cantillans* (4), *Chloris chloris* (4), *Oenanthe oenanthe* (7), *Passer montanus* (7), *Motacilla alba* (9), *Fringilia coelebs* (9)
Frequent	*Phoenicurus ochruros* (11), *Sylvia melanocephala* (11), *Saxicola torquatus* (16), *Emberiza calandra* (17), *Serinus serinus* (31), *Carduelis carduelis* (33)
Very frequent	*Columba livia* (46), *Pica pica* (63), *Galerida cristata* (79), *Petronia petronia* (97), *Passer domesticus* (108), *Carduelis cannabina* (112), *Sturnus unicolor* (244)

Scarce use: 1–10 observations; frequent: 11–40 observations; very frequent: >40 observations

Table 8.3 Birds' use of HSR structural elements

Species	Catenary	Passages	Ballast	Poles	Embankments	Fence
Phoenicurus ochruros		−	−		+	−
Sylvia melanocephala					+	−
Saxicola torquatus					+	+
Emberiza calandra	+				−	−
Serinus serinus					+	
Carduelis carduelis		−			+	−
Columba livia	−	+				
Pica pica	−	−	−	+	+	−
Galerida cristata			+	−	+	−
Petronia petronia	−	+			−	−
Passer domesticus	−	−	−		+	+
Carduelis cannabina	−		−		+	−
Sturnus unicolor	+	−		−	−	−
Average across species (%)	30.9	12.2	7.3	3.5	32.3	13.9

Only species with >10 observed individuals are included; +, intensive use (≥ 30% of observations); −, casual use (<30% of observations). Empty cells reflect no apparent use

fences (20 species; 11.3 birds km^{-1} h^{-1}), built structures (5 species; 9.9 birds km^{-1} h^{-1}), ballast (9 species; 5.9 birds km^{-1} h^{-1}), and, finally, catenary poles (5 species; 3.0 birds km^{-1} h^{-1}).

Table 8.3 summarizes HSR use by species observed more than 10 times. The embankments are used by most species as a feeding substrate and nesting place. Such nesting takes place both in bushes used for plant restoration as well as on the ground protected by them. For these species, embankments are islands with permanent and green vegetation year-round, contrary to the crop fields and fallows in the surrounding areas. The power lines are used as vantage and singing points by the corn bunting, in addition to the spotless starling. In the case of passages and bridges, the only note-worthy use was by the rock dove and the rock sparrow; both species rest and nest in holes within these structures. The ballast is only used by the crested lark, while the catenary poles are occupied as vantage points by the common magpie. It is noteworthy that during the monitoring presented in this chapter, it was discovered that these poles act as pitfall traps that cause the death of birds that fall inside (Malo et al. 2016). The spotless starling was the most affected species by this, but magpies and kestrels also suffered this fate. Finally, although the fences are used by many species, their use is only remarkable in the case of the house sparrow and European stonechat (*Saxicola torquatus*).

In conclusion, the HSR provides new structural elements for birds, which are used in accordance with the particular requirements of the species (Morelli et al. 2014). However, among the species that use these elements, some were found to present their maximum densities in close proximity to the railway (common linnet, house sparrow). More frequently, however, species were repelled by the railway (spotless starling, rock dove, crested lark). Therefore, the direct expectation that species that make an active use of the new opportunities provided by the HSR would increase their densities in proximity to the railway (Benítez-López et al. 2010), and in the area as a whole, was not fully met.

Bird Flight Over the Railway

In a sequential approach to the presentation of findings in the previous sections, to understand the potential magnitude of the HSR impact on birds, it is key to know to what extent flying birds face the risk of being overrun by circulating trains. Mortality from roadkill is one of the most serious direct impacts of operating transportation infrastructures (Loss et al. 2014). The potential relevance of the HSR is conditioned by the fact that the great speed of the trains, in most cases, precludes birds from avoiding train collisions (DeVault et al. 2015). If birds could fully avoid running trains by flying above their height, the railway would remain responsible only for effects on habitats and for increases in energy expenditure by individuals that cross it in flight.

The collision risk area associated with the HSR corresponds basically to the section framed by the catenary and the rails. In a simplified view, for twin tracks, it corresponds to a rectangle with an 8 m-wide base that is 5.3 m high (height of the power wire) where the trains run. Additionally, it is hazardous for birds to fly above the train collision risk area where they may come in contact with the catenary, suspenders, power wire, feeder, earth cable, and tensors (5.3–8.5 m above the ground). Although there is a constant risk of collision with these elements even when trains are not running, the risk of collision with the catenary may be increased by the passage of trains due to the potential for turbulence generated by the moving train to destabilize the normal flight of birds. Thus, it is reasonable to assume that birds that cross the railway by flying between or below the catenary wires face the mortality risk from train-kill.

In the HSR stretch analyzed here, surveys of birds flying over the infrastructure were carried out between 2010 and 2015, with a total of approximately 270 h of observation at nearly 1100 observation points. Sampling structure varied, depending on the objectives of particular studies, but in all cases observations were made from fixed stations for 10 min or 20 min. Bird flight was recorded from these points, with HSR crossing flight height estimated with respect to the rails. Therefore, for all crossing events over the railway (N = 3313 records and 10,776 individuals), it is known whether the birds crossed within the collision risk area (train path or catenary), allowing a risk index to be calculated for each species. Additionally, data obtained within 120 m of the observer have been used to estimate HSR crossing rates throughout the year (birds h^{-1} km^{-1}). Furthermore, spatial and temporal sampling patterns are informative for elucidating interactions of various factors, such as whether track crossover frequency is independent of HSR geometry, astronomical season, or environmental conditions such as wind speed. Complementarily, data available on birds from surrounding areas, such as parallel sampling of flight behavior, can be used to determine whether birds change their usual flight trajectories in the presence of the HSR. To ensure consistency of the results, the data presented are restricted to species for which, in at least one sampling station, 10 crossing records were obtained.

Our analysis showed that birds cross the HSR frequently, with an average HSR crossing frequency of 246.7 birds km^{-1} h^{-1} (experiment range, 160.2–514.0 birds km^{-1} h^{-1}). No consistent seasonal pattern was observed across the trials. However, our data analysis did reveal alterations in bird flight patterns in the vicinity of the HSR. An experiment conducted between June and July of 2011 at 237 observation stations along the HSR and 85 remote control sites showed that birds of any species fly, on average, almost 5 m higher near the HSR than in control sites (9.9 vs. 5.2 m over the ground). This increase in flight heights reflects some avoidance of the infrastructure, with an added energy cost. The average flight height of birds over the HSR was lower in areas where the railway runs along embankments relative to sections of railway that are level with the surrounding landscape. In fact, several experiments showed that birds reduced their crossing railway height between 1 m and 4 m for every 10 m of embankment height. However, these general patterns were found to vary greatly across different species, and all of them combine observations of birds both inside and outside the cross-section of collision risk.

Although the birds fly higher over the infrastructure than over an open field, the frequency with which they cross the railway within the risk area is high. For the 33 analyzed species, almost half of railway crossings (combined average, 46.5%) occurred less than 8.5 m above the ground (Table 8.4). More than two-thirds of these low crossings were under the catenary, and the remainder were through the catenary wires. In fact, birds of 21 species crossed within (rather than over) the risk area for at least 33% of their HSR crossings, and 14 species did so for at least 50%.

The partridge (100%), Eurasian tree sparrow (*Passer montanus*, 90%), hoopoe (*Upupa epops*, 88.5%), and northern wheatear (*Oenanthe oenanthe*, 83%) stand out

Table 8.4 Species for which it was possible to define flight behavior patterns relative to the collision risk area of the HSR (i.e. under or between the wires of the catenary)

Flying risk category	Species
Infrequent flight through risk area	*Otis tarda, Milvus migrans, Apus apus, Columba palumbus, Phalacrocorax carbo, Columba livia, Sturnus unicolor, Circus pygargus, Riparia riparia, Falco naumanni, Buteo buteo*
Frequent flight through risk area	*Anas platyrhynchos, undet. Passeriforme, Tetrax tetrax, Anthus pratensis, Fringilia coelebs, Falco tinnunculus, Corvus corone, Merops apiaster, Circus aeruginosus, Hirundo rustica, Melanocorypha calandra, Petronia petronia, Pica pica*
Usual flight through risk area	*Carduelis cannabina, Serinus serinus, Carduelis carduelis, Galerida cristata, Passer domesticus, Oenanthe oenanthe, Upupa epops, Passer montanus, Alectoris rufa*

The species are grouped into three categories of flying frequency through risk area defined as follows: infrequent, if they cross through it in less than a third of cases; frequent, between one-third and two-thirds of the cases; and usual, in more than two-thirds of the cases. Within each category the species are ordered by increasing values of risk values

as species with high-frequency risks of colliding with running trains. Conversely, the great bustard (0.0%), black kite (5.7%), common swift (*Apus apus*, 5.9%), wood pigeon (*Columba palumbus*, 8.7%), and great cormorant (*Phalacrocorax carbo*, 8.1%) showed low-risk flying behavior. However, the fact that 12 great bustard carcasses have been observed within the HSR study area indicates that great bustards must fly under or through the catenary at least occasionally (unpublished data, LIFE+ Impacto 0 2016). The collision risk was elevated on embankments, with the risk of collisions in one study being 14% greater on embankments over 5 m high than in nearby flat sections of the HSR. Birds from virtually all of the protected species present in the area crossed the risk area at least occasionally. Notably, 40.0% of little bustard crossings, 28.6% of lesser kestrel (*Falco naumanni*) crossings, and 24.3% of Montagu's harrier crossings were observed to be below 8.5 m.

Finally, another important factor that determines the height of flight over the HSR is wind. Because birds tend to fly lower in the presence of strong winds, percentages of railway crossings through the risk area increase under windy conditions. The presence of fog would be expected to affect birds' mortality risk in the vicinity of the HSR as well. However, foggy days were not sampled specifically. Generally, the potential influence of atmospheric conditions on bird behavior with respect to the HSR should be clarified to better understand the risk faced by birds crossing the railway.

In conclusion, although birds show some avoidance of the railway and catenary, many birds fly within the collision risk area. This risky behavior is also engaged in by protected species, although it is especially prevalent among species that make use of specific elements of the railway (e.g., common linnet, European serin, Eurasian skylark). There are also some species, such as the red-legged partridge and hoopoe, that tend to fly close to the ground naturally and thus face a high risk of collision with trains and the catenary.

Bird Behavior Towards Approaching Trains

A more detailed scale of bird interactions with the HSR is the behavior of individual birds facing an approaching train. Note that if birds are able to react before the arrival of a train, even high rates of crossing the risk area would be harmless. However, it is methodologically complex to analyze this issue. Data presented in the previous sections were in all cases collected with traditional direct observation methodologies, which involve time-space limitations (e.g., length of sections under scrutiny, difficulty or impossibility of accessing the infrastructure, and frequency of trains) that limit data quantity and quality (Wells et al. 1999). Such limitations are especially noticeable for relatively rare animals and exacerbated by the difficulty of observing the process in real time, given the velocity of HSR trains (Rodríguez et al. 2008).

To overcome these limitations, data in this section were obtained in the framework of a project currently under development (LIFE+ Impacto 0 2016) that uses high-speed video recording from the cockpits of running trains. This study is the first time that this method was applied in a railway study, previous studies used direct observations by train conductors (SCV 1996; SEO/BirdLife 1997; Wells et al. 1999), and video recordings have begun to be used to assess the interactions of wildlife with powerlines (Carlton and Harness 2001), wind turbines (Desholm et al. 2006; Cryan et al. 2014), airplanes (Doppler et al. 2015), and cars (Legagneux and Ducatez 2013; DeVault et al. 2015). Cockpit video data provide accurate information on location, time, conditions (i.e., speed of the train, type of stretch), and the behavior (i.e., distance of flight, type of reaction) of animals detected in front of the vehicle in events of crossing, infrastructure use, and collision. Also, observations can be made continuously along the train path for tens or even hundreds of kilometers, and include sections inaccessible to walking personnel due to viaducts and tunnels. Hence the cockpit video approach increases the number and types of observations that can be made.

Video recordings included in the present analysis were made with forward-facing, high-speed cameras (120–240 frames per second, GoPro Hero 3+ Black Edition) through the cockpit windshield. A technician located in the cabin operated the recording equipment together with GPS register systems and recorded all bird observations (perching, crossings, collisions) on standardized forms. A series of basic data were thus used to simplify later analysis (e.g., recording time, relative size or bird species, position in the image, type of observation). These forms served as a reference for the systematic review of recordings, focusing attention on direct observations. Technicians' observations, species, numbers of individuals, and behaviors were confirmed in the videos. In parallel, data from the navigation system, such as train speed, location (geographic coordinates and mile mark), stretch configuration, and surrounding habitat were extracted. High-speed recording allowed repeated and frame-to-frame review of each event to facilitate the analysis of the observations.

The train trips with high speed recording were between the Madrid and Albacete stations (321.7 km) of the Madrid-Levante line, which includes the section in which all field experiments were developed. A total of 66 trip recordings were collected over different seasons (14–20 per season). There were five additional trips during which the technical equipment was set up and tested (data not included). In all, 59 h and 55 min of recording (average trip, 55.3 min) along 14,700 km of accumulated train motion (average trip, 226.2 km; range: 78.3–288.7 km) were analyzed.

These recordings yielded 1090 confirmed bird observations, including 39 collisions (3.6% of observations), yielding a mortality risk of 0.0026 killed birds per km (1 hit per 406.1 km). Direct observation from the cockpit provided a new perspective in the analysis of the reactions of birds facing an approaching train. Train speed during bird collisions (mean ± standard deviation, 265.8 ± 39.2 km/h; range 175–305 km/h; N = 20) was similar to that during under-catenary crossings (251.5 ± 58.9 km/h; range 0–305 km/h; N = 183). However, in all cases, bird collisions occurred with trains travelling at high speeds. Collision rates varied seasonally, consistent with prior observations (SEO/BirdLife 1997; Frías 1999; Carvalho and Mira 2011; Bishop and Brogan 2013; Loss et al. 2014). The roadkill rate varied between 0.0018 and 0.0032 birds/km, with a noteworthy constant percentage of under-catenary crossings in front of the train resulting in collisions (∼ 12.2 ± 0.01%). These data suggest that train speed determines mortality risk independent of particular species' characteristics, with the risk being a result of the fact that birds are not adapted to avoiding objects approaching at such high velocities (Martin 2011; DeVault et al. 2015; Lima et al. 2015). Previous studies have pointed out that birds have difficulty avoiding vehicles travelling over 80–90 km/h (Pallag 2000; DeVault et al. 2014, 2015). The lowest speeds of collision recorded in our study were ∼ 180 km/h, similar to the findings of DeVault et al. (2015), who suggested that such speeds exceed birds' abilities to escape successfully.

Given these data and the infeasibility of reducing train speeds, information obtained from recordings with respect to birds' infrastructure use and behavior when facing a train, particularly in cases of collision, is of great interest. Of all recorded birds, 29.4% flew below the catenary (i.e. through the risk area), a datum similar to that registered in previous field studies (33.1, see above section). Cockpit recordings enabled us to determine that 37.7% of the birds that crossed the risk area were using some element of the infrastructure moments before their train encounters. That is, they commenced flight (most likely) due to the approach of the train, but crossed in front of it. Among the data obtained from the train, it is remarkable that 40.1% of the detected birds were initially observed in the infrastructure, and 28% of them crossed below the catenary. In addition, 5 of 29 collisions recorded from the cockpit (17.2%) involved birds resting on or in the infrastructure. Therefore, a high percentage of the birds that are finally exposed to the risk of collision were using the railway as a place to rest or feed. Hence, corrective measures, such as barriers designed to raise the flight of birds above the catenary (LIFE+ Impacto 0 2016), should also deter birds from HSR structures in order to reduce bird mortality risk in the HSR.

Table 8.5 Flight initiation distance (mean ± SD) of birds facing and approaching high speed trains

Bird group	Number of observations	Flight initiation distance (m)
Passerines	12	59.6 ± 33.5
Pigeons and doves	89	67.8 ± 36.8
Corvids	21	106.2 ± 42.2
Mid-sized raptors	40	136.0 ± 49.1

Data correspond to bird taxa with a sample size larger than 10 observations

On the other hand, our analysis of the flight initiation distance of different groups of birds showed that they do not seem to respond to trains as if they were predators, but rather human-associated elements (Møller et al. 2013; Neumann et al. 2013). Thus, species that are more tolerant of people or even anthropophilic (e.g., passerines and pigeons) show flight initiation distances that are significantly shorter than those of crows and birds of prey, which are more predatory (Table 8.5). Some species that are frequently observed around railways (e.g., the crested lark, doves, and pigeons) do not take flight in response to incoming trains, putting themselves at risk of being hit or dragged by the turbulence of the passing train.

In short, it is evident that at least some of the birds living in the surroundings of the HSR do not treat trains as a natural threat, and exhibit some habituation to HSR train circulation. However, HSR train speed often proves fatal when birds enter the collision risk area. Both factors could explain the high mortality rates near the HSR relative to rates near other linear infrastructures, though further research on other HSR stretches should be done to determine if this pattern of findings is representative of broader HSR areas.

Conclusions

Our results provide a first picture of the avian response to the presence and operation of HSR trains. Although the observed trends are not particularly prominent and are diluted by seasonal and inter-annual variability, they are consistent with the notion that train circulation could be responsible for a decrease in bird abundance in the traversed area. Accordingly, it can be concluded that the presence of the HSR has led to a process of ruderalization characterized by the increased numbers of species taking advantage of the railway. Some bird species representative of open and agricultural areas were found to avoid the HSR, whereas species that use man-made structures were found to be attracted to it. In fact, we documented the active use of these new elements, mainly the embankments and the catenary, by birds. In addition, birds alter their flight patterns near the railway, including showing some avoidance of it. However, birds risk train collision because such avoidance is not complete. Moreover, the species that are most exposed to train collisions are those that respond favorably to the railway and use its associated elements. The cockpit recordings confirmed bird habituation to moving trains and that some portion of over-run specific birds were actively using the railway moments before they faced the approaching train. Nevertheless, the flight characteristics of each species are also a determinant, and species that fly close to the ground are exposed to a disproportionate collision risk even if they do not make active use of the railway. These findings led us to conclude that changes in the community of birds in the surroundings of the HSR are relevant and related to structural landscape changes and the responses of bird species to the new infrastructure. In addition, bird mortality may have a significant impact on the populations of some species, given the fact that birds cross the collision risk area frequently. The results may be contentious from an environmental impact perspective given the limited opportunities for mitigation (Rodríguez et al. 2008). Firstly, changes in the avian community due to the presence of the HSR and the use of HSR structures are probably difficult to avoid because birds use elements that are fundamental to the infrastructure, thus, only minor design changes can be implemented to minimize HSR impacts on birds. Secondly, even if the installation of some devices can reduce flight paths across the railway, their implementation will be feasible only in short stretches of high ornithological interest due to cost limitations. Finally, reducing train speed in poder to reduce the risk of collisions with birds would conflict with the core objective of a HSR, making such a strategy again unlikely along long stretches of railway.

Acknowledgments Jorge Hernández Justribó, Mª José Pérez Sobola, Diego Sánchez Serrano, and safety personnel from Centro de Estudios y Experimentación of the Ministry for Public Works (CEDEX) and Ineco helped with fieldwork. Staff from CEDEX and Adif provided kind help with project logistics. This study was carried out as part the LIFE+ Impacto 0 research project (LIFE 12 BIO/ES/000660) and a research agreement between UAM and CEDEX; both focused on the mitigation of High Speed Railway impacts on bird populations. The Comunidad de Madrid, together with the European Social Fund, supports the Terrestrial Ecology Group research group through the REMEDINAL-3 Research Network (S-2013/MAE-2719).

References

Alonso, J. C., Palacín, C., & Martín, C. A. (Eds.). (2005). *La avutarda común en la Península Ibérica: población actual y método de censo*. Madrid: SEO/BirdLife.

Benítez-López, A., Alkemade, R., & Verweij, P. A. (2010). The impacts of roads and other infrastructure on mammal and bird populations: A meta-analysis. *Biological Conservation, 143*, 1307–1316.

Bibby, C. J., Burgess, N. D., & Hill, D. A. (2000). *Bird census techniques* (2nd ed.). London: Academic Press.

Bishop, C. A., & Brogan, J. M. (2013). Estimates of avian mortality attributed to vehicle collisions in Canada. *Avian Conservation and Ecology, 8*, 2.

Campos, J., & de Rus, G. (2009). Some stylized facts about high-speed rail: A review of HSR experiences around the world. *Transport Policy, 16*, 19–28.

Carlton, R. G., & Harness, R. E. (2001). Automated systems for monitoring avian interactions with utility structures and evaluating the effectiveness of mitigative measures. In: *Power Engineering Society winter meeting* (Vol. 351, pp. 359–361). IEEE.

Carvalho, F., & Mira, A. (2011). Comparing annual vertebrate road kills over two time periods, 9 years apart: A case study in Mediterranean farmland. *European Journal of Wildlife Research, 57,* 157–174.

Clavel, J., Julliard, R., & Devictor, V. (2011). Worldwide decline of specialist species: Toward a global functional homogenization? *Frontiers in Ecology and the Environment, 9,* 222–228.

Cryan, P. M., Gorresen, P. M., Hein, C. D., Schirmacher, M. R., Diehl, R. H., Huso, M. M., et al. (2014). Behavior of bats at wind turbines. *Proceedings of the National Academy of Sciences, 111,* 15126–15131.

Desholm, M., Fox, A. D., Beasley, P. D. L., & Kahlert, J. (2006). Remote techniques for counting and estimating the number of bird-wind turbine collisions at sea: A review. *Ibis, 148,* 76–89.

DeVault, T. L., Blackwell, B. F., Seamans, T. W., Lima, S. L., & Fernández-Juricic, E. (2014). Effects of vehicle speed on flight initiation by turkey vultures: Implications for bird-vehicle collisions. *PLoS ONE, 9,* e87944.

DeVault, T. L., Blackwell, B. F., Seamans, T. W., Lima, S. L., & Fernández-Juricic, E. (2015). Speed kills: Ineffective avian escape responses to oncoming vehicles. *Proceedings of the Royal Society of London B, 282,* 20142188.

Doppler, M. S., Blackwell, B. F., DeVault, T. L., & Fernández-Juricic, E. (2015). Cowbird responses to aircraft with lights tuned to their eyes: Implications for bird–aircraft collisions. *The Condor, 117,* 165–177.

Dorsey, B., Olsson, M., & Rew, L. J. (2015). Ecological effects of railways on wildlife. In R. van der Ree, D. J. Smith, & C. Grilo (Eds.), *Handbook of road ecology* (pp. 219–227). West Sussex: Wiley.

Frías, O. (1999). Estacionalidad de los atropellos de aves en el centro de España: número y edad de los individuos y riqueza y diversidad de especies. *Ardeola, 46,* 23–30.

Fu, X., Zhang, A., & Lei, Z. (2012). Will China's airline industry survive the entry of high-speed rail? *Research in Transportation Economics, 35,* 13–25.

Legagneux, P., & Ducatez, S. (2013). European birds adjust their flight initiation distance to road speed limits. *Biology Letters, 9,* 20130417.

LIFE+ Impacto 0. (2016). *Development and demonstration of an anti-bird strike tubular screen for High Speed Rail lines.* LIFE+ Biodiversity project (LIFE12 BIO/ES/000660). http://www.lifeimpactocero.com. Accessed October 14, 2016.

Lima, S. L., Blackwell, B. F., DeVault, T. L., & Fernández-Juricic, E. (2015). Animal reactions to oncoming vehicles: A conceptual review. *Biological Reviews, 90,* 60–76.

Loss, S. R., Will, T., & Marra, P. P. (2014). Estimation of bird-vehicle collision mortality on U.S. roads. *Journal of Wildlife Management, 78,* 763–771.

Madroño, A., González, A., & Atienza, J. C. (2004). Libro Rojo de las aves de España. Dirección General para la Biodiversidad-SEO/BirdLife, Madrid.

Mainwaring, M. C. (2015). The use of man-made structures as nesting sites by birds: A review of the costs and benefits. *Journal of Nature Conservation, 25,* 17–22.

Malo, J. E., García de la Morena, E. L., Hervás, I., Mata, C., & Herranz, J. (2016). Uncapped tubular poles along high-speed railway lines act as pitfall traps for cavity nesting birds. *European Journal of Wildlife Research, 62,* 483–489.

Mammen, U., Klammer, G., & Mammen, K. (2002). Greifvogeltod an Eisenbahntrassen—ein unterschaetztes problem [Dead of raptors at railways—An underestimated problem]. Paper presented at the 5. Internationales Symposium "Populationsökologie von Greifvogel- und Eulenarten" ["Population Ecology of Raptors and Owls"], Meisdorf/Harz.

Martin, G. R. (2011). Through birds' eyes: Insights into avian sensory ecology. *Journal of Ornithology, 153,* 23–48.

Møller, A. P., Grim, T., Ibáñez-Álamo, J. D., Markó, G., & Tryjanowski, P. (2013). Change in flight initiation distance between urban and rural habitats following a cold winter. *Behavioral Ecology, 24,* 1211–1217.

Morales, M. B., Traba, J., & Carriles, E. (2008). Sexual differences in microhabitat selection of breeding little bustards *Tetrax tetrax*: Ecological segregation based on vegetation structure. *Acta Oecologica, 34,* 345–353.

Morelli, F., Beim, M., Jerzak, L., Jones, D., & Tryjanowski, P. (2014). Can roads, railways and related structures have positive effects on birds?—A review. *Transportation Research Part D Transport and Environment, 30,* 21–31.

Neumann, W., Ericsson, G., Dettki, H., & Radeloff, V. C. (2013). Behavioural response to infrastructure of wildlife adapted to natural disturbances. *Landscape Urban Plan, 114,* 9–27.

Ninyerola, M., Pons, X., & Roure, J. M. (2005). *Atlas Climático Digital de la Península Ibérica. Metodología y aplicaciones en bioclimatología y geobotánica.* Bellaterra: Universidad Autónoma de Barcelona.

Pallag, O. (2000). COST 341. Habitat fragmentation due to transportation infrastructure. Hungarian State of the Art Report. Technical and Information Services on National Roads (ÁKMI), Budapest.

Petts, J. (Ed.). (1999). *Handbook of environmental impact assessment* (2 Vols.). Oxford: Blackwell Science.

Rodríguez, J. J., García de la Morena, E. L., & González, D. (2008). *Estudio de las medidas correctoras para reducir las colisiones de aves con ferrocarriles de alta velocidad.* Ministerio de Fomento, CEDEX, Madrid.

S. C. V. (1996). *Mortalidad de vertebrados en líneas de ferrocarril.* Documentos Técnicos de Conservación SCV. Sociedad Para la Conservación de los Vertebrados, Madrid.

SEO/BirdLife. (1997). *Mortalidad de aves en un tramo de línea de ferrocarril.* SEO/BirdLife, Grupo local SEO-Sierra de Guadarrama.

Spencer, K. J. (1965). Avian casualties on railways. *Bird Study, 12,* 257.

Todorovich, P., Schned, D., & Lane, R. (2011) *High-speed rail.* International Lessons for U.S. Policy Makers. Lincoln Institute of Land Policy, Cambridge, Massachusetts, USA.

Torres, A., Jaeger, J. A. G., & Alonso, J. C. (2016). Assessing large-scale wildlife responses to human infrastructure development. *Proceedings of the National Academy of Sciences, 113,* 8472–8477.

Torres, A., Palacín, J., Seoane, J., & Alonso, J. C. (2011). Assessing the effects of a highway on a threatened species using Before-During-After and Before-During-After-Control-Impact designs. *Biological Conservation, 144,* 2223–2232.

Traba, J., Sastre, P., & Morales, M. (2013). Factors determining species richness and composition of steppe bird communities in Peninsular Spain: Grass-steppe vs. shrub-steppe bird species. In M. Morales & J. Traba (Eds.), *Steppe ecosystems: Biological diversity, management and restoration* (pp. 29–45). New York: Nova Publishers.

Tryjanowski, P., Sparks, T. H., Jerzak, L., Rosin, Z. M., & Skórka, P. (2013). A paradox for conservation: Electricity pylons may benefit avian diversity in intensive farmland. *Conservation Letters, 7,* 34–40.

Wells, P., Woods, J. G., Bridgewater, G., & Morrison, H. (1999). Wildlife mortalities on railways: Monitoring methods and mitigation. In Proceedings of the Third International Conference on Wildlife Ecology and Transportation. ICOET, Missoula, Montana.

Wiącek, J., Polak, M., Filipiuk, M., Kucharczyk, M., & Bohatkiewicz, J. (2015). Do birds avoid railroads as has been found for roads? *Environmental Management, 56,* 643–652.

Wilson, J. R. U., Proches, S., Braschler, B., Dixon, E. S., & Richardson, D. M. (2007). The (bio)diversity of science reflects the interests of society. *Frontiers in Ecology and the Environment, 5,* 409–414.

Chapter 9
Relative Risk and Variables Associated with Bear and Ungulate Mortalities Along a Railroad in the Canadian Rocky Mountains

Benjamin P. Dorsey, Anthony Clevenger and Lisa J. Rew

Abstract Train-wildlife collisions can impact wildlife populations as well as create human and resource management challenges along railways. We identified locations and railroad design features associated with train-wildlife collisions (strikes) on a 134 km section of the Canadian Pacific Railroad (CPR) that travels through the Banff and Yoho National Parks. A 21-year dataset of train strikes with elk (*Cervus elaphus*), deer (*Odocoileus* spp.), American black bears (*Ursus americanus*) and grizzly bears (*U. arctos*) were compared to relative abundance estimates, and nine train and railroad variables. Train strikes and relative abundance varied spatially for elk, deer and bears. Hotspots and relative risk estimates were used to identify potential problem locations. Hotspots were defined as segments of the train line where strike counts were above the 95% confidence interval based on a Poisson distribution, and could be identified for elk and deer but not bears. Relative risk was estimated as the ratio of strike counts to that expected based on relative abundance. High relative risk locations, where more strikes occurred than were expected, were identified for elk, deer, and bears. Relative abundance was positively correlated with strikes for elk and deer but not bears. Train speed limit was positively associated with strikes for elk and deer. For bears, the number of structures (e.g., overpasses, tunnels, snow sheds and rock cuts) and bridges were positively correlated to strikes. To reduce the risk of train strikes on wildlife, our management recommendations include train speed reduction, habitat modifications and railroad design alterations.

Keywords Animal collision mortality · Grain spillage · Railroad · Relative risk · Transportation corridor

B.P. Dorsey (✉)
Parks Canada, Revelstoke, BC V0E 2S0, Canada
e-mail: bpdorsey@gmail.com

A. Clevenger
Western Transportation Institute, College of Engineering, Montana State University, 334 Leon Johnson, Bozeman, MT 59717, USA

L.J. Rew
Land Resources and Environmental Sciences, Montana State University, 334 Leon Johnson, Bozeman, MT 59717, USA

© The Author(s) 2017
L. Borda-de-Água et al. (eds.), *Railway Ecology*,
DOI 10.1007/978-3-319-57496-7_9

135

Introduction

Like roads, railroads affect wildlife through direct mortality, habitat loss and habitat fragmentation (van der Grift 1999; Forman et al. 2003; Davenport and Davenport 2006). Direct mortalities result when trains strike wildlife. Strikes can be a significant source of mortality for some wildlife populations and have been reported for decades (Child 1983; Gundersen et al. 1998; Bertch and Gibeau 2010a). Studies have reported strike rates for large mammals such as grizzly bears (*Ursus arctos*) (Bertch and Gibeau 2010a, b) and moose (*Alces alces*) (Child 1983, 1991; Modafferi 1991).

In Canada's Rocky Mountain National Parks, train strikes are a leading source of mortality for grizzly bears (0.35 year^{-1}) and black bears (1.95 year^{-1}), and the second largest source of mortality for deer (*Odocoileus* spp.), elk (*Cervus elaphus*) and moose (Bertch and Gibeau 2010a, b) in the Banff and Yoho National Parks. It is likely that true mortality rates due to train strikes are higher than reported. For example, as few as 50% of strikes with large mammals were reported by standard observers (train engineers) along the Canadian Pacific Railroad (CPR) during a six-year period (Wells et al. 1999). In other cases, strikes may not be reported except when large groups (>450) of wildlife are killed (Chaney 2011) or when strikes occur within protected areas (Waller and Servheen 2005). Long-term data of train strikes along the CPR exist because strikes have been reported to Parks Canada for at least 30 years. However, other railroads may not report or record strikes with such consistency.

Studies on roads have analyzed the spatial pattern of road-kills, which showed that these occurred in clusters (Finder et al. 1999; Clevenger et al. 2003; Malo et al. 2004). The spatial pattern of road-kills has been explained by landscape, environmental and infrastructure variables (Finder et al. 1999; Hubbard and Danielson 2000; Gunson et al. 2006; Kassar 2005). These studies have helped inform management actions targeted at reducing road-kills (Clevenger et al. 2001; Grilo et al. 2009). At least five general factors are thought to affect the spatial pattern of road-kills and train strikes (Seiler and Helldin 2006). Along railroads, these include: animal (e.g., wildlife abundance and behavior); train (e.g., train speed and frequency); railroad design (e.g., curvature or alignment); and landscape (e.g., vegetation type) variables (Huber et al. 1998; Bashore et al. 1985; Finder et al. 1999; Seiler and Helldin 2006). Driver behavior variables are largely removed from train strike analyses because trains generally cannot stop or swerve to avoid animals.

Landscape variables derived from land cover data have been found to be the best predictors of road-kill rates (Bashore et al. 1985; Finder et al. 1999; Roger and Ramp 2009). We suggest that estimates of relative abundance be used to assess the spatial pattern of road-kills and train strikes. The importance of including wildlife abundance in the analysis of factors contributing to wildlife strikes was demonstrated for moose: strikes coincided with locations of high moose abundance in wintering areas and on migration routes (Gundersen et al. 1998; Ito et al. 2008). Train variables and railroad design, as well as animal abundance, vary along the CPR and may also affect strike rates. If train or railroad variables altered the probability of a strike, the rate of strikes would vary spatially along with differences in these variables, even if wildlife abundance were constant. There are at least three theoretical mechanisms that could alter the spatial probability of strike occurrence. These mechanisms, based on wildlife behavior are: (1) constrained flight paths—some design features, such as bridges, may restrict wildlife movements on the rail bed surface instead of out of the path of oncoming trains; (2) reduced detectability—some design features may impede the sight and/or sound of oncoming trains, so that wildlife are less likely to detect trains resulting in increased strike rates; and (3) reduced reaction time—higher train speeds may reduce the time available to wildlife to successfully cross or flee before being struck. These mechanisms likely interact. For example, with increasing train speeds both the time to detect and flee may be reduced.

There is evidence that these mechanisms affect the rate of train strikes with ungulates and bears along railroads. Previous studies have shown that bears were struck at bridges and rock cuts (constrained flight paths) and in locations where the sound of oncoming trains was reduced (reduced detectability) (Van Why and Chamberlain 2003; Kaczensky et al. 2003). Moose have been observed to not flee off the railroad tracks in Alaska, with the suggestion that deep snow alongside the railroad restricted their ability to flee down the track (Andersen et al. 1991; Modafferi 1991). Similarly for bears, engine-mounted video footage suggested that some bears were struck while fleeing from trains where track-side slopes or other design variables appeared to obligate flight paths on the tracks instead of off and out of danger. Other associations are unclear, for example bighorn sheep (*Ovis canadensis*) strikes have been associated with rock cuts and may be a result of fine-scale habitat selection or multiple design mechanisms acting concurrently (Van Tighem 1981). As a result, it may be difficult to determine the exact mechanism(s) driving strike rates. However, identifying the animal, train or railroad variables associated with strikes and the direction of the effect (e.g., increased mortality at a particular design feature) may be a more tractable approach and could indicate ways to reduce strikes along the CPR and railways more broadly.

The purpose of this research was to identify locations where species-specific mitigation solutions are needed, and to identify the variables associated with those locations along the CPR. Four ungulate species (elk, mule deer [*Odocoileus hemionus*], white-tailed deer [*O. virginianus*] and two bear species (grizzly and black bear) were used for this analysis). We tested whether strike rates were correlated with the relative abundance and nine railway variables that could alter the flight path, detectability or reaction time of animals present.

Methods

Study Area

The study was conducted along a 134 km section of the CPR running from the western boundary of Yoho National Park, British Columbia (116° 39′W, 51° 14′N) to the

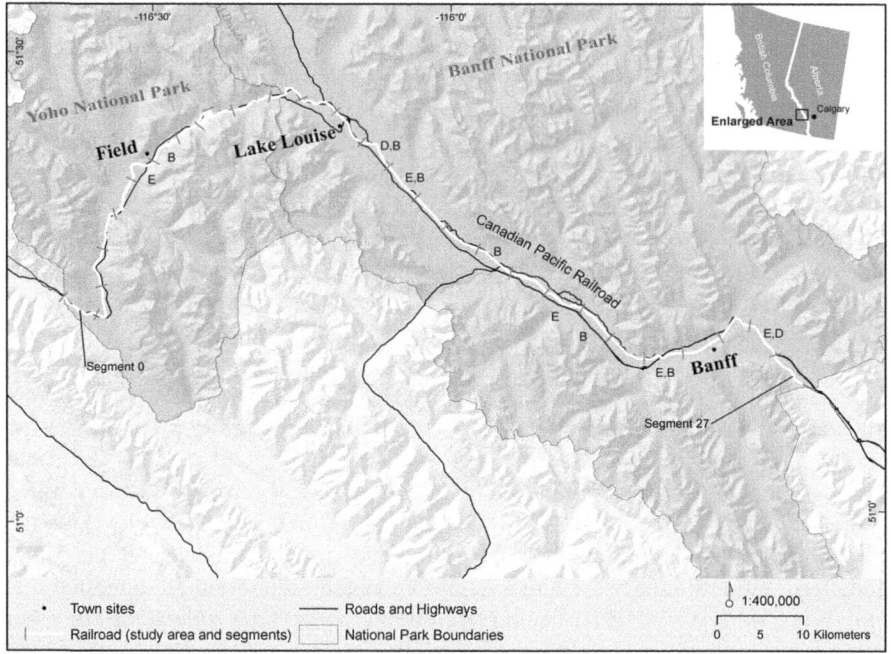

Fig. 9.1 Map of the 134 km of the Canadian Pacific Railroad study area that traverses the Banff and Yoho National Parks in the Canadian Rocky Mountains. Each analysis segment (4.86 km) is shown denoted by a small perpendicular bar along the railroad. Segments with high strike risk are labeled for (*E*) elk, (*D*) deer, and (*B*) bears during the 21-year period (1989–2009) along the Canadian Pacific Railroad through the Banff and Yoho National Parks

eastern boundary of Banff National Park, Alberta (115° 25′W, 51° 8′N) (Fig. 9.1). An average of 19 trains a day^{-1} pass through the study area at an average speed of 60 km h^{-1}. The Trans-Canada Highway runs parallel to the CPR throughout the study area, on average 416 m ($\pm$325 m) away. To prevent wildlife-vehicle collisions, there is a 2.4 m high wildlife exclusion fence along the highway for 48 km, starting at the eastern border of the study area (Clevenger et al. 2001). There were 27 wildlife crossing structures over or under the highway at the time of this study, one of the structures was a shared wildlife crossing and railroad underpass. An additional section of fencing (28 km) was completed in 2010 along with wildlife crossing structures and a shared wildlife railroad underpass (Clevenger et al. 2009).

The CPR crosses the Western Continental Divide at 1670 m above sea level 9.0 km west of Lake Louise, Alberta. The CPR consists of one track except for an 8.3 km long section near Lake Louise, where two tracks diverge to a maximum of 500 m. Due to large elevation gradients in the area (>2,000 m), a number of different vegetation types exist, and the average annual rainfall decreases from west to east. Approximately 39% of the land area in Banff and Yoho National Parks is at or above tree line ($\sim$2,300 m). Where vegetation was present, it consisted of closed and open canopy forest, dominated by lodgepole pine (*Pinus contorta*), white spruce (*Picea glauca*), and subalpine fir (*Abies lasiocarpa*) (Holland and Coen 1983). Vegetation within the CPR right-of-way (ROW) was kept in an early successional stage by mechanical treatments and consisted primarily of herbaceous vegetation (grasses, sedges and rushes), dandelions (*Taraxacum* spp.), bearberry (*Arctostaphylos uva-ursi*) and horsetail (*Equisetum* spp.). Numerous species of shrub including buffaloberry (*Sheperdia canadensis*) and tree species including lodgepole pine and white fir were present at reduced heights (<2 m). Vegetation between the rails was almost non-existent due to herbicide treatments and the rocky substrate. Where there was vegetation between the rails there was small herbaceous growth of native and non-native grasses or young sprouts of train-spilled agricultural products (cereal grains).

Data Collected

Wildlife Strike Records

Counts of train-struck ungulates and bears were obtained from Parks Canada for a 21-year period (1989–2009) (Parks Canada, unpublished data, 2010). These records included all strikes reported by CPR and Parks Canada personnel. Strikes were visited and removed only by Parks Canada personnel to avoid double counting. Records after 2001 were spatially referenced with a Garmin® global positioning system (GPS), but before 2001, records were referenced to permanent mile-marker posts or to geographic features along the CPR. As a result some strike records were spatially inaccurate by as much as 1.6 km (1.0 mile); thus, an error of plus or minus 1.6 km was assumed for all records. A spatial data layer representing the railroad was acquired from CPR and then imported into a geographic information system (GIS). The spatial layer was separated into 30 roughly equal segments 4.86 km (3 miles) in long. The segment length was based on a conservative estimate of the minimum mapping accuracy of the wildlife strike records, which were tallied to the appropriate 4.86 km (3 mile) segment in a GIS.

Relative Abundance Data

Relative abundance data were collected by sampling transects oriented parallel and perpendicular to the CPR (hereinafter called "rail bed transects" and "perpendicular transects," respectively). Both methods were binned to 4.86 km long rail segments, ensuring that data were analyzed at the same spatial scale. Sampling was carried during the summers of 2008–2009 by the same observer. A total of 313 km of rail bed transects and 129 perpendicular transects were surveyed in 2008 and 2009.

The relative abundance of bears was sampled using both methods. However, ungulates were sampled only using perpendicular transects that allowed for an accurate representation of ungulate distribution throughout the year.

Parallel transects were centered directly on the rail bed. Segments were treated as a strip transect 3 m wide and 4.86 km long. Locations of bear scats were recorded using a GPS unit (Trimble Navigation Ltd., Sunnyvale, USA). Black and grizzly bear scats were classified as "bear" because they could not be differentiated in the field. When the diameter, age and contents were visibly similar for scats found within 5 m, only one scat was recorded. All scats detected were removed from the CPR to avoid double counting. Where multiple tracks were encountered, e.g., sidings, all tracks were surveyed and the mean number of scats detected per track was used in the analysis.

Perpendicular transects were used to estimate relative abundance for bears and ungulates within the railroad corridor (<250 m). Two random points were generated within each rail segment in a GIS for each sampling year. Each point was used as the starting location for a perpendicular transect. At each point, a transect extended out perpendicularly away from the CPR in both directions for 250 m; the transects were 3 m wide. Pellet groups and scat found more than 1.5 m from the transect center were not recorded. Perpendicular transects were truncated at the edge of impassable features such as rock cliffs, rivers and open water. Along each transect ungulate pellet groups were classified into those of elk or deer based on differences in size and shaped as described by Elbroch (2003). White-tailed deer and mule deer pellets were combined into one class of "deer" because they could not be reliably differentiated in the field. The number of pellet groups was counted for each transect. Bear scats were recorded using the same criteria as the rail bed transects. Pellets and scat that appeared to be more than a year old-based on visible signs of decay (bleached color or unconsolidated) were excluded as were amorphous ungulate or bear feces (Elbroch 2003).

The count of bear scats were weighted based on the date they were detected, as they are detectable throughout the short summer season (Dorsey 2011). The first day that bears have been historically sighted in the study area is April 15; although the date of bears' emergence varies each year, it was used as a starting date for when bears could have been foraging on grain and left scats. As each day passed after April 15, there were cumulatively more days when bears could have visited and left scats. Therefore, each scat was given a weight inversely proportional to the day of year it was detected after April 15, which is the 105th Julian calendar day of the year. For example, the weighted value of one scat detected on September 13th, (day 273 of the year) would be $((273-105)^{-1} = 0.006)$. The final value was the sum of all weighted scats detected for each track segment stored as $RA_{rail\ bed}$.

The relative abundance for each species was defined as the number of pellet groups or weighted value km^{-1} within each 4.86 km long segment (i) (Eq. 1). For example, the relative abundance of elk, $RA_{rail\ corridor}$ (i) was the number of elk pellet groups km^{-1} averaged over n transects (t) within each rail segment (i).

$$RA_{rail\ corridor}(i) = \frac{\sum \left(\frac{pellet\ groups_t}{transect\ length_t} \right) * 1000}{\sum_{t=1}^{n} t} \tag{1}$$

Railroad Design and Train Spill Grain Data

Three additional variables were measured at the starting location of each perpendicular transect. The mean right-of-way width (ROW_{mean}) was measured using a Bushnell® Yardage Pro® 1,000 range finder (Bushnell Corporation, Denver, CO), which extended to the forest edge or open water. A maximum distance of 500 m was imposed in locations where open habitat and level ground resulted in an unidentifiable ROW corridor. Topography adjacent to the railroad was classified into three categories similar to those used to assess wildlife-vehicle collisions on the parallel Trans-Canada Highway (Clevenger et al. 2003; Gunson et al. 2006) (VERGESLOPE): (0) flat; (1) raised and partly raised; and (2) buried, buried-raised, and partly buried. Train-spilled grain was also measured at each transect starting location on the rail bed. Grain density was estimated visually by counting the wheat and barley seeds inside a frame (10 cm^2) placed randomly on the rail bed three times within a 5 m segment of track (Fig. 9.2). The sampling process was repeated for each track where sidings were present. Grain density was calculated as the average of all subsamples within a rail segment ($GRAIN_{mean}$); the density was affected by the date it was measured because grain spill decreased over the course of a summer (Dorsey

Fig. 9.2 Grain spill sampling method used on the rail bed. The sampling frame (10 × 10 cm^2), was randomly thrown three times to estimate the mean density of wheat and barley seeds within a 5 m zone

2011). Therefore, grain density was weighted by the week of year it was measured. For both years, the weighted value was calculated as the mean seed count multiplied by the average rate of decrease (3%) for each additional week after April 15th (data not shown). For example, a weighted seed count of 80 wheat and barley seeds measured on June 20th (4 weeks after April 15th) was [80 × (0.97 × 4) = 310.4]. The grain density data were binned to the appropriate 4.86 km long segment in a GIS.

The GIS layer representing the railroad contained additional information on train speed limits, bridges, sidings and track grades. The highest posted train speed limit (SPEED$_{max}$) and mean track grade (GRADE$_{mean}$) were calculated for each segment. The variable SINUOSITY was calculated using the "sinuosity" function in Hawth's analysis tools extension for ArcGIS 9 (ESRI Redlands, CA 2004). The number of bridges were counted and summed for the variable, hereinafter termed "BRIDGE$_c$". A count of vehicle overpasses, tunnels, snow sheds or rock cuts occurring along each segment were each given a value of "1" and summed for the variable hereinafter called "BARRIER$_c$". The lengths of track inside two tunnels and one secondary track west of Lake Louise were omitted from analyses; in all, 28 segments were used in the analysis.

Data Analysis

Analyses were conducted independently for elk, deer and bears. Strikes that occurred over the 21-year period were compared to relative abundance estimated using data covering a 2-year period (2008 and 2009). This comparison assumes that relative abundance across the study area remained relatively stable during the preceding years. To assess this assumption, we looked for changes in the distribution of strikes over time. A better approach would be to assess changes in relative abundance over time, but these data were not available for enough of the study area or species. Changes in the distribution of strikes over time were assessed by conducting ANOVA on year and rail segments. More variability in year than rail segment may mean that substantial shifts occurred in the relative abundance, therby invalidating further analysis. Next, to determine if strikes and/or wildlife abundance were evenly distributed along the CPR, chi-squared tests were used. To identify hotspots the upper 95% confidence interval of strikes per segment was used, which assumed that the counts followed a Poisson distribution (Bivand et al. 2008; Malo et al. 2004).

Risk estimates for each rail segment were developed through three steps. First, the RA$_{rail\ corridor}$ or RA$_{rail\ bed}$ estimates for each segment (i) were converted to a percentage of the total from all rail segments. Next, the expected number of strikes for each rail segment (Expected$_{(i)}$) was calculated by multiplying the percent of total scat on that segment by the overall mortality rate for that species (Eq. 2).

$$Expected_i = \frac{\sum STRIKES_i}{\sum RA_{rail\ corridor_i}} * RA_{rail\ corridor_i} \qquad (2)$$

Risk was estimated by assessing the ratio of the number of strikes to the expected number of strikes for each segment (Eq. 3). Both values were assumed to equal at least 1.0 for all rail segments because all species are included in the study area. In cases where no strikes or no pellets were detected the risk estimate was set to 1.0 without confidence intervals. Otherwise the risk estimate would have equaled zero or infinity in these cases.

$$Risk_i = \frac{Strikes_i}{Expected_i} \tag{3}$$

The risk estimate, evaluates relative risk between segments. It has been widely used in spatial epidemiology studies and is referred to as a standardized mortality ratio where the ratio is expected to equal 1.0 because internal standardization was used (Mantel and Haenszel 2004; Banerjee et al. 2004; Bivand et al. 2008). In these data, the average risk (risk = 1.0) meant that the number of strikes was proportional to the relative abundance for that species. When the estimated ratio was above 1.0 it indicated relatively high risk (more strikes occurred than expected). Internal standardization reduced the number of degrees of freedom to $n - 1$ which was used in chi-squared tests (Kim and Wakefield 2010).

To test for non-constant risk, the number of strikes observed was compared to the number expected using χ^2 tests (Bivand et al. 2008) and the relationship between observed and expected was evaluated with generalized linear models assuming a Poisson count distribution and a log link function. To determine which sections had an unusually high risk, a bootstrap function was used to resample the wildlife abundance estimates from scat counts for each rail segment. A new relative risk estimate was computed for each of 1,000 resampled abundance estimates; then confidence intervals were assessed from the distribution of the 1,000 estimates. High-risk segments were defined as those where 95% of bootstrapped estimates of risk were above 1.0. All analyses were performed in R (R Development Core Team 2009).

A set of variables hypothesized to affect train strikes were tested using generalized linear mixed models with a log link function (Table 9.1). Models were evaluated for each species by fitting the count of strikes per segment (summed over the 21-year period) to 10 predictors, including the appropriate wildlife abundance estimate (which included two for bears, due to evaluation at the rail bed and landscape scales). Initial fits used restricted maximum likelihood estimation and likelihood ratio tests to assess whether a negative binomial distribution was needed (Zuur et al. 2009). Each model was subjected to a "drop one approach," where the least significant parameter was dropped until all remaining parameters were significant. If models were reduced to a single predictor, the model with the minimal residual deviance was selected as the final model. Final models were refit with maximum likelihood and inspected using standard diagnostic plots. Semivariograms and residuals from the final model were plotted to assess the remaining spatial trends. Prior to performing the analysis, co-variates were tested for collinearity (Menard 1995). When correlated variables (r > 0.6) were found, the

Table 9.1 Description of field-collected and GIS-derived spatially varying train and railroad design variables and the hypothesized correlation to strike rates

Variable	Description	Hypothesis	Source
Continuous variables			
RA$_{rail\ corridor}$	**Mean number of weighted pellets or scats standardized by distance searched along perpendicular strip transects**	+	Field
RA$_{rail\ bed}$	Total of weighted bear scats detected on the CPR track bed	+	Field
GRAIN$_{mean}$	Mean number of train-spilled grains	+	Field
GRADE$_{mean}$	Mean percent grade of the railroad bed	+	GIS
ROW$_{mean}$	**Mean right-of-way width**	−	GIS
SINUOSITY	Ratio of rail segment length to the distance between the start and end locations	+	GIS
SPEED$_{max}$	**Maximum speed limit in miles per hour**	+	GIS
Categorical variables			
BARRIERc	**Count of overpass, snow shed, rock cut, or tunnel**	+	Field
BRIDGEc	Count of bridges	+	Field
SIDING	Presence/absence	+	Field
VERGESLOPE[a]	Railroad verge slope classes[a]	+	Field

(+) indicates a positive correlation and (−) indicates a negative correlation hypothesized for both ungulates and bears. Those in bold provided significant general linear mixed-effects model relationships see Table 9.3
[a](0) flat (1) at least one down slope (2) at least one up slope or water

one with a higher correlation to the response remained in the analysis (Guisan and Zimmermann 2000). Based on this cut-off, track grade was significantly correlated with train speed and was removed from the analysis.

Results

Strike Rates

Over the 21-year period (1989–2009), 862 strikes were recorded along 134 km of the CPR, consisting of 579 elk, 185 deer, 69 black bears, 9 grizzly bears, and 1 unidentified bear species. The bear data were combined into one class for further analysis. The spatial distribution of strikes per rail segment were non-uniform for elk $\chi^2(1, 27) = 834.4$, $p < 0.001$, deer $\chi^2(1, 27) = 252.5$, $p < 0.001$, and bears $\chi^2(1, 27) = 37.1$, $p = 0.03$ (Fig. 9.3).

The strike rates also varied temporally for elk, deer, and bears (Fig. 9.4). A decreasing trend was detected for elk strikes [$\beta_{elk} = -1.04$, t(19) $= -2.29$, $p = 0.03$], and conversely an increasing trend for deer [$\beta_{deer} = 1.13$, t(19) $= 8.18$,

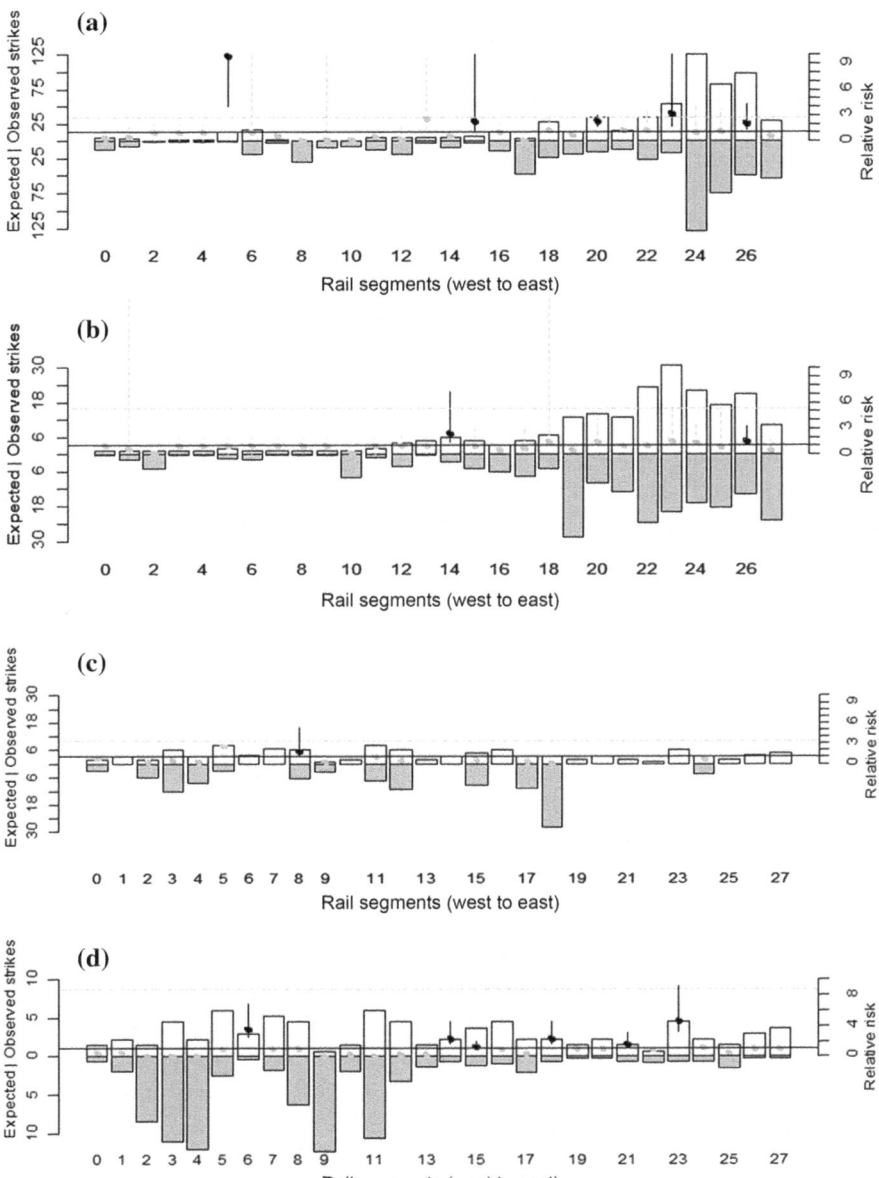

Fig. 9.3 The number of train strikes (*white bars*) for **a** elk; **b** deer; and **c** bear, compared to the expected number of strikes (*gray bars,* a positive value) based on the abundance of wildlife signs along each 4.86 km segment using perpendicular transects; **d** compares on-track relative abundance for bears to number of strikes. Hotspots are segments with a strike count above the 95% confidence interval (*gray dashed line*) and high-risk segments are those with a risk estimate significantly above 1.0 (*black points* and *error bars*). *Gray points* are risk estimates not significantly different from 1.0 (the number killed was close to the number expected based on wildlife abundance). Segments proceed west to east, where 0 is the first 4.86 km inside the western boundary of Yoho National Park. The field town site is located at segment 6, the Lake Louise town site corresponds to segment 13, and the Banff town site segment 24

$p < 0.001$] and bears [β_{bear} = 1.12, t(19) = 4.58, $p < 0.001$] over the 21-year period. Although annual mortality rates fluctuated for all three species, the spatial distribution of strikes accounted for more of the variability in strikes for both elk and bear, but not for deer (Table 9.2).

Relative Abundance

A total of 341 bear scats were detected on the CPR, resulting in a detection rate of 0.46 km^{-1} year^{-1}. On perpendicular transects, 599 elk pellet groups (5.98 km^{-1} year^{-1}), 212 deer pellet groups (2.23 km^{-1} year^{-1}), and 39 bear scats (0.33 km^{-1} year^{-1}) were detected. Relative abundance varied for elk [$\chi^2(27)$ = 940.12, $p < 0.001$] and deer [$\chi^2(27)$ = 294.2, $p < 0.001$]. Bear scats were unevenly distributed both on the rail bed [$\chi^2(27)$ = 263.3, $p < 0.001$] and on perpendicular transects [$\chi^2(27)$ = 125.4, $p < 0.001$] (Fig. 9.3c, d). The rail bed transects likely better represented relative risk for bears, because few bear scats were detected along the perpendicular transects (n = 39) relative to the 341 scats detected directly on the rail bed (Fig. 9.3d). Therefore, further analyses for bears were based on rail bed scats.

Hotspots and Relative Risk

Six elk hotspots were identified (21% of the CPR segments), which were segments with 29 or more strikes (Fig. 9.3a). Likewise, five deer hotspots (17% of the CPR segments) were identified, which averaged 6.54 ± 1.73 per segment (Fig. 9.3b). Bears were struck on average 2.86 ± 0.44 per segment, but no segment was identified as a hotspot because no single segment incurred more than seven strikes.

After estimating the ratio of strikes to wildlife abundance for each segment, overall non-constant risk was apparent for elk (χ^2 = 286.1, df = 27; $p < 0.0001$), deer (χ^2 = 182.0, df = 27; $p < 0.0001$), and bears (χ^2 = 7.4 × 10^6, df = 27; $p < 0.0001$). High-risk segments were identified for each species (Figs. 9.1 and 9.3). Five, two and six high-risk segments were determined for elk, deer and bears, respectively, and hotspots and high-risk segments differed in number and location (Fig. 9.3).

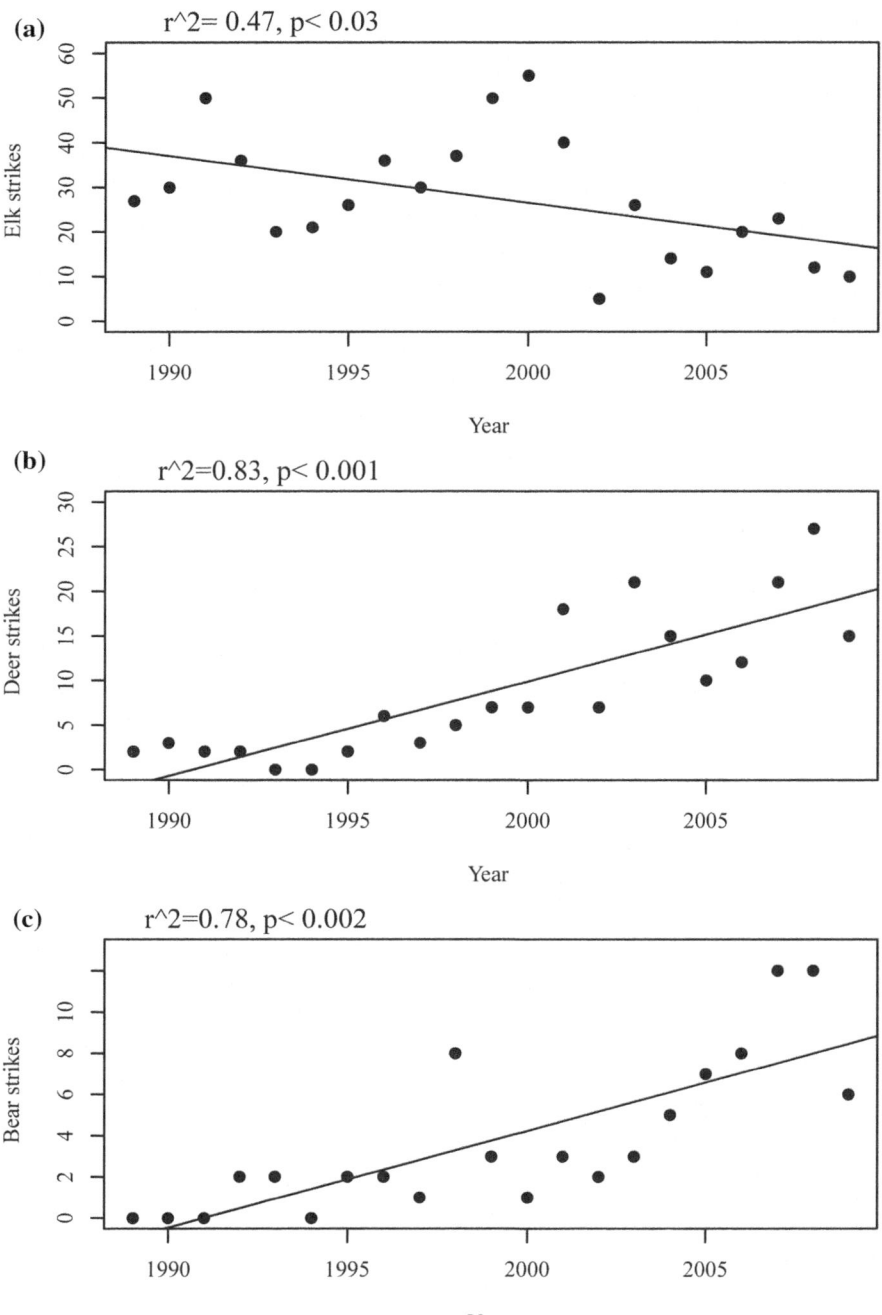

Fig. 9.4 Annual train strikes for **a** elk, **b** deer and **c** bears along 134 km of the Canadian Pacific Railroad through the Banff and Yoho National Parks, 1989–2009

Table 9.2 Variation between years and railroad segments for (a) elk, (b) deer, and (c) bears along the Canadian Pacific Railroad within the Banff and Yoho National Parks, 1989–2009

Factor	df	Sum Sq.	Mean Sq.	F	p
(a) Elk					
Year	20	311.31	15.57	2.4	<0.001
Railroad segment	26	760.81	29.26	4.6	<0.001
Residuals	234	1489.85	6.37		
(b) Deer					
Year	20	74.19	3.71	3.7	<0.001
Railroad segment	26	67.65	2.60	2.6	<0.001
Residuals	234	233.37	1.00		
(c) Bear					
Year	20	17.19	0.86	3.2	<0.001
Railroad segment	26	22.34	0.86	3.2	<0.001
Residuals	234	63.27	0.27		

Table 9.3 Significant ($p < 0.05$) parameter estimates explaining train strike rates with elk, deer, and bears

	Elk			Deer			Bears		
	β	SE	p	β	SE	p	β	SE	p
Explanatory variables									
Y-intercept	−1.063	0.782	<0.001	−1.917	0.751	<0.001	0.825	0.153	<0.001
RA$_{rail}$ corridor	0.024	0.007	<0.001	0.069	0.014	<0.001			
ROW$_{mean}$				0.009	0.003	0.005			
SPEED$_{max}$	0.080	0.020	<0.001	0.055	0.017				
Categorical variables									
BARRIERc							0.458	0.170	0.007
Dispersion (θ)	1.5			3.8			20.0		
Deviance explained (%)	41.7			28.3			82.042		

Model-fit statistics are provided from maximum likelihood fits

Variables Associated with Strikes

The variability in train strikes to relative abundance and nine train or railroad design variables (Table 9.1) indicated a negative binomial distribution for all species models. Initial models did not indicate that autocorrelation was present between rail segments and the most parsimonious models are reported (Table 9.3).

Elk

Train speed and relative abundance explained the variability in elk strikes (Log likelihood = -113.0 with 2 df, $\chi^2 = 26.3$, $p < 0.001$). Elk strikes increased on average $e^{0.07} = 1.07$ per segment, with each 1.0 mph increase in maximum posted train speed when elk relative abundance was held constant (Table 9.3). Likewise when the $SPEED_{max}$ was held constant strikes increased $e^{0.03} = 1.03$ for each one unit increase in elk $RA_{rail\ corridor}$. The variable $SPEED_{max}$ averaged 37.5 ± 1.85 mph (60.3 kmph) and ranged from 20 to 50 mph.

Deer

The variables deer relative abundance ($RA_{rail\ corridor}$), $SPEED_{max}$ and ROW_{mean} best explained the variability in deer strikes using the model selection process (Log likelihood = -78.46 with 5 df, $\chi^2 = 48.89$, $p < 0.001$). The parameter coefficients from a maximum likelihood fit indicated that $e^{0.06} = 1.06$ additional deer strikes were observed on average with an 1.0 mph increase in the posted train speed limit when deer abundance and ROW width were held constant. Likewise, $e^{0.009} = 1.009$ additional deer strikes were observed for each 1.0 m increase in ROW width when speed and deer abundance were held constant. The variable ROW_{mean} was on average 79.78 ± 7.68 m across all segments.

Bears

A single predictor model, including the variable $BARRIER_c$, best explained bear strikes (Log likelihood = -57.233 on 4 df, $\chi^2 = 14.686$, $p < 0.001$). The parameter estimated for $BARRIER_c$ ($\beta_{barrier} = e^{0.458} = 1.58$) revealed that for each additional barrier feature per segment bear strikes increased on average 1.58 (95% CI 1.13 to 2.20) (Table 9.3). A second single predictor model has almost equal explanatory power. This model included the variable $BRIDGE_c$. However, $BRIDGE_c$ and $BARRIER_c$ were not correlated (r = 0.23, t = 1.1968, df = 26, p = 0.24). The variable $BRIDGE_c$ was positively correlated to bear strikes (r = 0.44, $p < 0.02$).

For bears, neither estimate of relative abundance was significant; therefore, the final model did not include either variable describing bears relative abundance. This changes the modeling results from assessing risk to incidence rate. For this reason, the model selection procedure was repeated with the variable $RA_{rail\ bed}$ log transformed and held in the model as an offset to assess relative risk (Zuur et al. 2009). This model selection procedure resulted in a single predictor model that included $SPEED_{max}$, indicating that as posted train speed increased, so did the "risk" of bear strikes ($\beta = e^{0.053} = 1.054$). However, train speed explained only 17% of the deviance, compared to 82% explained by the $BARRIER_c$ model (Table 9.3).

Discussion

This study is the first attempt to account for unevenly distributed wildlife populations when assessing strikes along a railroad; its information was used to help
understand the spatial pattern of train strikes along the CPR. Hotspots were segments
of the CPR that had strike rates significantly higher than the overall mean, and were
identified for elk and deer, but not bears. Hotspots were generally associated with
higher relative abundance for elk and deer at the analysis scale (3 miles). Although
there was disagreement between bears' relative abundance and strike rates at this
scale (Fig. 9.3c, d), general correlation was apparent at a larger scale. For example,
bear strikes and relative abundance were relatively higher in the western half of the
study area (Fig. 9.3d segments 0–3), compared to the eastern half (Fig. 9.3d segments 14–27). High-risk segments had significantly higher kill rates than expected,
based on estimates of relative wildlife abundance, and they indicated potential
problem areas that may have been overlooked using strike data alone.

There was a significant relationship for elk, deer and bears with at least one train
or railroad design variable. These relationships indicated higher abundance and
train speeds (elk and deer), and larger right-of-way widths (deer) were associated
with increased strike rates. For bears, the number of barriers and presence of bridges
was also positively correlated with strike rates. These results supported the
hypothesis that there are at least three general variables that affect the spatial pattern
of train strikes: 1) the relative abundance of wildlife either on the rail bed or on the
adjacent landscape; 2) train speed; and 3) railroad designs such as highway overpasses, rock-cuts, tunnels, snow sheds and bridges. Other railroad studies have
shown similar associations with bears. In Slovenia, Eurasian brown bears
(*U. arctos*) were struck at rock cuts and on bridges (Kaczensky et al. 2003). At least
one road study has noted the effect of bridges on strikes with bears (Van Why and
Chamberlain 2003). Huber et al. (1998) studied locations where brown bears were
struck by trains in Croatia, and used variables similar to those of this study. They
found no difference in verge slope or longitudinal or perpendicular visibility but did
detect a difference in the presence of bear foods at strike locations compared to
random locations. They noted a slight difference in the distance at which trains were
first audible at strike locations (Huber et al. 1998). Data on train noise was not
collected in this study, but field observations suggested that train volume varied
depending on train direction within the study area, and the direction in which the
trains were moving when each animal was struck was not recorded.

There are at least four possible explanations for the lack of correlation between
bear strikes and relative abundance found in this study. The first is based on the
strength of non-constant risk phenomena, and three others deal with limitations of
the methods and analysis. First, railroad design or other variables may have strongly
affected strike probabilities. If these variables had strong causal or probabilistic
effects, the spatial pattern of strikes would be a function of the spatial pattern of
these variables and not the bears' relative abundance. However, it is unclear
whether this is the case because the sample size is relatively low (n = 80) and no

validation of these results has been attempted. Therefore, other explanations for the lack of agreement between strikes and bears' relative abundance could be due to the field and/or analysis methods used. The survey design has been shown to affect the estimation of roadkill hotspots for small-bodied species (Santos et al. 2015).

The field data collected may have failed to adequately represent the true distribution of bears along the CPR, since bears may have been moving or utilizing the railroad without leaving scat, which would have been undetectable using these methods. Additionally, scat sampling methods may have only represented animals that regularly encountered the CPR and have learned to coexist, to some degree, with the railroad. Bears that utilized grain as a food item in this study area may have developed behaviors that allowed them to forage on grain with reduced strike risk (learned behavior). Lastly, the analysis scale may have been too small or too large. Bears have been documented to have large home ranges thus a larger analysis scale may have been more appropriate. General large-scale agreement between relative abundance and strikes are apparent in Fig. 9.3d. However, it may be difficult to implement management solutions at this scale other than those generally aimed at population distribution, size, or health.

Similarly, the data collected may have failed to fully describe the deer distribution. This may have been why right-of-way width (ROW-width) was positively correlated to strikes for deer. The hypothesized influence of ROW width was that narrower ROW's (a negative relationship) would increase strike risk to wildlife. This variable may have accounted for spatial or temporal variability in deer abundance that the transect data failed to capture.

An animal's behavior when a strike occurs is only partially influenced by location-specific variables. Other variables that are likely to affect an animal's behavior are difficult to measure over long periods of time and large geographic areas. These variables include: fine-scale temporal variables and behavioral heterogeneity between individual animals. For this study, strikes spanning 21 years was used to provide adequate sample sizes for comparison to railroad variables that have not changed during that time; however, it is clear that strikes have changed over this period (Fig. 9.4), and thus relative abundance may have also changed during this period. To address this concern, analyses were repeated using a 10 year period. Results showed some shifts in the locations of hotspots and high-risk segments, but they these did not alter the general conclusions. The longer term 21-year analysis was likely better because of the larger sample size, which provided more information for hotspots, multivariate modeling, and more stable risk estimates.

Management Implications

This study found the wildlife abundance, train speed, barriers, bridges, and right-of-way widths affect strike rates for ungulates and/or bears. Based on these results, a variety of approaches could be used to reduce strike rates. These include those that affect wildlife abundance near railroads, reduced train speeds, or modified railway designs. Much of the CPR is co-aligned with a productive montane habitat, hence altering the abundance of deer or other species near the CPR may not be feasible. A two- step approach may be needed that both reduces forage quality and availability along the CPR and increases habitat quality away from it.

Two studies have documented a decrease in strike rates for moose (Jaren et al. 1991; Andreassen et al. 2005). One developed ungulate habitat away from a railroad (Jaren et al. 1991), and the other both modified habitat and provided supplemental forage away from the railroad (Andreassen et al. 2005). The feasibility of such an approach will be site-specific, but is likely to be beneficial to large mammal populations.

Train speed was also clearly important. Reduced train speeds may be an effective measure to reduce strike risk, particularly in problem areas. Train speed is a spatially and temporally explicit management solution, and likely interacts with the other mechanisms described above; thus, it could be implemented where or when strikes are most likely to occur. However, reduced speeds may be ineffective if other mechanisms are the primary driver in strike occurrences. For example, a study in Alaska observed a possible mechanism (constrained flight paths) resulting in strikes with moose (Becker and Grauvogel 1991). The study then evaluated reduced train speeds (reaction time mechanism) on reducing moose strike rates. The authors found reduced train speeds to be ineffective in this case but suggested that at some levels of speed reduction, the approach may have been effective; since these speeds are surely economically cost-prohibitive (Becker and Grauvogel 1991), and so variables (such as snow removal) enabling moose to move off the railroad, vegetative or other terrain modifications should be evaluated. Although moose were not analyzed in this study, speed was positively associated with increasing strike risks for elk and deer along the CPR. Speed reduction in hotspots and high-risk areas should be empirically evaluated in the future.

For bears, it may be most important to evaluate design modifications or mitigation solutions targeted at barriers including: highway vehicle overpasses, tunnels, snow sheds, rock cuts and bridges. However, multiple approaches are likely warranted in high-risk segments, including those that decrease the probability the bears will be exposed to strikes, increase the detectability of trains, increase the opportunities for safe flight paths off-track, and increase the time bears have to successfully avoid trains.

Acknowledgements B. Dorsey received support from the Western Transportation Institute (WTI) at Montana State University and Parks Canada to complete this project, as part of his Master's degree.

References

Andersen, R., Wiseth, B., Pedersen, P. H., & Jaren, V. (1991). Moose-train collisions: Effects of environmental conditions. *Alces, 27*, 79–84.

Andreassen, H. P., Gundersen, H., & Storaas, T. (2005). The effect of scent-marking, forest clearing, and supplemental feeding on moose-train collisions. *Journal of Wildlife Management, 69*, 1125–1132.

Banerjee, S., Carlin, B. P., & Gelfand, A. E. (2004). *Hierarchical modeling and analysis for spatial data*. London: Chapman & Hall.

Bashore, T. L., Tzilkowski, W. M., & Bellis, E. D. (1985). Analysis of deer-vehicle collision sites in Pennsylvania. *The Journal of Wildlife Management, 49*, 769–774.

Becker, E. F., & Grauvogel, C. A. (1991). Relationship of reduced train speed on moose-train collisions in Alaska. *Alces, 27*, 161–168.

Bertch, B., & Gibeau, M. (2010a). *Grizzly bear monitoring in and around the Mountain National Parks: Mortalities and bear/human encounters 1980–2009*. Parks Canada report.

Bertch, B., & Gibeau, M. (2010b). *Black bear mortalities in the Mountain National Parks: 1990–2009*. National Parks. Parks Canada report.

Bivand, R., Pebesma, E., & Gómez-Rubio, V. (2008). *Applied spatial data analysis with R*. New York: Springer.

Chaney, R. (2011, April 28). Antelope roam west to Clinton. *Missoulian*.

Child, K. (1983). Railways and moose in the central interior of BC: A recurrent management problem. *Alces, 19*, 118–135.

Child, K. (1991). Moose mortality on highways and railways in British Columbia. *Alces, 27*, 41–49.

Clevenger, A., Chruszcz, B., & Gunson, K. (2001). Highway mitigation fencing reduces wildlife-vehicle collisions. *Wildlife Society Bulletin, 29*, 646–653.

Clevenger, A., Chruszcz, B., & Gunson, K. (2003). Spatial patterns and factors influencing small vertebrate fauna road-kill aggregations. *Biological Conservation, 109*, 15–26.

Clevenger, A. P., Ford, A. T., & Sawaya, M. A. (2009). *Banff wildlife crossings project: Integrating science and education in restoring population connectivity across transportation corridors*. Final report to Parks Canada Agency.

Davenport, J., & Davenport, J. L. (Eds.). (2006). The ecology of transportation: Managing mobility for the environment. In *Environmental pollution*. Dordrecht: Springer.

Dorsey, P. B. (2011). *Factors affecting bear and ungulate mortalities along the Canadian Pacific Railroad through Banff and Yoho National Parks*. MS thesis, Montana State University.

Elbroch, M. (2003). *Mammal tracks and signs: A guide to North American species*. Mechanicsburg, PA: Stackpole Books.

Finder, R. A., Roseberry, J. L., & Woolf, A. (1999). Site and landscape conditions at white-tailed deer/vehicle collision locations in Illinois. *Landscape and Urban Planning, 44*, 77–85.

Forman, R. T. T., Sperling, D., Bissonette, J. A., Clevenger, A. P., Cutshall, C. D., Dale, V. H., et al. (2003). *Road ecology: Science and solutions*. Washington, DC: Island Press.

Grilo, C., Bissonette, J. A., Santos-Reis, M. (2009). Spatial–temporal patterns in Mediterranean carnivore road casualties: Consequences for mitigation. *Biological Conservation, 142* (2009), 301–313.

Guisan, A., & Zimmermann, N. (2000). Predictive habitat distribution models in ecology. *Ecological Modelling, 135*, 147–186.

Gundersen, H., Andreassen, H. P., & Storaas, T. (1998). Spatial and temporal correlates to Norwegian moose-train collisions. *Alces, 34*, 385–394.

Gunson, K., Chruszcz, B., & Clevenger, A. (2006). What features of the landscape and highway influence ungulate vehicle collisions in the watersheds of the Central Canadian Rocky mountains: A fine-scale perspective? In K. Mcdermott, C. L. Irwin, & P. Garrett (Eds.), *Proceedings of the 2005 international conference on ecology and transportation* (pp. 545–556).

Holland, W. D., & Coen, G. M. (1983). *Ecological (Biophysical) Land classification of Banff and Jasper National Parks* (Vol. 1), Edmonton, AB.

Hubbard, M., & Danielson, B. (2000). Factors influencing the location of deer-vehicle accidents in Iowa. *The Journal of Wildlife Management, 64*, 707–713.

Huber, D., Kusak, J., & Frkovic, A. (1998). Traffic kills of brown bears in Gorski Kotar, Croatia. *Ursus, 10*, 167–171.

Ito, T. Y., Okada, A., Buuveibaatar, B., Lhagvasuren, B., Takatsuki, S., & Tsunekawa, A. (2008). One-sided barrier impact of an International Railroad on Mongolian Gazelles. *Journal of Wildlife Management, 72* (4), 940–943.

Jaren, V., Andersen, R., Ulleberg, M., Pedersen, P., & Wiseth, B. (1991). Moose-train collisions: The effects of vegetation removal with a cost-benefit analysis. *Alces, 27,* 93–99.

Kaczensky, P., Knauer, F., Krze, B., Jonozovic, M., Adamic, M., & Gossow, H. (2003). The impact of high speed, high volume traffic axes on brown bears in Slovenia. *Biological Conservation, 111,* 191–204.

Kassar, C. (2005). *Wildlife—Vehicle collisions in Utah: An analysis of wildlife road mortality hot spots, economic impacts, and implications for mitigation and management.* Thesis: Utah State University, Logan, Utah.

Kim, A., & Wakefield, J. (2010). *R data and methods for spatial epidemiology: The SpatialEpi Package.*

Malo, J., Suárez, F., & Díez, A. (2004). Can we mitigate animal-vehicle accidents using predictive models. *Journal of Applied Ecology, 41,* 701–710.

Mantel, N., & Haenszel, W. (2004). Statistical aspects of the analysis of data from retrospective studies of disease. *The Challenge of Epidemiology: Issues and Selected Readings, 1,* 533–553.

Menard, S. W. (1995). *Applied logistic regression analysis.* Thousand Oaks, CA: Sage.

Modafferi, R. (1991). Train moose-kill in Alaska: Characteristics and relationship with snowpack depth and moose distribution in lower Susitna Valley. *Alces, 27,* 193–207.

R Core Team. (2009). *R: A language and environment for statistical computing.* R Foundation for Statistical Computing, Vienna, Austria. http://www.R-project.org/.

Roger, E., & Ramp, D. (2009). Incorporating habitat use in models of fauna fatalities on roads. *Diversity and Distributions, 15,* 222–231.

Santos, S. M., Marques, J. T., Lourenço, A., Medinas, D., Barbosa, A. M., Beja, P., et al. (2015). Sampling effects on the identification of roadkill hotspots: Implications for survey design. *Journal of Environmental Management, 162,* 87–95.

Seiler, A., & Helldin, J. O. (2006). Mortality in wildlife due to transportation. In J. Davenport & J. L. Davenport (Eds.), *The ecology of transportation: Managing mobility for the environment* (pp. 165–189). Berlin: Springer.

van der Grift, E. (1999). Mammals and railroads: Impacts and management implications. *Lutra, 42,* 77–91.

Van Tighem, K. (1981). *Mortality of bighorn sheep on a railroad and highway in Jasper National Park, Canada.* Parks Canada report.

Van Why, K., & Chamberlain, M. (2003). Mortality of black bears, *Ursus americanus,* associated with elevated train trestles. *Canadian Field Naturalist, 117,* 113–114.

Waller, J., & Servheen, C. (2005). Effects of transportation infrastructure on grizzly bears in northwestern Montana. *Journal of Wildlife Management, 69,* 985–1000.

Wells, P., Woods, J., Bridgewater, G., & Morrison, H. (1999). Wildlife mortalities on railways; monitoring methods and mitigation strategies. In G. Evink, P. Garrett, & D. Zeigler (Eds.), *Proceedings of the third international conference on wildlife ecology and transportation* (pp. 237–246), Tallahassee, FL.

Zuur, A., Ieno, E., Walker, N., Saveliev, A., & Smith, G. (2009). *Mixed effects models and extensions in ecology with R. statistics.* New York: Springer.

Chapter 10
Railways and Wildlife: A Case Study of Train-Elephant Collisions in Northern West Bengal, India

Mukti Roy and Raman Sukumar

Abstract The extensive network of the Indian Railways cuts through several forested landscapes, resulting in collisions of trains with a variety of wildlife species, including the largest land mammal–the elephant. In India, railway lines cross elephant habitats in several states, with accidents that resulted in more than 200 elephant deaths between 1987 and 2015. As the 161-km Siliguri–Alipurduar track in the northern West Bengal state witnesses train–elephant collisions frequently, we developed a case study there with the objectives of mapping locations of collisions and generating a susceptibility map showing locations prone to accidents. We mapped elephant crossing points and movement paths along this railway track, as well as accident locations. Between 1974 and 2015, collisions occurred throughout the line, although there were several hotspots where elephant deaths were concentrated. A disproportionate number of accidents occurred during the night. Crop raiding in villages and train elephant accidents seem to be closely related, probably due to an increased frequency of elephant movement near or across this railway track during the cultivation season. Male elephants were much more prone to accidents, possibly because of behavioural characteristics that make them cross railway tracks more frequently. To reduce the frequency of accidents in this region, we recommend reducing the speed of trains, limiting the operation of trains during at night, provisioning overpasses and underpasses, using communications technology, realigning a portion of the track, and fencing the track except for corridor areas.

Keywords Train elephant accidents · Elephant deaths · Crop raiding · Accident susceptibility map · Northern West Bengal

M. Roy (✉)
Asian Nature Conservation Foundation, c/o Centre for Ecological Sciences, Indian Institute of Science, Bangalore 560012, Karnataka, India
e-mail: muktiroy@rediffmail.com

R. Sukumar
Centre for Ecological Sciences, Indian Institute of Science, Bangalore 560012, Karnataka, India
e-mail: rsuku@ces.iisc.ernet.in

© The Author(s) 2017
L. Borda-de-Água et al. (eds.), *Railway Ecology*,
DOI 10.1007/978-3-319-57496-7_10

157

Introduction

Many animals range widely across landscapes in the quest to meet their daily, seasonal and annual biological needs of food, water, shelter and mates (Caughley and Sinclair 1994). Habitat connectivity within a landscape helps dispersal and re-colonization, thus maintaining regional metapopulations and minimizing risks of inbreeding within populations (Newmark 1987; Wilcove et al. 1998). Historically, the spread of agriculture probably contributed the most to loss and fragmentation of their natural habitat. In an increasingly industrialized world, however, transport networks (railways, roads, and waterways) or so-called "linear infrastructures," restrict the movement of wildlife populations by fragmenting their habitat, increasing edge effects, constricting ecological corridors, blocking animal movement, and increasing the risk of mortality due to direct collisions with motorized vehicles (Forman and Deblinger 2000; Trombulak and Frissell 2000; Inell et al. 2003; Van der Ree et al. 2011). These processes hinder the persistence of species in human-dominated landscapes because small and isolated populations are more vulnerable to extinction from stochastic demographic processes and loss of genetic variation (Soule and Wilcox 1980).

Roads are perhaps the most widespread and pervasive form of "linear infrastructure" that have greatly impacted wildlife populations in the more developed parts of the world through habitat loss, restriction of animal movements, alteration of animal behaviour, and directly injuring or killing very large numbers of animals in collisions with vehicles (Trombulak and Frissell 2000; Seiler and Helldin 2015). A variety of animals are involved in road–kills, from large-bodied ones such as moose to smaller creatures such as frogs (Fahrig et al. 1995; Formann et al. 2003; Dorans et al. 2012). One estimate of road–kills suggested that 1,000,000 vertebrates were killed in the USA every day (Lalo 1987), while Conover et al. (1995) estimated that the number of deer–vehicle collisions in the USA exceeds 1,000,000 annually, causing approximately 29,000 human injuries and 200 fatalities, apart from animal fatalities. Given the magnitude of animal road–kills and the occurrence of human fatalities, there has been much attention paid to mitigating wildlife road kills through the appropriate design of roadways in many developed countries (Clarke et al. 1998; Inell et al. 2003). In contrast, only limited attention has been paid to wildlife–train collisions, although this has also been happening on significant scales in many countries. For instance, 200 moose train collisions were estimated in the province of British Columbia, Canada, during 1988–1990 (Child et al. 1991), with 266 collisions in a 92 km section of railways in Norway during 1980–1988 (Andersen et al. 1991), and between 9 and 725 collisions annually in Alaska (Modafferi 1991). Similarly, 69 roe deer train collisions in the Czech Republic were reported during 2009 (Kusta et al. 2014). Therefore, there is an urgent need for more information on train–wildlife collisions, and how these may be mitigated.

India is one of the countries with serious problems of train–wildlife collisions. This is because the Indian Railways (IR) is one of the world's largest railway networks, comprising 115,000 km of tracks over a route of about 65,000 km and

7500 stations (Indian Railways 2014). In 2015–2016, IR transported more than 22,000,000 passengers and 3,000,000 tons of freight daily (Indian Railways 2015). Furthermore, the Indian Railway network cuts across various forested landscapes, and its impact on wildlife and their habitats has been a matter of increasing concern, although there are few studies on the subject. Direct as well as indirect impacts of the railways have exacerbated over the years with the expansion of the rail network, gauge conversion, and increases in frequency and speed of trains to meet the needs of a modernizing society and an increasing population. A number of larger mammalian species including gaur (*Bos gaurus*), rhinoceros (*Rhinoceros unicornis*), sambar deer (*Rusa unicolor*), nilgai (*Boselaphus tragocamelus*), sloth bear (*Melursus ursinus*), leopard (*Panthera pardus*), lion (*Pathera leo*) and tiger (*Panthera tigris*) have been killed in collision with trains (MOEF 2010; Raman 2011; Dasgupta and Ghosh 2015).

Ironically, the mascot of the Indian Railways, the Asian elephant (*Elephas maximus*), has itself been a frequent victim of train collisions (Singh et al. 2001; Sarma et al. 2006; Roy et al. 2009), which is a cause of conservation and safety concerns. Railway tracks pass through elephant habitat in several Indian states, including Assam, West Bengal, Uttarakhand, Jharkhand, Odisha, Kerala and Tamil Nadu, with accidents resulting in more than 200 elephant deaths between 1987 and 2015. Train–elephant collisions have been common along railway lines such as the Siliguri–Alipurduar line in northern West Bengal (Roy et al. 2009), Guwahati–Lumding line in Assam and Meghalaya (Sarma et al. 2006), Haridwar–Dehradun line in Uttarakhand (Singh et al. 2001), and Coimbatore–Thrissur line in Tamil Nadu and Kerala (Jha et al. 2014). Railway–elephant collisions have also occurred along the Delhi–Howrah line within Jharkhand state (see news report, The Times of India, 02.08.2013), and the Kharagpur–Adra railway line in southern West Bengal (West Bengal Forest Department 2013). To give an idea of the numbers of accidents involving elephants along these tracks, 18 elephant deaths were recorded at Rajaji National Park, Uttarkhand, between 1987 and 2001 (Singh et al. 2001), 35 deaths in Assam from 1987 to 2006 (Sarma et al.2006), 16 in Odisha (Palai et al. 2013) and 13 elephant deaths in Tamil Nadu between 2002 and 2013 (Jha et al. 2014). However, the Siliguri Alipurduar track in northern West Bengal has witnessed the highest numbers of elephant deaths in train collisions; while 27 deaths were recorded between 1974 and 2002, the figure rose to 65 between 2004 and 2015 (Roy et al. 2009; Dasgupta and Ghosh 2015; Roy and Sukumar 2016).

As northern West Bengal is a "hotspot" of train–elephant collisions, we carried out a case study of this phenomenon with the following objectives: (1) mapping the locations of collisions and generating a susceptibility map showing locations prone to accidents; (2) understanding patterns in accidents involving elephants; (3) mapping elephant crossing points and their movement paths along the railway track; and (4) providing suggestions for reducing the incidence of accidents.

Methods

Study Area

This case study was carried out along the 161-km long railway line between the Siliguri Junction (latitude 26.70758°N, longitude 88.42761°E) and the Alipurduar Junction (latitude 26.49099°N, longitude 89.52644°E) in northern West Bengal state, India. The Eastern Bengal Railway (EBR) was one of the pioneer railway companies in the former Bengal and Assam provinces of British India from 1857 to 1942. The Siliguri Alipurduar track was built during 1910–1911, using non-standard meter gauge (MG). Major tracts of tropical moist forests dominated by sal (*Shorea robusta*) were cleared to establish the railway line and to expand tea (*Camellia sinensis*) plantations. Other forests were also cut to make way for establishing agricultural land and to provide the wood to make the railway sleepers. These forests supported a large elephant population from historical times (GoWB 1957) and even supplied elephants for use in armies of Sultanate and Mughal rulers, and later in logging operations by the British (Sukumar 2011). Post-independence, the Eastern Bengal Railway operating in Assam and northern Bengal came under the jurisdiction of the North-East Frontier Railways.

This railway line passes through an elephant range that is part of the Eastern Himalayan biodiversity hotspot and the Eastern Dooars Elephant Reserve (Roy et al. 2009). The line cuts through a mixture of forests, tea gardens, towns, crop lands, human settlements, army establishments, streams, rivers, and elephant corridors within the districts of Darjeeling, Jalpaiguri, and Alipurduar. Established in the early nineteenth century primarily to serve the tea industry (Ray 2002) and the extraction of timber from this region, the line initially had a "meter-gauge track" distance between two rails = 1000 mm/3 ft 3 3/8 in.); at that time, only a few trains used the line and did so at maximum speeds of 60 kph (37 mph). However, given the increasing demand to improve connectivity between the states of north-eastern India and the rest of the country, this railway line was converted from meter-gauge to broad-gauge (distance between two rails 1676 mm or 5' 6") in 2003. Broad-gauge lines can carry heavier loads (including more people) and withstand higher train speeds; consequently, the number of freight and passenger trains on the track has almost doubled. The speed of trains has also increased to more than 100 kph (62 mph). Over 74 km (46 mi) of this line run through forested areas, including several protected areas (Mahananda Wildlife Sanctuary, Gorumara National Park, Chapramari Wildlife Sanctuary, Jaldapara National Park and Buxa Tiger Reserve), as well as through more than 40 tea gardens, 4 army establishments, and agricultural land (mainly paddy and maize, cultivated seasonally).

Data Collection

We collected data on elephant casualties due to collisions with trains between 1974 and 2015, using records maintained by the Government of West Bengal (2012), the Alipurduar Divisional Railway Manager's office, several offices of the West Bengal Forest Department, tea association offices, scientific reports and publications (Roy et al. 2009; Dasgupta and Ghosh 2015), and newspaper reports. These records include date, time, location, number of animals injured or killed, and the gender and age groups of these animals. We also collected first-hand information on some of the more recent accidents during our field research on elephant ecology in this region.

To estimate the distribution and relative abundance of elephants, we conducted surveys along this railway track during three seasons in 2015: non-crop (Jan–Mar), maize cultivation/harvest (Apr–Jul), and paddy cultivation/harvest (Aug–Dec). The survey was done along a 150 km stretch of the 161-km railway track, excluding a part of the Siliguri Junction–Gulma station and Rajabhatkhawa station–Alipurduar junction, as these portions fall within human habitation areas with no elephant presence. The surveys were conducted along the track on foot by a team consisting of one researcher and two assistants, aiming to detect elephant signs (dung, tracks, footprints, feeding signs), as well as direct sightings, on both sides of the track, up to a distance of 10 m. Every sign was counted separately and thus footprints, dung, and tracks at a particular location were counted as three separate pieces of evidence, as in general it was not possible to determine whether these signs, most of which were old, came from the same elephant or different ones. The paddy season survey also included the mapping of elephant corridors, which mostly pass through tea gardens in this region. All the evidence of elephant evidence that we detected were GPS-referenced and plotted on Arc GIS 10.3.1. (ESRI 2014). The non-crop season survey took 33 days, the maize season survey 22 days, and the paddy season survey 41 days to complete.

Data Analysis

Using data on the spatial distribution of elephant signs (dung, footprints, track signs, feeding signs), we generated a "susceptibility map" to estimate the locations along the railway track where accidents were more likely to occur due to the concentration of elephant activity. This map was computed using the point density tool of Arc GIS 10.3.1 software, and the GPS locations of elephant signs. The point density tool calculates the density of point features around a given neighborhood of each raster cell by dividing the number of points that fall in the neighborhood by the area of the neighborhood. In our computations, we used an output cell size of 0.001375 km^2 and search radius of 0.11458 km, based on values provided automatically by the GIS software. We also generated a "kernel density map" in ARC GIS 10.3.1, taking into account all indirect evidence of elephants (dung,

Table 10.1 Number of elephant signs (see text for list) recorded in different sectors along the railway track in northern West Bengal (Siliguri Junction–Alipurduar Junction; see Fig. 10.7 for locations of stations)

From (sector station)	To (sector station)	Distance (km)	Signs/km
Siliguri Junction	Bagrakote	33	7.7
Bagrakote	Chalsa	22	0.8
Chalsa	Dalgaon	46	4.4
Dalgaon	Hasimara	25	3.5
Hasimara	Alipurduar Junction	35	5.9

footprints, track signs, feeding signs) found along the railway track. For point data, the kernel density is a nonparametric method of estimating the probability density function of a random variable that calculates the density of point features around each output raster cell with a smoothly curved surface fitted over each point. Estimates were based on the Silverman (1986) quartic kernel function. The output cell size and search radius were taken as for the "susceptibility map."

To understand temporal patterns of elephant casualties, we used nonparametric statistical tests to determine whether the number of accidents changed between before gauge conversion (1974–2002) and after it (2004–2015). We also evaluated temporal patterns in accidents according to season of the year and time of day for records after gauge conversion. For analysis, accidents were aggregated within the three seasons described above, and in two daily periods, i.e., day (06:00–18:00) vs night (18:00–06:00).

Results

Elephant Distribution

During the three seasonal field surveys for elephants, we found 589 indirect signs during the dry season, 599 signs during the maize cultivation/harvest season, and 1,124 signs during the paddy cultivation/harvest season along the 150 km stretch of the railway track. Overall, most signs were found in the Siliguri–Bagarkote (Mahananda–Monpong stretch) sector in the western region (8.0 signs/km), followed by the Hasimara–Alipurduar (Buxa Tiger Reserve West) sector in the eastern region (6.0 signs/km) (Table 10.1).

Elephant Fatalities

Eighty–nine (89) elephant deaths were reported from 61 accidents over a 41-year period (1974–2015), with a marked increase over time (Fig. 10.1) in the number of

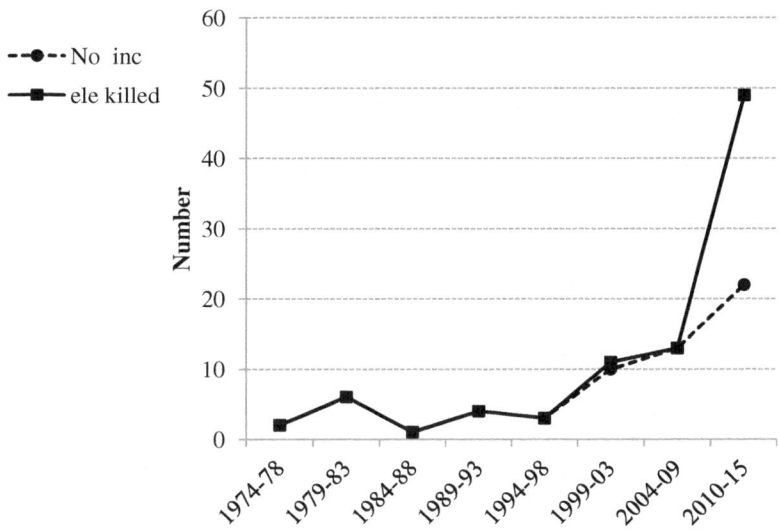

Fig. 10.1 Temporal variation between 1974 and 2015 in the number of train–elephant collision accidents (No inc) and in the number of elephants killed (ele killed) per 5-year period, in the Siliguri–Alipurduar railway, northern West Bengal, India

accidents (Pearson's r = 0.82, P < 0.05) and the number of elephants killed per year (r = 0.96, P < 0.0001). Between 1974 and 2002, when meter gauge (MG) was in operation, 27 elephants died in 26 accidents, corresponding to 0.90 accidents/year ±1.11 (SD) and 0.93 ± 1.19 (SD) elephants/accident. In the 12 years (2004–2015) after conversion to broad gauge (BG), these figures jumped to 62 fatalities in 35 accidents, corresponding to 2.92 accidents/year ±1.98 (SD) and 5.17 ± 5.89 (SD) elephants/accident. There were statistically significant differences between the time periods before and after gauge conversion, for both the annual number of accidents (Mann–Whitney U = 283.5, Z = −3.13, P < 0.001) and the annual number of animals killed (Mann–Whitney U = 290.5, Z = −3.33, P < 0.001).

We also analyzed data in more detail for accidents occurring from 2004 to 2015 (after gauge conversion). Females accounted for most of the animals killed in train collisions, while adults (>15 years) constituted nearly 60% of all mortality recorded (Table 10.2). Also noteworthy is the relatively high mortality of calves (Table 10.2). Comparing these figures with available data on the age structure of elephant populations in northern West Bengal (Sukumar et al. 2003), we found that adult elephant males were killed far more frequently (31% or more than 2.5–fold) than what would have been expected from their representation in the population (13%; Table 10.2). A proportions test also showed that adult males (Z = 3.831, P < 0.001), and calves (Z = 2.136, P = 0.037) were involved to a significantly greater extent in train accidents (n = 62) than their representation in the population (n = 650).

Table 10.2 Age structure of the elephant population in northern West Bengal (based on Sukumar et al. 2003) and those killed in train elephant collisions in the Siliguri–Alipurduar railway between 2004 and 2015

Age class	% Females in population	% Females killed in train accidents	% Males in population	% Males killed in train accidents
Adult	33.3	29.0	12.9	30.6
Sub-adult	11.1	11.3	12.3	8.1
Juvenile	13.5	4.8	8.8	0.0
Calves	4.1	8.1	4.1	8.1
Total	62.0	53.2	38.0	46.8

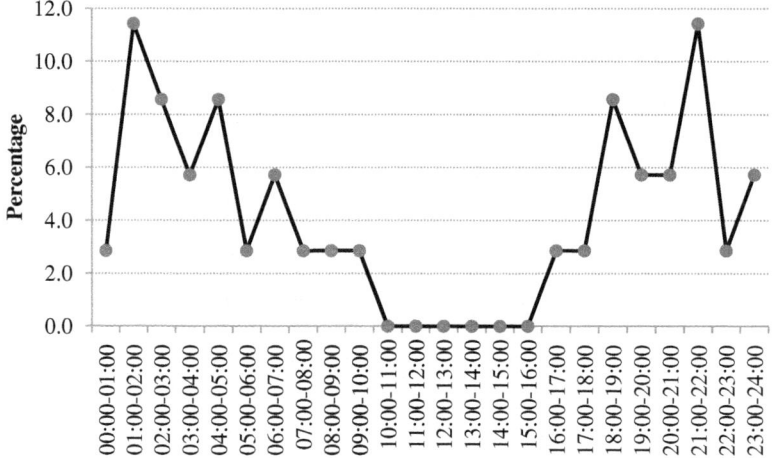

Fig. 10.2 Temporal distribution of train–elephant collisions during the daily cycle from 2004 until 2015 (after broad gauge conversion), in the Siliguri–Alipurduar railway, northern West Bengal, India

Also in 2004–2015, we found that the occurrence of elephant–train collisions varied markedly during the daily cycle. In fact, most accidents occurred between midnight and 6:00 and 16:00–24:00, with peaks at 01:00–2:00 and 22:00–23:00 (Fig. 10.2). More accidents occurred during the night (80%), and this was significantly higher than during the day ($Z = 3.198$, $P = 0.002$). There were also significant seasonal patterns in the number of accidents (Kruskal–Wallis $H = 12.69$, $P = 0.05$) and the number of elephant deaths (Kruskal–Wallis $H = 39.70$, $P = 0.05$), with both occurring most frequently during the paddy cultivation season, followed by maize and non-crop seasons (Fig. 10.3). The association of accidents with elephants with the crop cultivation seasons is further emphasized by the strong correlation between the monthly proportion of elephants killed and the frequency of crop raiding by elephants in villages (all seasons: $r = 0.38$, $P = 0.11$; maize season: $r = 0.81$, $P = 0.025$; paddy season: $r = -0.037$, $P = 0.47$). In fact, there were

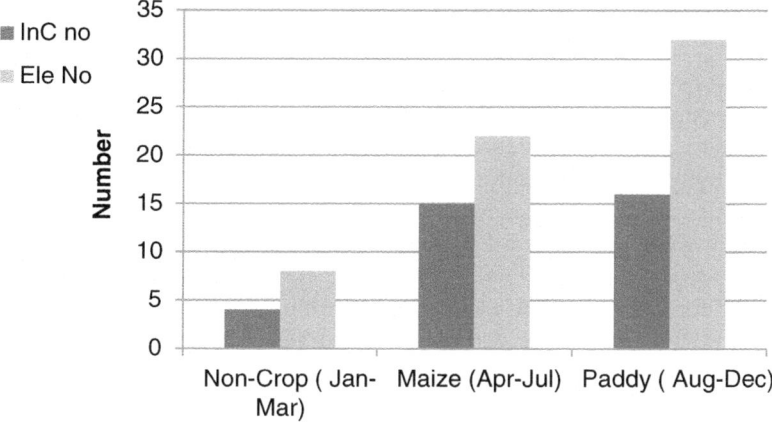

Fig. 10.3 Variation in relation to season in the number of accidents (InC no) and number of elephants killed (ele no) in collisions with trains during 2004–2015 (after broad gauge conversion) in the Siliguri–Alipurduar railway, northern West Bengal, India

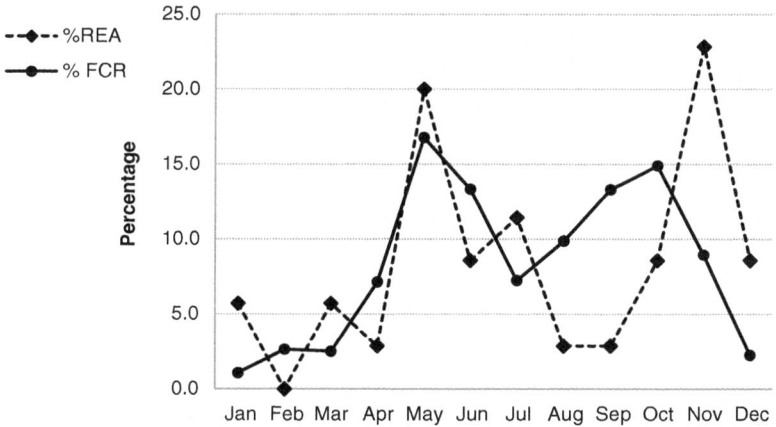

Fig. 10.4 Monthly variation of railway elephant accidents (REA) between 2004 and 2015 (after broad gauge conversion), in the Siliguri–Alipurduar railway, and in the frequency of crop raiding (FCR) by elephants in villages. Data on crop raiding are based on Sukumar et al. (2003) and Roy (2010)

simultaneous peaks in elephant mortality and crop raiding during the year, corresponding to the maize (May–June) and paddy-harvesting season (Oct–Dec) (Fig. 10.4).

Spatial Distribution of Collision Risk

There was a strong spatial variation in the number of accidents both before and after gauge conversion (Fig. 10.5), with many accidents occurring in the same areas and none in others. The concentrations of accidents matched some areas with high elephant activity as estimated from the densities of elephant signs (Fig. 10.6), but there were also areas with large densities of elephant signs but with no or just a few accidents. However, when we divided the entire railway line surveyed into segments of 2 km each (n = 76) and plotted the elephant signs found (dung, tracks, foot prints, feeding signs) in each segment with the number of accidents that occurred, we found a strong correlation (Spearman's rank correlation r = 0.80, $P \leq 0.001$).

During 2004–2015, after gauge conversion, the Buxa Tiger Reserve West Division region recorded the highest number (29%) of the accidents that took place (n = 35), followed by the Mahananda Wildlife Sanctuary (26%), the Jalpaiguri Forest Division (17%), the Jaldapara Wildlife Division (14%), the Kalimpong Forest Division (9%), and the Gorumara Wildlife Division (6%). When we consider the number of elephants killed (n = 62), the highest figures were observed in Jalpaiguri (31%), followed by the Buxa Tiger Reserve West (26%), Gorumara Wildlife (16%), Mahananda Wildlife Sanctuary (15%), Jaldapara Wildlife Division (8%), and Kalimpong Division (5%). More elephant groups were involved in accidents in Gorumara and Jalpaiguri resulting in multiple fatalities per accident.

Discussion

Our study describes the spatial and temporal mortality patterns of elephants due to collisions with trains along a 150-km stretch of the 161-km railway line between the Siliguri and Alipurduar Junctions. Collisions occurred throughout the study period (1974–2015), but they greatly increased after 2004, when a new schedule of trains began operating, after the line was converted from meter gauge to broad gauge. Most collisions occurred during the night, and adult male elephants and calves appeared particularly susceptible to this type of mortality. There was an uneven distribution of accidents during the annual cycle, with peaks in May–June and September–October, during the crop cultivation seasons of maize and paddy rice, respectively. The spatial distribution of accidents was also uneven, occurring mostly in a few well-defined hotspots. Overall, our study provides useful information to mitigate elephant–train accidents in the Siliguri–Alipurduar railway, and elsewhere in India.

One reason for the increase that was seen in the frequency of accidents was probably the increase in the number of circulating trains, from about 10 trains/day during the pre-conversion period, to estimates of between 17 (Ghosh 2001) and 22–26/day (Das 2013) during the post-conversion period, thereby increasing the

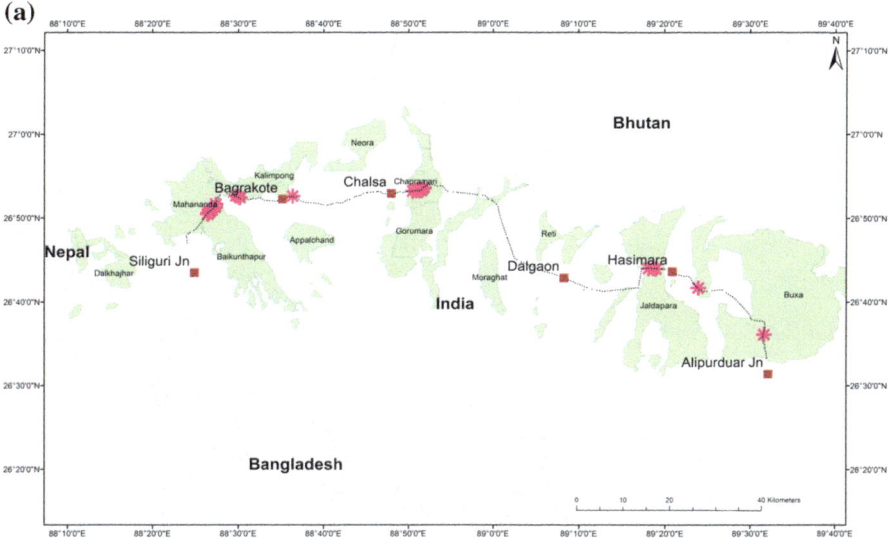

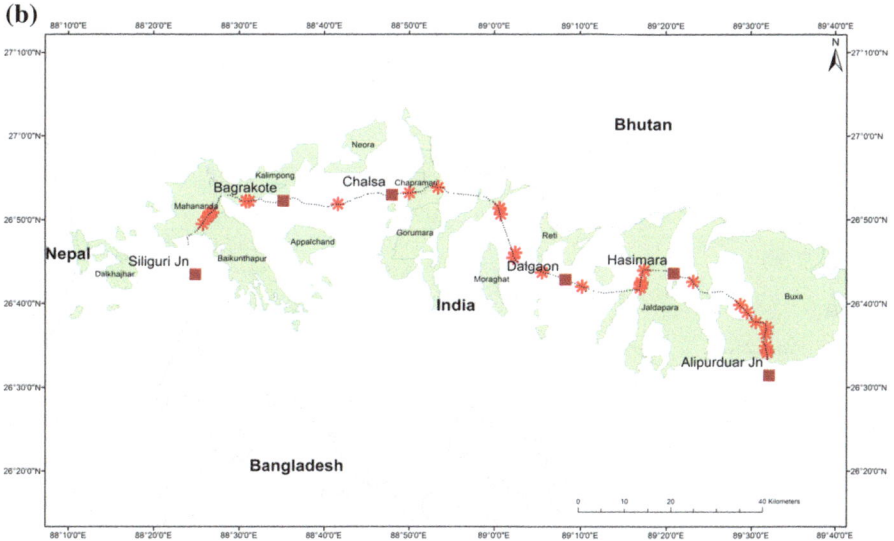

Fig. 10.5 Locations of elephant–train accidents in northern West Bengal between 1974 and 2002 (*upper panel*, **a**) when meter gauge was in operation (*pink stars* indicate locations where accidents occurred), and during 2004–2015 (*lower panel*, **b**), after broad gauge conversion (*red stars* indicate locations where accidents occurred)

chances of an elephant encountering a train when crossing the railway. Also, broad gauge allows trains to reach higher velocities, making it harder for elephants to avoid a moving train. In fact, after gauge conversion, the maximum speed of trains

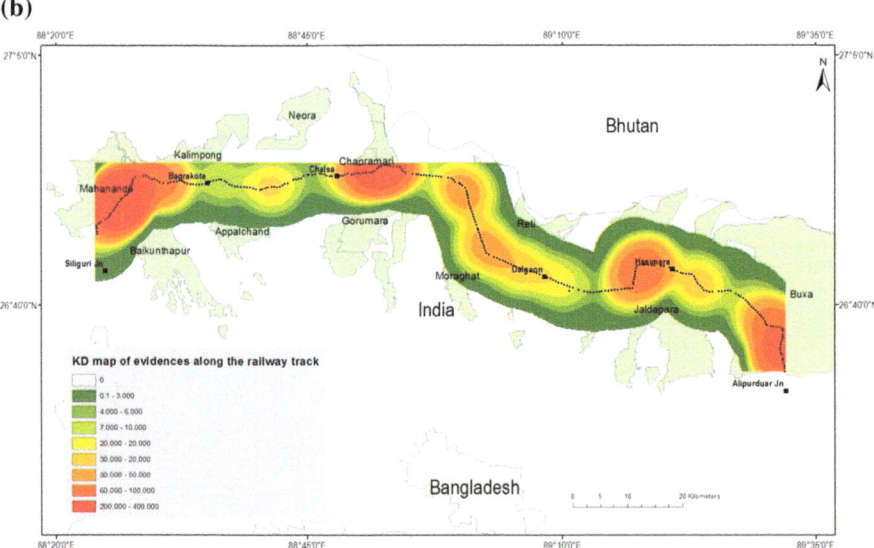

Fig. 10.6 Maps estimating the spatial variation in the risk of collision of elephants with trains along the railway track in northern West Bengal based on point densities (susceptibility map, *upper panel* **a**) and kernel densities (*lower panel* **b**)

increased from about 60 kph (37 mph) to over 100 kph (62 mph). A second possible reason for the increasing number of accidents was the increase in the elephant population of this region; the population more than doubled or even tripled here when 150 elephants were reported in 1980 from this region (Roy 2010).

In terms of age and gender, we found that adults were killed more often than other age groups, but this was roughly proportional to their representation in the population. What is striking, however, is that the proportion of adult males killed was nearly three times higher than that expected from their representation in the population. The reasons for this are uncertain, but it is possible that adult males range over larger areas, raid crops in villages more frequently, and thus cross railways more often than females do (Sukumar 1989; Sukumar et al. 2003; Roy 2010; Williams et al. 2001). It is also noteworthy that juveniles were killed to a lesser extent than expected, implying that they are alert and protected within a family group, whereas calves were killed more often than expected from their percentage of the population, probably because of their inability to respond quickly to dangerous situations.

The concentration of accidents at night was probably a consequence of elephant movement patterns, rather than resulting from variation in the number of circulating trains. In fact, about 80% of accidents occurred between 18:00 and 6:00, although only 35% of trains operate at this time; this is probably because in daytime, elephants tend to rest in more dense forests, moving towards cultivated fields at night (Sukumar 2003) and eventually crossing railway lines. Other studies have reported similar patterns, showing a matching between animal daily activity patterns and collisions with trains. For instance, Ando (2003) found that 69% of the collisions of Sika deer *Cervus nippon* with trains in the eastern Hokkaido line, Japan, were recorded in the evening, which is the period when animals move between resting and feeding sites.

The seasonal distribution of collisions may also be a consequence, at least partly, of elephant movements from forests to agricultural areas, where they forage for crops. This is supported by the close matching between accident peaks and the harvest seasons of maize and paddy rice, as well as the observed spatial and temporal patterns of elephant raiding of crops in the region. Similar patterns have been observed elsewhere in India, with Singh et al. (2001) reporting that 78% of collisions with trains in the Rajaji National Park occurred in January–June and peaked in May, which matches the wheat and sugar cane cultivation/harvest season. Likewise, Sarma et al. (2006) in Assam found one peak in June–July during the active monsoon period, when elephants move away from floodplains to higher ground through which railway tracks run, and another in November that coincides with the paddy crop harvesting season. Studies on other herbivores, such as Sika deer (Ando 2003), moose (*Alces alces*, Andersen et al. 1991), and roe deer (*Capreolus capreolus*, Kusta et al. 2014), also suggested that collisions with trains peaked during the seasons when animals moved widely in search of food.

The spatial distribution of train–elephant collisions was uneven, with the occurrence of a few well-defined hotspots. Reasons for this are uncertain, but it may be the result of a number of factors, such as a variation in animal abundance, activity, and behaviour, habitat distribution, landscape topography, and rail design, as described for other species elsewhere (Gundersen and Andreassen 1998; Inbar and Mayer 1999; Haikonen and Summala 2001; Joyce and Mahoney 2001; Mysterud 2004; Roy 2010). In our case, accidents seemed concentrated in areas where the railway track crossed forested habitats heavily used by elephants, and in areas where elephants cross from forests to agricultural areas during their foraging movements. There is probably also a high risk of collisions in corridors that are used when animals move/migrate among forest patches, including the Reti Reserved Forest (RF)-Moraghat RF, Reti RF-Diana RF, Neora RF-Apalchand RF (Roy and Sukumar 2015).

It is not certain whether the observed mortality from collisions with trains has any significant impact on the local elephant population, but it is worth noting that train accidents from 2004 to 2013 were responsible for causing about 25% of the 224 elephant deaths recorded in northern West Bengal. This compares with a population size of 674 elephants estimated for northern West Bengal in 2015 (West Bengal Forest Department 2015), suggesting that there may indeed be some negative effects. As mentioned, the elephant population in this region has been increasing over the past three decades. Although northern West Bengal accounts for only 2.3% of the free-ranging elephant population in India, it is part of a much larger elephant population (about 2,700–3,000 animals) of the North Bank landscape (of the Brahmaputra River) that shares a contiguous habitat with Bhutan in the north and Assam in the east (Chowdhury and Menon 2006).

The impact of these accidents is very significant, Trains have to stop, elephants have to be removed, carriages may be damaged, etc. Due to these railways–elephant accidents revenue loss was estimated all over India Rs. 69,675.97 cores from freight and Rs. 28,645.52 cores (Nayak 2013).

Management Implications

Together with other studies (Singh et al. 2001; Sarma et al. 2006), our work clearly shows that elephant mortality due to collision with trains is a significant problem that urgently needs to be resolved. Addressing this issue requires that collision risks be duly considered in "environmental impact assessments," both when developing new railway projects, and when improving the operational characteristics of extant ones (e.g., changes from meter to broad gauge). These studies should assess the spatial and temporal patterns of elephant abundance, movements, and activity, thereby identifying areas where the risk of collision is high and new railway lines should not be laid. Where new lines are unavoidable, or where lines are already in operation, efforts should be targeted to mitigating impacts by implementing measures such as fencing, bridges, under- or overpasses, and local topography adjustments, among others. These processes should be conducted by technically qualified agencies in close collaboration with local and regional stakeholders, including government, NGOs, and the general public. It is also important that monitoring be put in place–in both the construction and exploration phases–aiming at the early detection of unanticipated impacts, and for assessing the efficiency of mitigation and compensation measures (Inell et al. 2003).

Along railway lines where collisions with elephants are frequent, such as in our study area, measures should be put in place to reduce the risks and thus avoid elephant mortality to the extent possible. Below are a number of recommendations that need to be considered in the Siliguri–Alipurduar line, and that may also be useful in similar areas elsewhere:

1. **Reduce the speed of trains in the most vulnerable stretches of the track** This suggestion has been repeated by several committees that have addressed train–elephant collisions in this region, but there have been genuine technical difficulties faced by the railway companies in implementing it due to operational constraints. With the current train operations along the Siliguri–Alipurduar line, it seems that reduced speed can only be implemented along a few stretches.

2. **Limit the operation of trains during the night-time** Since the frequency of accidents is much higher during the night than during the day, the curtailment of train movement along this track at night might be a possible way to reduce accidents. The acceptance of this suggestion would again depend on the operational needs of the railways.

3. **Provision of overpasses and underpasses at vulnerable crossing points** This would require an engineering survey along the entire stretch in order to determine the need, feasibility, and costs of any structure at locations where elephants are expected to cross more frequently. Considerable expenses would be incurred in undertaking such constructions. A pilot project on landscaping at some critical places such as Chalsa-Nagrakata, where the track is laid through a narrow passage in a hillock with steep excavation on either side, can be undertaken to assess its success or otherwise in reducing train elephant accidents. It is also important to consider whether elephants would use such engineered passages for crossing the railway track or would be suspicious of them and avoid them.

4. **Patrolling along the track by forest/wildlife/railway staff** This is already being practiced in northern West Bengal but can perhaps be strengthened during crop–raiding seasons, when the frequency of elephants crossing the tracks increases. The patrolling staff can then pass on the information to the railway signalmen on the presence of elephants. These measures are not without their problems, as patrolling at night inside dense forest is hazardous. The provision of more watch towers along the track in forest areas may be useful.

5. **Implement garbage clearance and ensure better visibility along the track** Elephants are attracted by food items, including packaged ones that passengers may throw out of trains. Creating awareness among passengers and employing staff to remove such garbage along the track would help reduce the possibility of elephants lingering along the railway track in search of food. Vegetation along the track has to be cleared on a regular basis to ensure the best possible visibility to train drivers who have to be sensitized to the risks of elephants crossing the tracks (such awareness is already being instituted among the drivers). There have been instances of elephants, usually solitary bulls, feeding on an attractive plant (such as banana) that grows close to a railway track, and being hit by a train. It is important to remove such elephant forage plants along the railway tracks.

6. **Use of communications technology as an advance warning to train drivers and signalling staff** Electronic surveillance can be undertaken to assess its efficacy in detecting elephants close to the railway track. Various technologies have been suggested, including seismic sensors that measure ground vibrations to record elephant presence, infrared (thermal) sensors and cameras to provide images of elephants along railway tracks, infrared beams to detect the movement of an elephant, and even acoustic devices to scare elephants from tracks when trains are approaching. These require considerable R&D for field situations, extensive field trials, and evaluation of cost-effectiveness before being deployed on a large scale. GPS collaring of representative animals in family herds and bulls would help in the fine-grained analysis of their movement pattern along the railway track, and help in planning mitigation measures.

7. **Realignment of a portion of the track** The existing track can be realigned so as to avoid passing through the Buxa Tiger Reserve West Division and the Jaldapara National Park, where a large number of accidents has taken place. The Southern line between New Jalpaiguri and New Alipurduar can be doubled to handle the increased traffic. There are four possibilities for realignment of the Northern line (Fig. 10.7).

   (i)   From the Dalgaon Railway station the track can be diverted and/or linked to the southern line, namely, Salbari–Falakata station between the New Jalpaiguri Junction– and the New Alipurduar Junction line so that some trains can be diverted before the Jaladapara—Buxa Reserve forest areas to avoid accidents.

   (ii) From the Madarihat Station, the track can be diverted and/or linked to the southern line, namely, the Falakata station, along the New Jalpaiguri Junction–New Alipurduar Junction line so that some trains can be diverted before the Jaladapara–Buxa Reserve forest areas to avoid accidents.

   (iii) From the Madarihat Station the track can be diverted and/or linked along the southwest periphery of the Jaldapara National Park (NP) and southern periphery of Jaldapara and Buxa Tiger Reserve (BTR) linked to Alipurduar Junction.

   (iv) The fourth possibility is to realign the northern line from Hamiltanganj/ Kalchini to Damanpur by diverting the line along the south west border of the Buxa Tiger Reserve via Nimati and along south of highway NH31, which is linked to Damanpur, to avoid going through the Buxa Tiger Reserve.

Besides these technical recommendations, it should be kept in mind that limiting accidents to elephants from train collisions is extremely challenging, requiring the active participation of a number of stakeholders that include the Indian Railways, Ministry of Environment, Forest and Climate Change (Govt. of India), West Bengal Forest Department, NGOs, elephant biologists/conservationists, engineers and technologists, tea garden management, and the people living along the track. We suggest that various solutions be implemented at an experimental level, and then monitored over a reasonable time frame for testing their efficacy. The most cost-effective solutions should then be selected and implemented on a larger scale, thereby reducing collision risks and ultimately contributing to the conservation of elephants in India and elsewhere in Asia

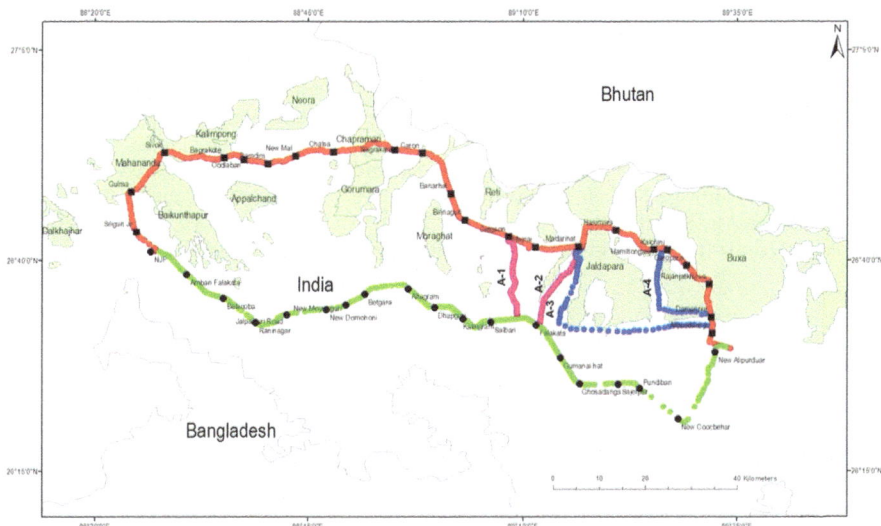

Fig. 10.7 Suggestions for realignment of the existing railway track from the Siliguri Junction to the Alipurduar Junction to reduce the risk of train–elephant collisions in northern West Bengal. *#Red line* Northern sector railway track (Siliguri Junction–Alipurduar Junction). *#Green line* Southern sector railway track (New Jalpaiguri–New Alipurduar Junction) that does not pass through any forest. *#Pink lines* Possible realignment of track (Dalgaon- Falakata (*A-1*), Madarihat–Falakata (*A-2*)). *#Blue dotted lines* Possible realignment of proposed track (Madarihat–Alipurduar Junction (*A-3*) and Hamlitonganj–Damanpur (*A-4*))

Photo credit @Debopratim Saha

Acknowledgements We thank the International Elephant Foundation, the primary agency that funded this project; the Asian Nature Conservation Foundation for the work place and administration of the project; the Centre for Ecological Sciences, Indian Institute of Science, for office space, Internet and report writing; the West Bengal Forest Department for permissions and help in the field to carry out the study; the Indian Railways (North Frontier Railway) for permissions to survey along the track; the Dooars Branch of the Indian Tea Association for permissions to survey elephant paths within their tea gardens; and the Environmental System Research Institute for providing online software as well as Arc GIS 10.3.1. We also thank Mr. Santosh Chhetri, Mr. Sapwan Chhetri, Mr. Sushil Manger, Mr. Rabin Chhetri, Mr Ratua Oran, Mr. Rohit Munda and Mr. Bikash Kairala for their help in the field, often under difficult conditions.

References

Andersen, R., Wiseth, B., Pedersen, P. H., & Jaren, V. (1991). Moose–train collisions: Effects of environmental conditions. *Alces, 27,* 79–84.

Ando, C. (2003). The relationship between deer–train casualties and daily activity of the sika deer, *Cervus nippon. Mammal Study, 28,* 135–143.

Caughley, G., & Sinclair, A. R. E. (1994). *Wildlife Ecology and Management.* Cambridge, MA: Blackwell Science.

Child, K. N., Barry, S. P., & Aitken, D. A. (1991). Moose mortality on highways and railways in British Columbia. *Alces, 27,* 41–49.

Chowdhury, A., & Menon, V. (2006). Conservation of Asian elephant in north-east India. *Gajah, 25,* 47–60.

Clarke, G. B., White, P. C. L., & Harris, S. (1998). Effects of roads on badger *Meles meles* populations in south-west England. *Biological Conservation, 86,* 117–124.

Conover, M. R., Pitt, W. C., Kessler, K. K., Dubow, T. J., & Sanborn, W. A. (1995). Review of human injuries, illnesses, and economic losses caused by wildlife in the United States. *Wildlife Bulletin, 23,* 407–414.

Das, K. (2013). *Man Elephant Conflict in Northern Bengal.* Unpublished report to TERI University. Retrieved on November 05, 2016, from http://www.teriuniversity.ac.in/mct/pdf/assignment/Kalyan-Das.pdf

Dasgupta, S., & Ghosh, A. K. (2015). Elephant–railway conflict in a biodiversity hotspot: Determinants and perceptions of the conflict in Northern West Bengal, India. *Human Dimensions of Wildlife, 20,* 81–94.

Dorans, R. A. P., Kindel, A., Bager, A., & Freitas, S. R. (2012). Avaliação da mortalidade de vertebrados em rodovias no Brasil. In A. BAGER (ed.), Ecologia de estradas: tendências e pesquisas. Lavras: UFLA, pp. 139–152.

ESRI. (2014). *ArcGIS Desktop: Release 10.3.1.* Redlands, CA: Environmental Systems Research Institute.

Fahrig, L., Pedlar, J. H., Pope, S. E., Taylor, P. D., & Wegner, J. F. (1995). Effect of road traffic on amphibian density. *Biological Conservation, 73,* 177–182.

Forman, R. T. T., & Deblinger, R. D. (2000). The ecological road-effect zone of a Massachusetts (USA) suburban highway. *Conservation Biology, 14,* 36–46.

Formann, R. T. T., Sperling, D., Bissonette, J. A., Clevenger, A. P., Cutshall, C. D., Dale, V. H., et al. (2003) *Road Ecology: Science and Solutions* (481 p). Washington: Island Press.

Fahrig, L., France, R., Goldman, C. R., Heanue, K., & Jone, I. A. (2003) *Road ecology: Science and solutions* (481 p). Washington: Island Press.

Ghosh, L. (2001). *Tracking death. India today.* Retrieved November 5, 2016 from, http://indiatoday.intoday.in/story/railway-expansion-project-increases-threat-to-wildlife-in-bengal-sanctuaries/1/232130.html)

Government of West Bengal, Forest Department (GoWB). (1957). Sixth working plan of the reserved forests of Jalpaiguri Division (1957–58 to 1966–67), Vol. 1, prepared by the Divisional Forest Officer, Jalpaiguri Division.

Government of West Bengal. (2012). *State forest reports: 2006–2012*. Kolkata: Government of West Bengal, Directorate of Forests, Office of the Principal Chief Conservator of Forests.

Government of West Bengal. (2013). *State forest reports: 2013*. Kolkata, India: Government of West Bengal, Directorate of Forests, Office of the Principal Chief Conservator of Forests.

Government of West Bengal. (2015). *Synchronized population estimation of the Asian elephant in forest divisions of Northern West Bengal-2014*. Jalpaiguri, West Bengal: West Bengal, Directorate of Forests, Office of Conservator of Forests.

Government of India. (2015). Indian railways lifeline of the nation (A White Paper) February 2015, Ministry of Railways New Delhi.

Gundersen, H., & Andreassen, H. P. (1998). The risk of moose *Alces alces* collision: A predictive logistic model for moose train accidents. *Wildlife Biology, 4,* 103–110.

Haikonen, H., & Summala, H. (2001). Deer-vehicle clashes: Extensive peak at 1 hour after sunset. *American Journal of Preventive Medicine, 21,* 209–213.

https://en.wikipedia.org/wiki/Track_gauge_in_India

http://indiatoday.intoday.in/story/railway-expansion-project-increases-threat-to-wildlife-in-bengal-sanctuaries/1/232130.html

Inbar, M., & Mayer, R. T. M. (1999). Spatio-temporal trends in armadillo diurnal activity and road-kills in central Florida. *Wildlife Society Bulletin, 27,* 865–872.

Indian Railways. (2014). *Statistical summary—Indian railways*. Retrieved November 05, 2016, from http://indianrailways.gov.in/railwayboard/uploads/directorate/stat_econ/IRSP_201314/pdf/Statistical_Summary/Summary%20Sheet_Eng.pdf

Inell, B., Bekker, G. J., Cuperus, R., Dufek, J., Fry, G., Hicks, C., et al. (Eds.). (2003). *Wildlife and traffic: A European handbook for identifying conflicts and designing solutions*. Brussels: European Co-operation in the Field of Scientific and Technical Research.

Jha, N., Sarma, K., & Bhattacharya, P. (2014). Assessment of elephant (*Elephas maximus*) mortality along Palakkad–Coimbatore railway stretch of Kerala and Tamil Nadu using Geospatial Technology. *Journal of Biodiversity Management & Forestry, 3,* 1–7.

Joyce, T. L., & Mahoney, S. P. (2001). Spatial and temporal distributions of moose–vehicle collisions in Newfoundland. *Wildlife Society Bulletin, 29,* 281–291.

Kusta, T., Hola, M., Keken, Z., Jezek, M., Zika, T., & Hart, V. (2014). Deer on the railway line: Spatiotemporal trends in mortality patterns of roe deer. *Turkish Journal of Zoology, 38,* 479–485.

Lalo, J. (1987). The problem of road kill. *American Forests, 50,* 50–52.

Modafferi, R. D. (1991). Train moose kill in Alaska: Characteristics and relationships with snowpack depth and moose distribution in Lower Susitna Valley. *Alces, 27,* 193–207.

Mysterud, A. (2004). Temporal variation in the number of car-killed red deer *Cervus elaphus* in Norway. *Wildlife Biology, 10,* 203–210.

Ministry of Environment and Forests (MOEF). (2010). Gajah: Securing the future for elephants in India. New Delhi, India

Nayak, S. (2013). A petition under article 32 of the Constitution of India in the nature of public Interest litigation challenging the inaction of ministry of Environment & Forest, Indian railways and many State Governments and under their failure to protect wealth like the elephants, In the Supreme Court of India, New Delhi.

Newmark, W. (1987). A land–bridge island perspective on mammalian extinctions in western American parks. *Nature, 325,* 430–432.

Palai, N. C., Bhakta, P. R., & Kar, C. S. (2013). Death of elephants due to railway accidents in Odisha, India. *Gajah, 38,* 39–41.

Raman, T. S. R. (2011). *Framing ecologically sound policy on linear intrusions affecting wildlife habitats (background paper for the National Board of Wildlife)*. Mysore: Nature Conservation Foundation.

Ray, S. (2002). *Transformations on the Bengal frontier: Jalpaiguri 1765–1948*. Curzon, NY: Routledge.

Roy, M., Baskaran, N., & Sukumar, R. (2009). The death of jumbos on railway lines in northern West Bengal. *Gajah, 31,* 36–39.

Roy, M. (2010). *Habitat use and foraging ecology of the Asian Elephant (Elephas maximus) in Buxa Tiger Reserve and adjoining areas of northern West Bengal.* Ph.D. thesis, Vidyasagar University, West Bengal, India.

Roy, M., & Sukumar, R. (2015). Elephant corridors in northern West Bengal. *Gajah, 43,* 26–35.

Roy, M., & Sukumar, R. (2016). *Survey of elephant movement paths/corridors across rail tracks in northern West Bengal.* Final Report to International Elephant Foundation, March 2016.

Seiler, A., & Helldin, J. O. (2015). Greener transport infrastructure—IENE 2014 international conference. In A. Seiler & J. O. Helldin (Eds.), *Proceedings of IENE 2014 International Conference on Ecology and Transportation*, Vol 11, pp. 5–12. Malmö, Sweden. Nature Conservation. doi:10.3897/natureconservation.11.5458

Sarma, U. K., Easa, P. S., & Menon, V. (2006). *Deadly lines: A scientific approach to understanding and mitigating elephant mortality due to train hits in Assam* (Occasional Report no. 24). New Delhi: Wildlife Trust of India.

Silverman, B. W. (1986). *Density estimation for statistics and data analysis.* New York: Chapman and Hall.

Singh, A. K., Kumar, A., Mookerjee, A., & Menon, V. (2001). *Jumbo express: A scientific approach to understanding and mitigating elephant mortality due to train accidents in Rajaji National Park* (Occasional Report no. 3). New Delhi: Wildlife Trust of India.

Soule, M. E., & Wilcox, B. A. (Eds.). (1980). *Conservation biology: An evolutionary-ecological perspective.* Sunderland, MA: Sinauer.

Sukumar, R. (2011). *The story of Asia's elephants.* Mumbai: Marg Foundation.

Sukumar, R. (1989). *The Asian elephant: Ecology and management.* Cambridge: Cambridge University Press.

Sukumar, R. (2003). *The living elephants: Evolutionary ecology, behavior and conservation.* New York: Oxford University Press.

Sukumar, R., Baskaran, N., Dharmarajan, G., Roy, M., Suresh, H.S., & Narendran, K. (2003). *Study of Elephants in the Buxa Tiger Reserve and adjoining areas of northern West Bengal and preparation of Conservation action plan.* Final Report submitted to West Bengal Forest Department. Centre for Ecological Sciences, Indian Institute of Science, Bangalore, India.

Trombulak, S. C., & Frissell, C. A. (2000). Review of ecological effects of roads on terrestrial and aquatic communities. *Conservation Biology, 14,* 18–30.

Van der Ree, R., Jaeger, J. A. G., Van der Grift, E. A., & Clevenger, A. P. (2011). Effects of roads and traffic on wildlife populations and landscape function: Road ecology is moving towards larger scales. *Ecology and Society, 16,* 48.

Wilcove, D. S., Rothstrin, D., Dubow, J., Philips, A., & Locos, E. (1998). Quantifying threats to imperiled species in the United States. *BioScience, 48,* 607–615.

Williams, A. C., Johnsingh, A. J. T., & Krausman, P. R. (2001). Elephant–human conflicts in Rajaji National Park, north-western India. *Wildlife Society Bulletin, 29,* 1097–1104.

Chapter 11
Assessing Bird Exclusion Effects in a Wetland Crossed by a Railway (Sado Estuary, Portugal)

Carlos Godinho, Luísa Catarino, João T. Marques, António Mira and Pedro Beja

Abstract Linear transportation infrastructures may displace wildlife from nearby areas that otherwise would provide adequate habitat conditions. This exclusion effect has been documented in roads, but much less is known about railways. Here we evaluated the potential exclusion effect on birds of a railway crossing a wetland of international importance (Sado Estuary, Portugal). We selected 22 sectors representative of locally available wetland habitats (salt pans, rice paddy fields, and intertidal mudflats); of each, half were located either close to (0–500 m) or far from (500–1500 m) the railway line. Water birds were counted in each sector between December 2012 and October 2015, during two months per season (spring, summer, winter, and autumn) and year, at both low and high tide. We recorded 46 species, of which the most abundant (>70% of individuals) were black-headed gull, greater flamingo, northern shoveler, black-tailed godwit, and lesser black-backed gull. Peak abundances were found in autumn and winter. There was no significant variation between sectors close to and far from the railway in species richness, total abundance, and abundance of the most common species. Some species tended to be most

C. Godinho (✉) · L. Catarino
LabOr—Laboratório de Ornitologia, ICAAM—Instituto de Ciências Agrárias e Ambientais Mediterrânicas, Universidade de Évora, Mitra, 7002-554 Évora, Portugal
e-mail: capg@uevora.pt

J.T. Marques · A. Mira
Unidade de Biologia da Conservação, Departamento de Biologia, Universidade de Évora, Mitra, 7002-554 Évora, Portugal

J.T. Marques · A. Mira
CIBIO/InBIO, Centro de Investigação em Biodiversidade e Recursos Genéticos, Universidade de Évora, Pólo de Évora, Casa do Cordovil 2° Andar, 7000-890 Évora, Portugal

P. Beja
CIBIO/InBIO, Centro de Investigação em Biodiversidade e Recursos Genéticos, Universidade do Porto, Campus Agrário de Vairão, Rua Padre Armando Quintas, 4485-661 Vairão, Portugal

P. Beja
CEABN/InBIO, Centro de Ecologia Aplicada "Professor Baeta Neves", Instituto Superior de Agronomia, Universidade de Lisboa, Tapada da Ajuda, 1349-017 Lisboa, Portugal

© The Author(s) 2017
L. Borda-de-Água et al. (eds.), *Railway Ecology*,
DOI 10.1007/978-3-319-57496-7_11

179

abundant either close to or far from the railway albeit not significantly so but this often varied across the tidal and annual cycles. Overall, our study did not find noticeable exclusion effects of this railway on wetland birds, with spatial variation in abundances probably reflecting habitat selection and daily movement patterns. Information is needed on other study systems to assess the generality of our findings.

Keywords Aquatic birds · Habitat loss · Human disturbance · Environmental impact · Transportation infrastructures · Zone of influence

Introduction

Linear transportation infrastructures such as roads and railways are increasing worldwide, and this is generally considered a serious threat to wildlife conservation (van der Ree et al. 2015). Transportation infrastructure networks are particularly dense in developed countries, with a recent study showing that in Europe about 50% of the land area is within 1.5 km of a paved road or railway line (Torres et al. 2016). This problem also increasingly affects less developed countries, where most infrastructure and urban development is expected to take place over the coming decades (Seto et al. 2012). Therefore, there is an urgent need to evaluate the impacts of linear infrastructure development on wildlife, and to devise planning and technical solutions to mitigate such impacts (DeFries et al. 2012).

The impacts of linear transportation infrastructures are generally associated with direct mortality caused by animal collisions with circulating vehicles (Loss et al. 2015; Santos et al. 2016), and with barrier effects and the consequent increases in habitat fragmentation and reduced landscape connectivity (Carvalho et al. 2016). However, there may also be widespread indirect habitat degradation, because many animal species avoid using areas close to transportation infrastructures (Forman and Deblinger 2000; Torres et al. 2016). For instance, a recent study showed that many bird species strongly avoided traffic noise, while certain others remaining close to noisy environments exhibited degradation in body condition (Ware et al. 2015). Another study showed that even unpaved roads reduced herbivore densities in a protected area (D'Amico et al. 2016). Most of these studies, however, have focused on roads, either ignoring railways or assuming that roads and railways have largely similar impacts (e.g. Torres et al. 2016). Furthermore, the few studies available are contradictory (see Chap. 6), with some pointing out negative effects of noise, lights and vibrations associated with circulating trains, while others suggest that wildlife ignore or adapt to railway disturbance (Owens 1977; Waterman et al. 2002; Gosh et al. 2010; Li et al. 2010; Mundahl et al. 2013; Wiacek et al. 2015). Thus, more information is needed on the exclusion effects of railways, particularly on the species and ecological conditions associated with the most negative effects.

If railways have strong exclusion effects, their consequences may be particularly serious when crossing wetland habitats. This is because wetlands are declining fast, they are among the most threatened habitats, and the species they support are among the most endangered taxa (e.g. Millennium Ecosystem Assessment 2005; Tittensor et al. 2014); therefore, further habitat degradation due to railways and other transportation infrastructures may have particularly negative consequences. Moreover, wetlands are very productive habitats that are home to large concentrations of wildlife, particularly aquatic birds such as herons, egrets, waders, waterfowl and gulls, among others, many of which are declining and are highly sensitive to habitat loss and degradation (e.g. Kleijn et al. 2014). Despite its importance, there is very little information on the eventual exclusion effects of railways crossing wetlands, with very few studies focusing on aquatic bird species. In one of these studies, Waterman et al. (2002) found significant negative effects of railways on densities of black-tailed godwit *Limosa limosa* and garganey *Anas querquedula*, for noise levels above 45–49 dB(A). In contrast, Owens (1977) observed that geese *Branta* spp. disregarded trains passing nearby, while the single study using a Before-After-Control-Impact (BACI) design found no exclusion effects of a new railway line on a diverse wetland bird assemblage throughout the annual cycle (see Chap. 16). Possibly, the type and magnitude of effects depend on a number of factors related to wetland and railway characteristics, as well as the behaviour and life history traits of the species involved. Therefore, the development of a number of case studies covering a wide range of socio-ecological conditions would be needed to find general patterns.

In this study, we addressed the eventual exclusion effects on birds of a new railway line crossing the upper sector of the Sado Estuary (Portugal). This estuary is a "wetland of international importance" under the Ramsar Convention, and it is included in a Natural Reserve and in the Natura 2000 Network of Special Protected Areas and Sites of Community Importance. The Sado Estuary is inhabited by thousands of wetland birds, particularly from early autumn to spring, with large numbers occurring in the area where the railway line started operating in December 2010 (Lourenço et al. 2009; Alves et al. 2011). The environmental impact assessment study raised the possibility of the railway causing the displacement of these birds, thereby causing potential losses in prime wetland habitat. The study was designed to clarify this issue, aiming at: (1) describing the wetland bird assemblages occurring in the vicinity of the railway, and estimating how wetland bird assemblages varied in relation to proximity to the railway line, in terms of (2) species richness, (3) total bird density, and (4) densities of the most common species. Results of our study were used to discuss the potential exclusion effects of railways crossing wetland habitats.

Methods

Our study was carried out in the Natural Reserve of the Sado Estuary (Portugal), in a 3-km section of the upper estuary crossed by the "Variante de Alcácer" railway. Specifically, we focused in the area known as *Salinas da Batalha* (38° 24′N, 8° 36W), where the railway crosses the Sado River and adjacent wetland habitats through a bowstring bridge (see details in Chap. 7). During the study period

Table 11.1 Number (N) and area (ha) of habitat types in relation to distance categories (<500 and >500 m) from a railway bridge, which were sampled to estimate the exclusion effects on aquatic birds in the Sado Estuary, Portugal

Habitat	Distance (m)	N	Mean area (ha) ± SD	Total area (ha)
Rice paddy fields	<500	6	2.7 ± 1.4	16.6
	>500	5	4.0 ± 1.5	19.9
Intertidal mudflats	<500	3	0.7 ± 0.1	2.2
	>500	2	0.9 ± 0.3	1.7
Salt pans	<500	2	3.2 ± 0.9	6.4
	>500	4	4.5 ± 2.0	18.0

(26 weeks in 2013–2015), the number of trains using the railway increased from 26 ± 4 [SD] trains/day in 2013 to 34 ± 6 trains/day in 2015, and they mostly crossed the line (70%) during the day. The railway is bordered by habitats used by wetland birds, including mainly rice paddy fields, salt pans and intertidal mudflats. Rice paddy fields are particularly important foraging habitats in autumn and winter, outside the crop growing season, when they are partly flooded and occupied by stubbles, which may be used by black-tailed godwits and a number of herons, egrets, ibis, waders and waterfowl species (Fasola and Ruiz 1996; Lourenço and Piersma 2009). Salt pans are also artificial habitats of utmost importance for wetland birds in the Mediterranean region, as they provide important foraging grounds and critical high-tide roost for species foraging mainly on intertidal areas (Paracuellos et al. 2002; Dias et al. 2006). Finally, intertidal flats are critical foraging habitats for waders and some waterfowl species, which in the Mediterranean region are mainly used by wintering birds, and as stopovers during the autumn and spring migration periods (Moreira 1999; Granadeiro et al. 2007). In late spring and summer these habitats tend to be much less used by wetland birds, although they are still important for a few breeding species, such as the purple heron *Ardea purpurea* and black-winged stilt *Himantopus himantopus*.

We counted birds in 11 wetland sectors close to (0–500 m) and 11 sectors far from (500–1500 m) the railway line between December 2012 and October 2015. These sectors were selected to encompass the variety of habitats represented in the study area (Table 11.1), including intertidal mudflats (3.9 ha), rice paddy fields (36.5 ha), and salt pans (24.4 ha). We tried to balance the number and area of habitat types per distance category, but this was only partially achieved due to the local distribution of habitats around the railway bridge (Fig. 11.1). To minimize this problem, we reported and analyzed the data in terms of bird densities, rather than using absolute counts.

In each of the three study years, counts were made during two months in each of the winter, spring, summer and autumn seasons. In each month, counts were carried out both at low and high tide, between two hours before and two hours after peak tide. This design was used to account for the movements of wetland birds during the tidal cycle between foraging and roosting areas, and among foraging areas (e.g. Dias et al. 2006). In each wetland sector, counts were made using the "scanning method," beginning in the plot limit and sequentially covering the entire plot, and

Fig. 11.1 Map of the study area in the Sado Estuary (Portugal), showing the location of the sectors close to (0–500 m) and far from (500–1500 m) the railway, where wetland birds were counted between December 2012 and October 2015

thus minimizing double counts (Bibby et al. 2005). We did not carry out counts on days with adverse weather conditions, such as heavy rain and strong winds.

We used Generalized Linear Mixed Models (GLMM; Zuur et al. 2009) to evaluate how species richness and wetland bird densities varied in relation to the railway line. In order to model species' richness, we used Poisson errors and a log link, while the modelling of bird densities was done on log-transformed data using Gaussian errors and the identity link (Zuur et al. 2009). As fixed effects, we used dichotomous variables coding the distance to the railway line (<500 vs. >500 m) and the stage of the tidal cycle (low vs. high tide), and the interaction between them. We considered random effects on the intercept, including as factors the sector, the sampling month, and the sampling year. Models were fit using the package lmer (Bates et al. 2015) for the R software 3.2.2 (R Core Team 2015).

Results

Wetland Bird Assemblages

Overall, we counted 32,647 individuals of 46 wetland bird species during the surveys ("Annex"). The most abundant species accounted for 71% of the birds observed, and included black-headed gull *Chroicocephalus ridibundus* (21.9% of birds observed), greater flamingo *Phoenicopterus roseus* (20.7%), northern shoveler *A. clypeata* (10.0%), black-tailed godwit (9.6%) and lesser black-backed gull *Larus fuscus* (9.3%). Many more birds were recorded in autumn (47.3%) and winter (22.0%), than in spring (11.9%) and summer (18.8%). This pattern of seasonal occurrence was also observed for the most abundant species, except in the case of flamingos, which were mainly recorded in winter and spring, and godwits, which peaked in autumn and summer.

The distribution of birds was highly concentrated, with about 76% of individuals counted on just eight of the 22 plots surveyed. Many more birds were counted in the tanks of salt pans (64.5%), than in either rice fields (27.2%) or intertidal areas (8.4%). This pattern was also found for most individual species, although there were also several species with peak counts (>75% of observations) in rice fields (e.g. cattle egret *Bubulcus ibis*, purple heron, white stork *Ciconia ciconia*, black stork *C. nigra*, glossy ibis *Plegadis falcinellus*, and lapwing *Vanellus vanellus*). Only a few species were most counted (>50%) in intertidal areas, including mallard *A. platyrhynchos*, grey plover *Pluvialis squatarola*, common sandpiper *Actitis hypoleucos*, and whimbrel *Numenius phaeopus*.

Effects of Distance to Railway

There were no statistically significant effects of distance to railway on species richness, total bird abundance and the abundance of the 12 most abundant species (overall counts >500 individuals) (Table 11.2). The observed patterns of bird distribution in relation to the railway varied markedly across species, and for each it often varied in relation to the tidal cycle and the season (Fig. 11.2). For instance, black-headed gull densities were highest close to the railway in winter, both at low and high tides, while in both autumn and summer, densities were highest far from the railway, particularly at high tide. Also, the densities of the shoveler were highest close to the railway in autumn irrespective of tide, and in winter at low tide, while the reverse was found in winter at high tide. A comparable variation in relation to tide was found for pied avocets *Recurvirostra avosetta*, which in autumn, showed the highest densities close to the railway at low tide, and far from the railway at high tide. In white storks, densities in autumn and winter were always higher close to the railway, while the reverse was found in summer, irrespective of the tidal cycle. Other species had more consistent patterns, albeit not statistically significant, with larger densities tending to be found either close to (e.g. flamingo, dunlin *Calidris alpina*) or far from (e.g. godwit, great cormorant *Phalacrocorax carbo*) the railway.

Discussion

The bird assemblages recorded around the railway line crossing the upper Sado Estuary were comparable to those occurring in other coastal wetlands in Portugal, including a diversity of species such as herons, egrets, ibis, flamingos, gulls, waders,

Table 11.2 Summary results of generalized mixed models (GLMM) relating bird species richness and abundances to distance (Dist) to the railway (<500 vs. >500 m) and tidal cycle (low vs. high tide), while controlling for the random effects of sampling month and sector

Species	Fixed effects				Random effects			
	Inter	Dist	Tide	Dist:Tide	Sector	Month	Year	Residual
Black-headed gull	0.17 (2.34)	−0.002 (−0.02)	−0.009 (−0.22)	0.09 (1.63)	0.031	0.014	<0.001	0.142
Black-tailed godwit	0.05 (1.68)	−0.02 (−0.60)	−0.02 (−0.87)	0.03 (0.84)	0.002	0.002	<0.001	0.049
Black-winged stilt	0.04 (1.44)	0.01 (0.41)	0.002 (0.11)	−0.02 (−0.64)	0.003	0.003	<0.001	0.035
Dunlin	0.02 (1.69)	−0.01 (−0.86)	0.005 (0.39)	−0.01 (−0.80)	<0.001	<0.001	<0.001	0.016
Flamingo	0.08 (1.72)	−0.002 (−0.03)	0.03 (0.94)	−0.04 (−0.92)	0.016	0.002	<0.001	0.084
Great cormorant	0.01 (0.42)	0.02 (0.58)	−0.002 (−0.20)	0.01 (0.73)	<0.001	<0.001	<0.001	0.011
Lesser black-backed gull	0.12 (2.49)	−0.001 (−0.03)	−0.4 (−1.36)	0.03 (0.72)	0.008	0.009	<0.001	0.070
Little egret	0.09 (3.49)	−0.02 (−0.85)	−0.02 (−1.04)	0.03 (1.29)	0.003	<0.001	<0.001	0.026
Pied avocet	0.05 (1.79)	−0.04 (−0.96)	−0.004 (−0.24)	0.03 (1.34)	<0.001	<0.001	<0.001	0.030
Shoveler	0.07 (1.86)	−0.05 (−1.12)	−0.01 (−0.55)	0.02 (0.66)	0.008	0.003	<0.001	0.051
Spotted redshank	0.009 (1.37)	−0.002 (−0.33)	−0.003 (−0.67)	0.004 (0.69)	<0.001	<0.001	<0.001	<0.001
White stork	0.029 (1.99)	0.005 (0.33)	0.01 (0.74)	−0.02 (−0.92)	<0.001	<0.001	<0.001	0.018
Total abundance	0.43 (4.96)	−0.09 (−0.91)	−0.002 (−0.07)	0.02 (0.38)	0.048	0.014	0.003	0.130
Species richness	0.48 (0.05)	0.08 (0.76)	−0.09 (0.29)	0.19 (0.08)	0.368	0.18	0.001	–

Abundance models were fitted on log-transformed data, using Gaussian errors and the identity link. The species richness model was fitted using Poisson errors and the log link. For each model we provide the regression coefficients, the t-value (Z-value in the case of species richness model), and the variance accounted by the random effects

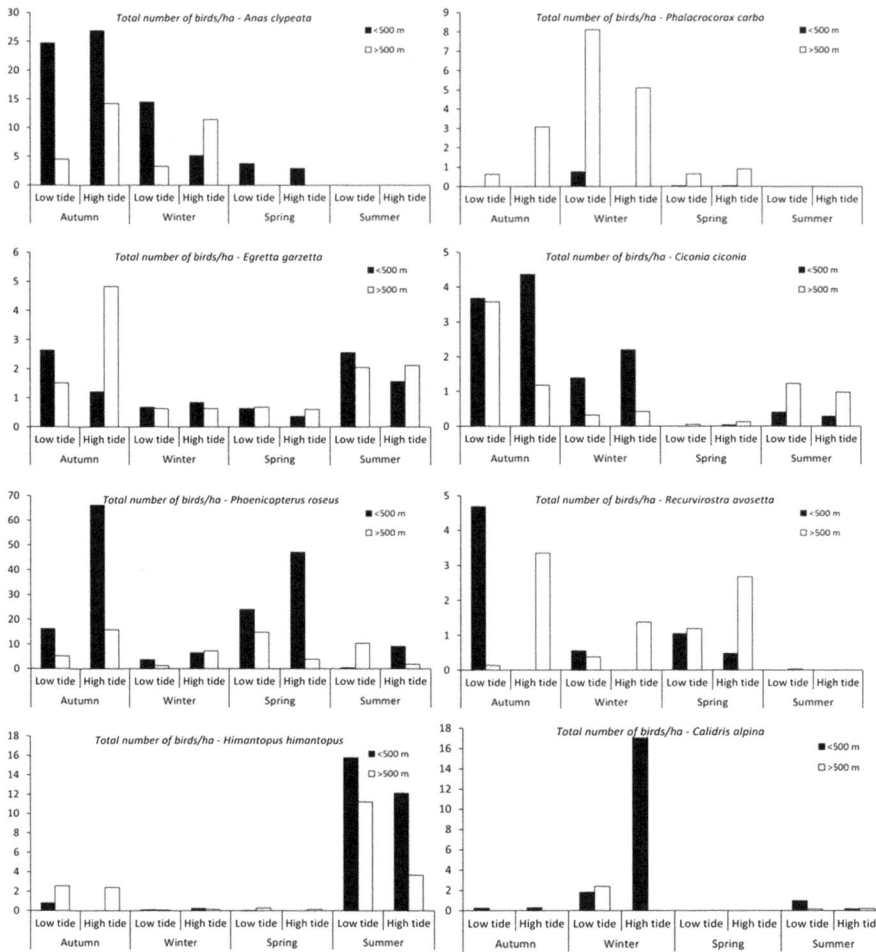

Fig. 11.2 Variation in mean densities (birds/ha) of the most abundant wetland birds counted in sectors close to (<500 m) and far from (>500 m) the railway line, in relation to season and level of the tidal cycle

and waterfowl (Lourenço et al. 2009; Alves et al. 2011). Also, the temporal patterns of species' composition were similar to those of other wetlands, with diversity and abundance of most species peaking during the autumn and, to a lesser extent, spring migration periods, and in winter. There was no evidence that the spatial patterns of bird assemblages were affected by proximity to the railway, with species richness and abundances not showing significant differences in areas close to and far from the railway. Although higher densities of some species were recorded far from the railway on some occasions, there was no consistent pattern due to strong variation across species and seasons, and through the tidal cycle. Overall, we could find no noticeable exclusion effects of this railway line on wetland birds, which is in line

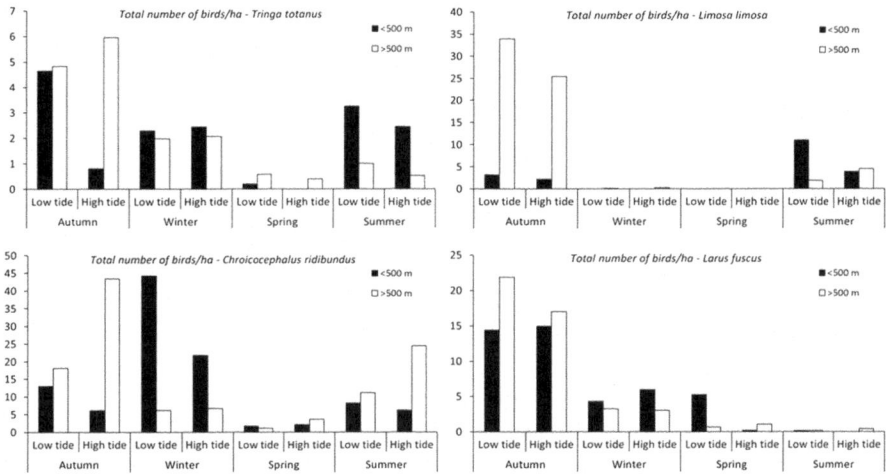

Fig. 11.2 (continued)

with some of the few studies addressing the same issue (Owens 1977, and Chap. 16), but in contrast with another study (Waterman et al. 2002).

Although our study had some limitations and potential shortcomings, it is unlikely that they significantly affected our key conclusions. The main problem is that no information on abundances and bird spatial distribution was available for the period before the construction of the railway line. Because of this, it was impossible to carry out a proper BACI study, which would be essential to fully demonstrate whether or not there were any significant impacts (e.g. Tarr et al. 2010). Nevertheless, our results are still useful to suggest that wetland bird abundances at present are not lower at <500 m from the railway than elsewhere in the estuary, though they cannot be used to make comparisons in relation to the situation before the construction of the railway. Another problem is that our study focused only on species richness and densities, although there might be more subtle impacts due, for instance, to changes in vigilance and feeding activity close to the railway (e.g. Yasué 2006). Although this issue was not dealt with in our study, non-systematic observations of 55 individuals of 18 species did not provide any evidence that birds are more alert close to than there are far from the railway. Finally, we considered a relatively wide area to represent proximity to the railway (<500 m), although exclusion effects may only be apparent at shorter distances from disturbance. This was a constraint required to obtain sufficiently large sample sizes, but observations during the study period suggest that this was not serious because large bird concentrations were often recorded very close to the railway.

The lack of obvious avoidance responses observed in our study may be a consequence of wetland birds tolerating well or eventually adapting behaviourally to human structures and vehicles such as circulating trains, and thus largely ignoring these potential sources of disturbances when selecting foraging and roosting habitats. Studies addressing this hypothesis are scarce, but there is some supportive evidence. For instance, a study showed that the displacement effects of wind farms on pink-footed geese *Anser brachyrhynchus* occurred over relatively short distances (<150 m), and that the negative effects declined markedly in the 10 years after wind farm construction (Madsen and Boertmann 2008). Also, Donaldson et al. (2007) found that waterbirds were more tolerant of human presence in developed than in undeveloped sites, probably reflecting some level of behavioural adaptation to potential anthropogenic disturbance (see also Lowry et al. 2013). In contrast to these studies, however, in Chap. 12, Múrias and colleagues, in a study akin to a BACI design, reported significantly lower shorebird abundances in saltpans close to a new railway that had recently begun to operate. Another study by Burton et al. (2002) reported reduced wader abundances in relation to footpaths, roads and railroads, although this greatly varied across species and human structures. Together, these studies suggest that the exclusion effects of railways on wetland birds may be small in some cases but not in others, and thus drawing generalities would require a larger number of studies covering a wide range of socio-ecological contexts.

Although our study did not find obvious negative effects, the densities of some species were sometimes higher far from (or close to) the railway, although this was not statistically significant and there were inconsistencies across species, seasons, and stages of the tidal cycle. Reasons for this are not certain, but they may be a consequence of spatial and temporal variations in habitat requirements and quality for foraging and roosting, rather than reflecting negative (or positive) effects of the railway. It is well known that waders and other aquatic birds undertake daily movements between foraging and roosting habitats, which are jointly affected by the circadian and tidal cycles, and which induce great variations in bird spatial distributions in estuaries and other wetlands (Dias et al. 2006; Granadeiro et al. 2006). Also, there are changes in the use of habitats through the annual cycle, due for instance, to differences in habitat selection between wintering and migrating birds (Martins et al. 2016), or temporal changes in habitat quality (Lourenço et al. 2009). These factors may underlie, for instance, the observation of avocets concentrating at high densities in intertidal mudflats close to the railway at low tide in autumn, but not in winter, while at high tide they were mainly found in a single salt pan far from the railway in both autumn and winter. Conversely, dunlins were found concentrated in a single salt pan close to the railway in winter at high tide, while at low tide during the same season they were mostly counted in intertidal areas close to the railway and in a salt pan far from the railway. Overall, these results suggest that care should be taken when evaluating the effects of railways on wetland birds, due to the confounding effects resulting from species-specific habitat requirements and preferences, the availability and quality of habitats, and the changes in these factors over the seasonal, circadian, and tidal cycles.

Acknowledgements We would like to thank to Pedro Salgueiro and Luís Gomes for their support in field work. This study was supported by Infraestruturas de Portugal with the contribution of the Portuguese Science Foundation through the Doctoral Grant SFRH/BD/81602/2011 (Carlos Godinho) doctoral grant.

Annex

Wetland birds counted in the study area (Sado Estuary, Portugal) between December 2012 and October 2015. For each species we present the number of birds counted per season, in 11 sectors close to (<500 m) and 11 sectors far from (>500 m) a railway line. We also present the proportion of birds counted in each of the three main habitats represented in the study area: rice fields (rice), intertidal mudflats (intertidal) and salt pans (tanks).

Species		Winter		Autumn		Spring		Summer		Total	Habitat		
		<500 m	>500 m	<500 m	>500 m	<500 m	>500 m	<500 m	>500 m		Rice (%)	Intertidal (%)	Tank (%)
Common shelduck	*Tadorna tadorna*	65		20			20	7	15	127		7	93
Ruddy shelduck	*Tadorna ferruginea*		2							2			100
Mallard	*Anas platyrhynchos*	142	6	6	27	37	54	6		278	9	67	24
Gadwall	*Anas strepera*			2						2			100
Northern shoveler	*Anas clypeata*	490	585	1289	747	166				3277		4	96
Eurasian teal	*Anas crecca*			1						1			100
Black-necked grebe	*Podiceps nigricollis*	105	29	20	45		1			200	3		97
Little grebe	*Tachybaptus ruficollis*	61	94	9	39		7		2	212	1		99
Great cormorant	*Phalacrocorax carbo*	19	528		148	2	62			759		3	97
Western cattle egret	*Bubulcus ibis*	7	14	13	61	10	3	60	320	488	93	1	6
Little egret	*Egretta garzetta*	38	50	96	254	25	51	103	167	784	64	11	25

(continued)

(continued)

Species	Winter <500 m	Winter >500 m	Autumn <500 m	Autumn >500 m	Spring <500 m	Spring >500 m	Summer <500 m	Summer >500 m	Total	Habitat Rice (%)	Habitat Intertidal (%)	Habitat Tank (%)
Great egret *Casmerodius albus*	6	10	1	13		6			36	64	3	33
Grey heron *Ardea cinerea*	58	48	49	51	9	8	26	42	291	62	15	22
Purple heron *Ardea purpurea*							2	9	11	100		
White stork *Ciconia ciconia*	90	30	201	190	1	7	17	88	624	75		25
Black stork *Ciconia nigra*		3							3	100		
Glossy ibis *Plegadis falcinellus*				116		1	260	30	407	100		0
Eurasian spoonbill *Platalea leucorodia*	18	42	50	97	16	10	10	44	287	18	18	64
Greater flamingo *Phoenicopterus roseus*	261	343	2061	843	1776	741	235	484	6744	33	0	67
Eurasian coot *Fulica atra*			4						4	100		
Common moorhen *Gallinula chloropus*							2	2	4	100		
Pied avocet *Recurvirostra avosetta*	14	70	117	139	38	155		1	534	1	32	67
Black-winged Stilt *Himantopus himantopus*	8	7	20	198	1	17	696	594	1541	20	4	76
Common ringed plover *Charadrius hiaticula*	34	1				1	1	8	45	44	31	24
Kentish plover *Charadrius alexandrinus*	19	2			5	6	14	8	54	37	9	54
Grey plover *Pluvialis squatarola*	37	67			8	1			113	14	54	32
Northern lapwing *Vanellus vanellus*	82	32							114	100		

(continued)

(continued)

Species		Winter		Autumn		Spring		Summer		Total	Habitat		
		<500 m	>500 m	<500 m	>500 m	<500 m	>500 m	<500 m	>500 m		Rice (%)	Intertidal (%)	Tank (%)
Sanderling	*Calidris alba*		1			1			2	4			100
Dunlin	*Calidris alpina*	470	96	13				28	14	621	72	5	23
Curlew sandpiper	*Calidris ferruginea*							2		2			100
Green sandpiper	*Tringa ochropus*	13	7	4	2	2		2	22	52	77	12	12
Common sandpiper	*Actitis hypoleucos*	25	15	4	4	4	1	5	7	65	23	66	11
Spotted redshank	*Tringa erythropus*	118	161	136	431	5	39	142	61	1093	8	15	77
Common Redshank	*Tringa totanus*	4	6		1			6	31	48	18	23	59
Common greenshank	*Tringa nebularia*	2	22	4	10	3	68		4	113	4	7	89
Black-tailed Godwit	*Limosa limosa*		11	132	2374			370	253	3140	2	8	90
Whimbrel	*Numenius phaeopus*								1	1		100	
Common snipe	*Gallinago gallinago*	12	2	2		7				23	100		
Ruff	*Philomachus pugnax*				17			1	10	28			100
Black-headed Gull	*Larus ridibundus*	1649	512	478	2456	95	189	356	1417	7152	20	14	66
Yellow-legged Gull	*Larus michahellis*		1							1			100
	Larus fuscus	256	249	733	1555	136	67	4	20	3020	63	7	30

(continued)

(continued)

Species	Winter		Autumn		Spring		Summer		Total	Habitat		
	<500 m	>500 m	<500 m	>500 m	<500 m	>500 m	<500 m	>500 m		Rice (%)	Intertidal (%)	Tank (%)
Lesser Black-backed Gull												
Little tern *Sternula albifrons*								16	16			100
Sandwich tern *Sterna sandvicensis*		7		5					12			100
Common tern *Sterna hirundo*						2			2			100
Caspian tern *Hydroprogne caspia*				1				3	4			100
Sterna spp.				2					2			100
Unidentified	3	1	7	5				1	17	12	18	71
Anas spp.			120	14					134	67	90	10
Charadrius spp.	2			1					3	67	33	
Calidris spp.	2								2	100		
Tringa spp.	2	15		9		29	1	93	149	3		97
Larus spp.						1			1		100	

References

Alves, J. A., Dias, M., Rocha, A., Barreto, B., Catry, T., Costa, H., et al. (2011). Monitorização das populações de aves aquáticas dos estuários do Tejo, Sado e Guadiana. Relatório do ano de 2010. *Anuário Ornitológico (SPEA), 8,* 118–133.

Bates, D., Maechler, M., Bolker, B., & Walker, S. (2015). Fitting linear mixed-effects models using lme4. *Journal of Statistical Software, 67,* 1–48.

Bibby, C. J., Burgess, N. D., Hill, D. A., Mustoe, S. H. (2005). *Bird census techniques* (2nd ed.). London: Elsevier Academic Press.

Burton, N. H., Armitage, M. J., Musgrove, A. J., & Rehfisch, M. M. (2002). Impacts of man-made landscape features on numbers of estuarine waterbirds at low tide. *Environmental Management, 30,* 0857–0864.

Carvalho, F., Carvalho, R., Mira, A., & Beja, P. (2016). Assessing landscape functional connectivity in a forest carnivore using path selection functions. *Landscape Ecology, 31,* 1021–1036.

D'Amico, M., Périquet, S., Román, J., & Revilla, E. (2016). Road avoidance responses determine the impact of heterogeneous road networks at a regional scale. *Journal of Applied Ecology, 53,* 181–190.

DeFries, R. S., Ellis, E. C., Chapin, F. S., Matson, P. A., Turner, B. L., Agrawal, A., et al. (2012). Planetary opportunities: A social contract for global change science to contribute to a sustainable future. *BioScience, 62,* 603–606.

Dias, M. P., Granadeiro, J. P., Lecoq, M., Santos, C. D., & Palmeirim, J. M. (2006). Distance to high-tide roosts constrains the use of foraging areas by dunlins: Implications for the management of estuarine wetlands. *Biological Conservation, 131,* 446–452.

Donaldson, M. R., Henein, K. M., & Runtz, M. W. (2007). Assessing the effect of developed habitat on waterbird behaviour in an urban riparian system in Ottawa, Canada. *Urban Ecosystems, 10,* 139–151.

Fasola, M., & Ruiz, X. (1996). The value of rice fields as substitutes for natural wetlands for waterbirds in the Mediterranean region. *Colonial Waterbirds, 19,* 122–128.

Forman, R. T., & Deblinger, R. D. (2000). The ecological road-effect zone of a Massachusetts (USA) suburban highway. *Conservation Biology, 14,* 36–46.

Gosh, S., Kim, K., & Bhattacharya, R. (2010). A survey on house sparrow population decline at Bandel, West Bengal, India. *Journal Korean Earth Science Society, 31,* 448–453.

Granadeiro, J. P., Dias, M. P., Martins, R. C., & Palmeirim, J. M. (2006). Variation in numbers and behaviour of waders during the tidal cycle: Implications for the use of estuarine sediment flats. *Acta Oecologica, 29,* 293–300.

Granadeiro, J. P., Santos, C. D., Dias, M. P., & Palmeirim, J. M. (2007). Environmental factors drive habitat partitioning in birds feeding in intertidal flats: Implications for conservation. *Hydrobiologia, 587,* 291–302.

Kleijn, D., Cherkaoui, I., Goedhart, P. W., van der Hout, J., & Lammersma, D. (2014). Waterbirds increase more rapidly in Ramsar-designated wetlands than in unprotected wetlands. *Journal of Applied Ecology, 51,* 289–298.

Li, Z., Ge, C., Li, J., Li, Y., Xu, A., Zhou, K., et al. (2010). Ground-dwelling birds near the Qinghai-Tibet highway and railway. *Transportation Research Part D, 15,* 525–528.

Loss, S. R., Will, T., & Marra, P. P. (2015). Direct mortality of birds from anthropogenic causes. *Annual Review of Ecology Evolution and Systematics, 46,* 99–120.

Lourenço, P. M., Groen, N., Hooijmeijer, J. C. E. W., & Piersma, T. (2009). The rice fields around the estuaries of the Tejo and Sado are a critical stopover area for the globally near-threatened black-tailed godwit *Limosa l. limosa*: Site description, international importance and conservation proposals. *Airo, 19,* 19–26.

Lourenço, P. M., & Piersma, T. (2009). Waterbird densities in South European rice fields as a function of rice management. *Ibis, 151,* 196–199.

Lowry, H., Lill, A., & Wong, B. (2013). Behavioural responses of wildlife to urban environments. *Biological Reviews, 88,* 537–549.

Madsen, J., & Boertmann, D. (2008). Animal behavioral adaptation to changing landscapes: Spring-staging geese habituate to wind farms. *Landscape Ecology, 23,* 1007–1011.

Martins, R. C., Catry, T., Rebelo, R., Pardal, S., Palmeirim, J. M., & Granadeiro, J. P. (2016). Contrasting estuary-scale distribution of wintering and migrating waders: The potential role of fear. *Hydrobiologia, 768,* 211–222.

Millennium Ecosystem Assessment. (2005). *Ecosystems and human well-being: Wetlands and water synthesis.* Washington, DC, USA: World Resources Institute.

Moreira, F. (1999). On the use by birds of intertidal areas of the Tagus estuary: Implications for management. *Aquatic Ecology, 33,* 301–309.

Mundahl, N. D., Bilyeu, A. G., & Maas, L. (2013). Bald Eagle Nesting Habitats in the Upper Mississippi River National Wildlife and Fish Refuge. *Journal of Fish and Wildlife Management, 4,* 362–376.

Owens, N. W. (1977). Responses of wintering brent geese to human disturbance. *Wildfowl, 28,* 5–14.

Paracuellos, M., Castro, H., Nevado, J. C., Oña, J. A., Matamala, J. J., García, L., et al. (2002). Repercussions of the abandonment of Mediterranean saltpans on waterbird communities. *Waterbirds, 25,* 492–498.

R Core Team. (2015). *R: A language and environment for statistical computing.* Austria: R Foundation for Statistical Computing.

Santos, S. M., Mira, A., Salgueiro, P. A., Costa, P., Medinas, D., & Beja, P. (2016). Avian trait-mediated vulnerability to road traffic collisions. *Biological Conservation, 200,* 122–130.

Seto, K. C., Güneralp, B., & Hutyra, L. R. (2012). Global forecasts of urban expansion to 2030 and direct impacts on biodiversity and carbon pools. *Proceedings of the National Academy of Sciences, 109,* 16083–16088.

Tarr, N. M., Simons, T. R., & Pollock, K. H. (2010). An experimental assessment of vehicle disturbance effects on migratory shorebirds. *The Journal of Wildlife Management, 74,* 1776–1783.

Tittensor, D. P., Walpole, M., Hill, S. L., Boyce, D. G., Britten, G. L., Burgess, N. D., et al. (2014). A mid-term analysis of progress toward international biodiversity targets. *Science, 346,* 241–244.

Torres, A., Jaeger, J. A., & Alonso, J. C. (2016). Assessing large-scale wildlife responses to human infrastructure development. *Proceedings of the National Academy of Sciences, 113,* 8472–8477.

van der Ree, R., Smith, D. J., & Grilo, C. (2015). *Handbook of Road Ecology.* Hoboken: Wiley.

Ware, H. E., McClure, C. J., Carlisle, J. D., & Barber, J. R. (2015). A phantom road experiment reveals traffic noise is an invisible source of habitat degradation. *Proceedings of the National Academy of Sciences, 112,* 12105–12109.

Waterman, E., Tulp, I., Reijnen, R., Krijgsveld, K., & Braak, C. (2002). Disturbance of meadow birds by railway noise in The Netherlands. *Geluid, 1,* 2–3.

Wiącek, J., Polak, M., Filipiuk, M., Kucharczyk, M., & Bohatkiewicz, J. (2015). Do birds avoid railroads as has been found for roads? *Environmental Management, 56,* 643–652.
Yasué, M. (2006). Environmental factors and spatial scale influence shorebirds' responses to human disturbance. *Biological Conservation, 128,* 47–54.
Zuur, A. F., Leno, E. N., Walker, N. J., Saveliev, A. A., & Smith, G. M. (2009). *Mixed effects models and extensions in ecology with R.* New York: Springer.

Chapter 12
Evaluating the Impacts of a New Railway on Shorebirds: A Case Study in Central Portugal (Aveiro Lagoon)

Tiago Múrias, David Gonçalves and Ricardo Jorge Lopes

Abstract In 2007, the Portuguese Railway Company (REFER) began to build a new railway connecting the commercial port of Aveiro (Central Portugal) to the national rail network, which extended for about 9 km and crossed three of the remaining saltpans of the Aveiro Lagoon through a viaduct. Due to the importance of these habitats for shorebirds, and because the railway crossed a Natura 2000 site, the national biodiversity conservation authority required the impact assessment of this infrastructure. The study extended from 2006 (pre-construction), through 2008 and 2009 (construction), to 2011 (post-construction), encompassing four breeding and four wintering seasons. We used a Before-After-Control-Impact (BACI) design, with three impacted (those under the viaduct) and six control saltpans. During the breeding season (April–July) we monitored the numbers and breeding parameters of Black-winged stilt (*Himantopus himantopus*), and the numbers of Kentish plover (*Charadrius alexandrinus*) and Little tern (*Sterna albifrons*). In the winter, we monitored the numbers and spatial distribution of all shorebirds, and the activity of Dunlin (*Calidris alpina*). There was evidence for reductions in the abundance of the three breeding species in impacted saltpans in 2011, while no significant negative effects were found on abundances during 2008–2009 and on the breeding parameters of Black-winged Stilt. There was also evidence for reductions in wintering shorebird abundances in impacted saltpans in 2011, with no significant effects on abundances in 2008–2009, and on shorebird spatial distribution and activity patterns. Overall, the results suggest that the operation of a new railway close to important wetland habitats had negative impacts on the abundance of breeding and wintering shorebirds.

T. Múrias · D. Gonçalves · R.J. Lopes (✉)
CIBIO/InBIO, Centro de Investigação em Biodiversidade e Recursos Genéticos, Universidade do Porto, Campus Agrário de Vairão, Rua Padre Armando Quintas, 4485-661 Vairão, Portugal
e-mail: riclopes@me.com

D. Gonçalves
Departamento de Biologia, Faculdade de Ciências, Universidade do Porto, Rua do Campo Alegre, 4169-007 Porto, Portugal

© The Author(s) 2017
L. Borda-de-Água et al. (eds.), *Railway Ecology*,
DOI 10.1007/978-3-319-57496-7_12

197

Keywords Shorebirds · Wetlands · Saltpans · Impact · Railway

Introduction

The assessment of ecological effects on habitats and species of the construction and posterior operation of railways is still in its infancy when compared with other human-made infrastructures, like such as roads or powerlines, as largely stressed in Chap. 1. Although mortality is the most obvious and direct ecological effect of railways (as with other linear structures), the railways may also have indirect effects-namely as barriers for wildlife (Chap. 4), as vectors for the transport of alien species (Chap. 5), and as sources of disturbance (Chap. 6). In this book, most examples refer to railways on land or aquatic inland ecosystems (e.g. lakes, rivers), which is rather logical, since for the most part, railway lines are built overland. However, they may also impact non-terrestrial habitats (and associated species), when they are built in or near commercial ports. These are almost invariably located on estuaries, one of the most sensible ecosystems in terms of functioning and conservation, and critical for the conservation of many bird species (McLusky and Elliott 2004).

The eventual negative impacts of railways in estuarine systems may be particularly relevant for shorebirds and related species (Charadriiformes), which are among the birds of highest conservation concern (International Wader Study Group 2003; van de Kamp et al. 2004), due to their migratory status, feeding specialization, and, as a result, dependency on a limited set of areas for breeding, staging, and wintering (McLusky and Elliott 2004). Most of these species are very dependent on estuarine habitats and neighboring inland wetlands, as saltmarshes (Prater 1981) and, in southern Europe, also saltpans (Luís et al. 2002; Masero 2003; Múrias et al. 1997; Rufino et al. 1984). Therefore, it is important to assess the impacts of railways on this group of species, thereby allowing the development of adequate mitigation measures.

Portugal is one of the countries in southern Europe that holds international important wetlands for shorebirds, and where the development of railway networks to serve harbors can impact on estuarine and other coastal habitats. In these wetlands, saltpans appear to play an important role for shorebird conservation, as they are used both as refuges for sheltering at high tides, and as foraging habitats, especially when the mudflats are covered. Therefore, saltpans provide important sources of supplementary energy for birds, particularly during the demanding migratory periods and in winter (Dias et al. 2014; Lopes et al. 2005; Múrias et al. 2002; Rufino et al. 1984). Moreover, saltpans provide suitable breeding locations to populations of some shorebird species, such as Black-winged Stilt *Himantopus himantopus*, Kentish plover *Charadrius alexandrinus*, Redshank *Tringa tetanus* and Little tern *Sterna albifrons* (Fonseca et al. 2005; Jardim 1984; Múrias et al. 1997; Rufino and Araújo 1987). For the Black-winged stilt they even represent the most important breeding areas in Portugal (Rufino and Neves 1992).

Despite their importance for shorebirds, saltpans in Portugal are fast disappearing due to abandonment or conversion to aquaculture. As a result, only a few artisanal and industrial salt works still persist in the wetlands of Aveiro, Mondego, Tagus, and Sado estuaries, along the western coast of the country, as well as in wetlands along the southern coast of Algarve. According to figures from the most recent national census in 2015, there were only 73 saltpans in the whole country, which produced a total of 117 tons of salt (INE 2016). Once a thriving industry in the majority of the estuaries and coastal lagoons of the Portuguese coast, the marine salt extraction declined with the onset of new food conservation technologies and the reduction or abandonment of the once-prosperous long-range fishing fleet, particularly since the early 1960s (Rodrigues et al. 2011; Vieira and Bio 2014).

In this chapter, we provide a case study developed in the coastal lagoon of Aveiro (Ria de Aveiro), evaluating the response of shorebirds using saltpans to the impacts associated with the construction and operation of a new railway, both during the breeding and the wintering seasons. Under this scope, our specific objectives were to evaluate whether the railway affected (a) the numbers of the main breeding shorebirds (Black-winged Stilt, Kentish plover and Little tern); (b) the phenology and breeding success of Black-winged Stilt; and (c) the numbers, distribution, and activity of wintering shorebirds.

Methodology

Study Area

The study was carried out in Ria de Aveiro (Fig. 12.1), which is a complex lagoon system located in western Portugal at the mouth of the Vouga River that covers an area of approximately 450 km^2. For centuries, marine salt extraction occupied a large part of this system (Bastos 2009; Rodrigues et al. 2011). By the early 1970s, there were still some 270 saltpans (1160 ha), but in subsequent decades, most of these saltpans were abandoned, flooded, and/or converted for other purposes (e.g., aquaculture production), with only nine saltpans (56 ha) persisting in 2007 (Bastos 2009; Rodrigues et al. 2011). In contrast, a busy trading and fishing harbor has grown in the last few decades, particularly since it was linked by a highway to the Spanish border. To connect the harbor to the national rail network, a new railroad connection about 9 km long started to be built in 2007 (REFER 2009). The engineering and environmental problems faced by the project were challenging, because this was a densely populated area, and Ria de Aveiro is a wetland included in the Natura 2000 network, established by the European Directive 92/43/EEC. Furthermore, the proposed rail path crossed important wetland areas, including eight of the last remaining untransformed saltpans, and there were no technically viable alternatives. For this reason, it was decided that the railway should cross the saltpans and adjoining areas through a 5.5 km viaduct, to minimize the loss of important shorebird habitats. Despite these efforts, three saltpans were directly affected by sustaining viaduct pillars (Fig. 12.1), or the construction of a service road (REFER 2009).

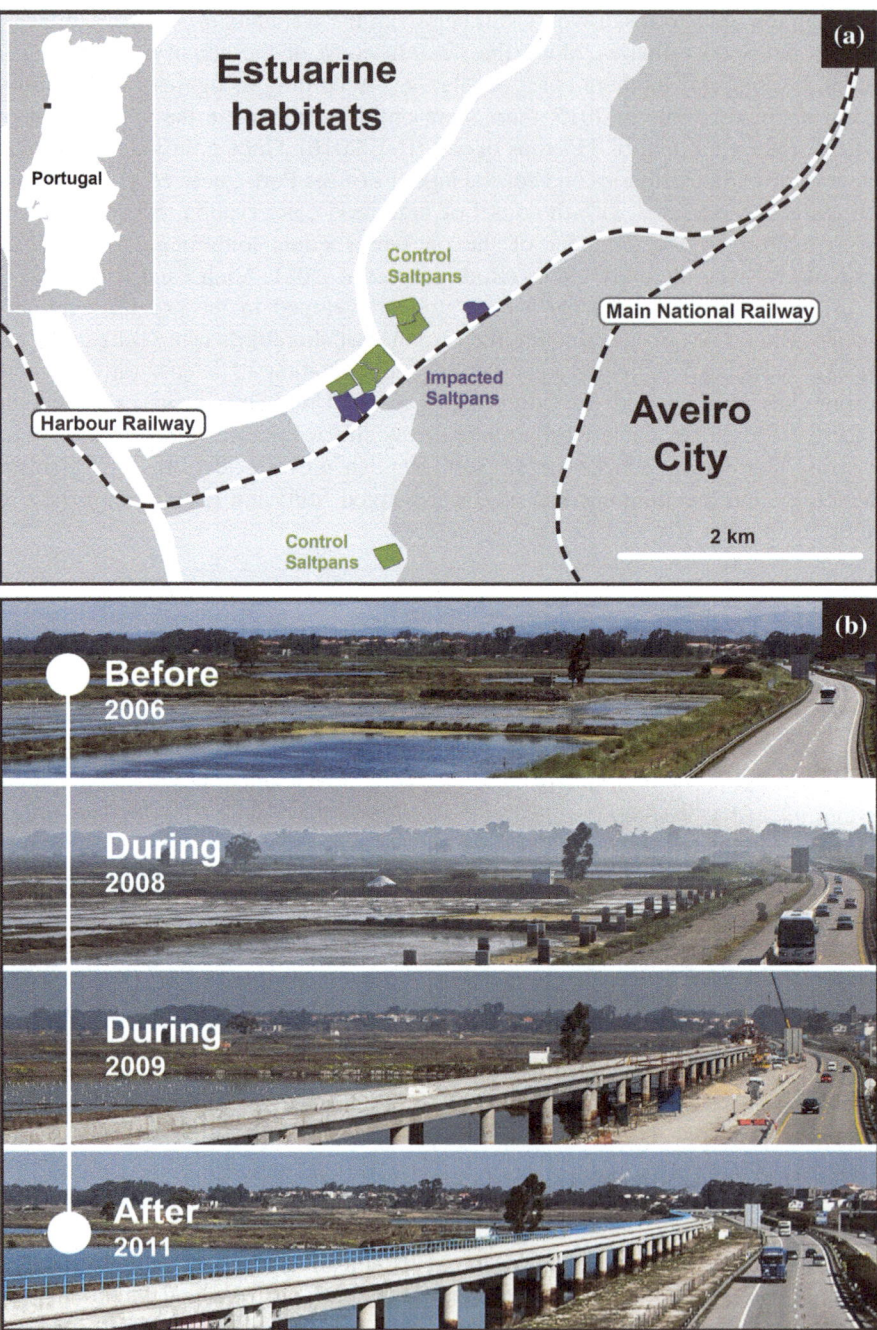

Fig. 12.1 **a** Map of the study area in Ria de Aveiro (Portugal), showing the location of the new railway line and the saltpans that were impacted by the construction and functioning of the railway (*blue*) and the saltpans that were used as control sampling units (*green*). **b** Timeline of the different phases, with photos of the same railway viaduct section, also showing one of the impacted saltpans

Study Design

The study was based on a Before-After-Control-Impact (BACI) designed to test for the impacts of railway construction and operation in shorebirds using saltpans. Sampling was carried out on three time periods (Fig. 12.1) six years: one control year before the construction (2006); two years during the phase of construction (2008 and 2009); and one year in the post-construction (i.e., operation) phase (2011). Each "treatment" phase (i.e., construction and post-construction) was compared with the control phase (i.e., pre-construction) using as sampling units three impacted (those under the viaduct) and six control saltpans (Fig. 12.1). The presence of significant interactions between year and type of pond (i.e., control vs. impact) was understood to indicate significant railway impacts-either positive or negative-depending on the sign of the interaction term. We used saltpans as sampling units because they were the most important shorebird habitats affected by the project, and because they form discrete units where it is possible to make absolute counts both of all birds and their nests. Moreover, they have easy access and they were the only nesting areas used by shorebirds close to the railway corridor. Periodical winter counts in the surrounding saltmarshes did not reveal a substantial usage of these areas by shorebirds, with the exception of a small number of Dunlin (*Calidris alpina*) and common sandpiper (*Actitis hypoleucos*), which seemed to be largely insensitive to the construction of the pillars (personal observation).

We focused the impact assessment on breeding and wintering seasons, which are the two most critical periods in the phenology of shorebirds. During the breeding season (April–July), data were gathered from visits carried out at weekly intervals to all saltpans. During the winter season (January–February) we made regular, but interspersed visits (at least 1-week intervals) to all saltpans to obtain a suitable number of sampling occasions. In each sampling visit, selected parameters (see below) were estimated in all saltpans based on pre-established transects and observation points, by a team of two researchers using binoculars and telescopes. Detailed maps of bird distribution in each saltpan were produced at each visit.

Statistical analysis was carried out in the R 3.2.6 statistical environment (R Core Team 2015), using packages lmer4 (Bates et al. 2015), glmmADMB (Fournier et al. 2012), and lsmeans (Lenth 2016; Bates et al. 2015).

Evaluating Effects on Breeding Shorebirds

Shorebirds are mobile animals and can thus avoid potential impacts by moving away from impacted saltpans (Cayford 1993). When they remain in the impacted saltpans, however, they may still change their spatial distribution (individuals or nests), which may eventually affect their phenology or breeding success, through reduced nest or chick survival (Sansom et al. 2016; Pearce-Higgins et al. 2007). We tested these ideas by focusing on the three main breeding shorebird species:

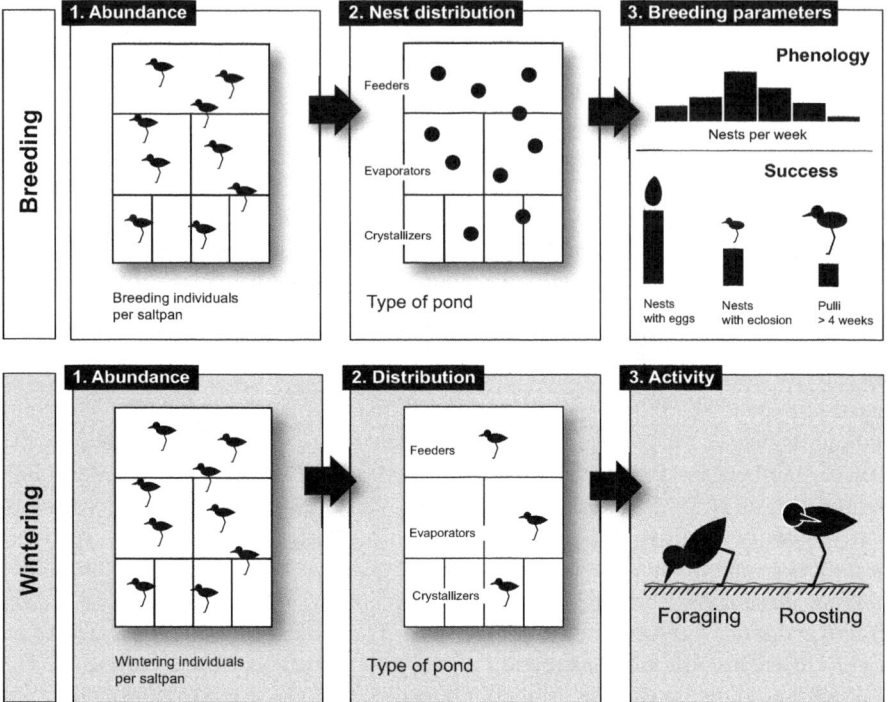

Fig. 12.2 Outline of the sampling rationale and parameters collected to test for the impacts of a new railway on shorebirds breeding and wintering in saltpans of Ria de Aveiro (Portugal)

Black-winged Stilt, Kentish plover, and Little tern. A more detailed study was made of the Black-winged Stilt, because it was the only species breeding on all saltpans in reasonable numbers, and it was the only one amenable to reliable surveys while minimizing disturbances, due to the relatively large size and conspicuousness of nests and chicks. Therefore, besides abundance, for this species we estimated for this species the distribution of nests, phenology and breeding success (Fig. 12.2). Overall, we tested the following hypothesis:

1. The overall abundance of each breeding species should decrease in the impacted saltpans.
2. The distribution of Black-winged Stilt nests should change within impacted saltpans.
3. For the Black-winged Stilt, the (a) nesting phenology should be delayed, and (b) breeding success should be lower in impacted saltpans in comparison to control saltpans.

Breeding Shorebird Abundance

For the three focal species, we recorded the number of individuals per saltpan on a weekly basis during the breeding season. We tested for the effects of time (year) and type of saltpan (impacted vs. control) on the weekly number of individuals using GLMM with negative binomial distribution, controlling for overdispersion (Crawley 2013). The saltpan and the week of census were used as random factors. An offset variable was always specified to account for variation in the area (hectares) of the saltpans. The models' fitness was assessed graphically (residuals inspection), and they were compared with null models to test their significance (Zuur et al. 2009).

Spatial Distribution of Black-Winged Stilt Nests

Each saltpan was divided into three pond categories (feeders, evaporators, and crystallizers), according to their position, size, and microhabitat characteristics, such as water level (Rodrigues et al. 2011). Black-winged stilt nests are not distributed randomly among these categories (Tavares et al. 2008), and we expected that the disturbance caused by the construction and operation of the railway would affect the distribution of nests. For this reason, we recorded the spatial distribution of all Black-winged stilt nests according to type of pond. We then tested for differences in the number of nests and the type of pond, time of year, and type of saltpan (impacted vs. control) using a GLMM with a negative binomial distribution, controlling for overdispersion and saltpan area (Crawley 2013), and specifying the saltpan and week of visit as random factors.

Black-Winged Stilt Nest Phenology and Breeding Success

Black-winged stilts incubate 3–4 eggs for 4 weeks, and chicks remain in the same saltpan while not yet able to fly and are protected by their parents. Five weeks after hatching, chicks are no longer restricted to the saltpan where they were born, because they can already fly and follow their parents (del Hoyo et al. 1996). We recorded the timing of deployment of nests during the breeding season on a weekly basis (see above) in all saltpans (impacted and control). We tested for differences in the temporal distribution of the nests with time of year and type of saltpan (impacted vs. control), using Kolmogorov–Smirnov tests (Griffin 1999) with the ks.test procedure (R Core Team 2015). Following nest identification (see above) we recorded, on a weekly basis, the number of chicks and their ages. This allowed estimation of two binomial parameters of breeding success: nest and fledgling success. A nest was considered successful if at least one chick up to one week old was observed, while fledging success was considered for each clutch when at least one chick reached the age of fledging (4–5 weeks). We tested for differences with time of year and saltpan (impacted vs. control) on nest and fledging success, using

GLMM with a binary distribution function, and controlling for saltpan area (Crawley 2013). The saltpan identity was specified as a random factor.

Evaluating Effects on Wintering Shorebirds

As in the breeding season, wintering shorebirds can avoid potential impacts by moving away from impacted saltpans (Cayford 1993). If they stay, they may change their spatial distribution on impacted saltpans and their activity patterns, which may eventually affect their winter survival, mainly by affecting energy intake (Yasué et al. 2008; Robinson and Pollitt 2002). We monitored the number of all wintering shorebirds, and assessed their distribution and patterns of activity (Fig. 12.2), thereby testing the following hypothesis:

1. Shorebird abundance should decline in impacted saltpans in comparison to control saltpans.
2. The distribution of shorebirds within impacted saltpans should change after the beginning of the construction and operation of the railway.
3. The distribution and activity of shorebirds within impacted saltpans should change after the beginning of the construction and operation of the railway.

Wintering Shorebird Abundance

We estimated the number of individuals of all shorebird species in the saltpans, from counts carried out at regular intervals during the months of January and/or February (3–5 sampling occasions). Given the small number of counts per species and the associated overdispersion, we pooled the counts of all species. We tested for differences in the total number of shorebirds with time of year and type of saltpan (impacted vs. control), using a GLMM with a negative binomial distribution, controlling for overdispersion and using the saltpan area as an offset (Crawley 2013). Saltpan identity and survey week were specified as random variables.

Shorebird Spatial Distribution and Activity Within Saltpans

Using the same rationale established for the breeding season, we recorded the spatial distribution of shorebirds in each saltpan according to the type of pond (feeders, evaporators, and crystallizers). We then tested for variation in the total number of shorebirds in relation to the type of pond, time (year) and type of saltpan (impacted vs control) using a GLMM with a negative binomial distribution, controlling for over-dispersion and saltpan area (Crawley 2013). Saltpan identity and the survey week were specified as random variables.

We also recorded the activity (foraging vs roosting) of each shorebird. Because activity budgets differ between species, we used only for Dunlin data for this analysis, because this was the most abundant species. We tested for differences in activity with time (year) and type of saltpan (impacted and control) using a GLMM with a binomial distribution function, controlling for saltpan area (Crawley 2013).

Results

Effects on Breeding Shorebirds

Shorebird Abundances

The average abundance of breeding shorebirds presented large variations, which ranged from the complete absence in some saltpans, in a breeding season, to maximum average values per saltpan of 36.2 Black-winged stilts (9.3 birds/ha), 26.0 Kentish plovers (7.9 birds/ha) and 89.7 Little terns (23.0 birds/ha) in 2006. For the three focal species, there was a significant interaction term between time and type of pond (Table 12.1), thus pointing out an impact of the construction and operation of the railway viaduct. Regarding the construction phase, there were significant increases in the abundance of Kentish plover (2008) and Little tern (2008 and 2009) in impacted saltpans in relation to the expectation from temporal trends at control saltpans (Table 12.1). However, during the operation phase (2011), the numbers of Black-winged Stilt, Kentish plover, and Little tern declined in impacted saltpans in relation to the expectation from temporal trends at control saltpans (Table 12.1).

Black-Winged Stilt Nest Spatial Distribution Within the Saltpans

The total number of nests at the saltpans monitored showed a steady negative temporal trend: annual average = 81.2 (density=22.3 nests/ha), 2006 = 97 (26.6 nests/ha), 2008 = 89 (22.6 nests/ha), 2009 = 81 (21.1 nests/ha), 2011 = 58 (17.5 nests/ha). Considering the four years, it was found that most of the Black-winged Stilt nests were built in feeders (annual average = 32.2 nests) and evaporators (45.7 nests), while few were found in crystallizers (3.2 nests). There was no evidence of the railway affecting nest numbers, as the full GLMM did not differ significantly from the model without the interaction term (Table 12.1).

Black-Winged Stilt Breeding Phenology and Success

Overall, the nesting phenology of Black-winged stilts was clearly unimodal, with the first nest detected in week 16 (late April) and the last on week 27 (early July), with a peak number of nests in week 21 (end of May) (Fig. 12.3). There were no significant differences in phenology between impacted and control in any phase, but

Table 12.1 Summary of the results of generalized linear mixed models (GLMM) used to evaluate the impact of a railway on breeding and wintering shorebirds in saltpans of Ria de Aveiro (Portugal)

GLMM	AIC		Model comparison			Interaction terms				
	Full model	Null model	Dev	df	p	Contrast	Coef	SE	p
Breeding shorebirds									
Abundance									
Black-winged Stilt[a]	2290	2298	13.63	3	<0.010	2006 versus 2008	−0.24	0.25	0.325
						2006 versus 2009	0.19	0.25	0.438
						2006 versus 2011	−0.73	0.26	<0.05
Kentish plover[a]	1887	1930	49.00	3	<0.001	2006 versus 2008	0.64	0.26	<0.05
						2006 versus 2009	0.23	0.26	0.386
						2006 versus 2011	−1.46	0.30	<0.001
Little tern[a]	2040	2058	24.21	3	<0.001	2006 versus 2008	1.34	0.46	<0.01
						2006 versus 2009	1.09	0.42	<0.01
						2006 versus 2011	−0.63	0.48	0.189
Breeding parameters (Black-winged Stilt)									
Nest distribution[b]	455	447	7.38	8	0.50	Null model accepted			
Nest success[a]	154	154	6.45	3	0.09	Null model accepted			
Fledging success[a]	90	86	1.87	3	0.60	Null model accepted			
Wintering shorebirds									
Abundance[a]	1293	1299	12.49	3	<0.01	2006 versus 2008	−2.91	1.59	0.067
						2006 versus 2009	1.52	1.48	0.304
						2006 versus 2011	−3.01	1.52	<0.05

(continued)

Table 12.1 (continued)

GLMM	AIC		Model comparison				Interaction terms			
	Full model	Null model	Null model	Dev	df	p	Contrast	Coef	SE	p
Spatial distribution[b]	2078		2088	22.00	6	<0.01	All interaction terms with $p > 0.05$			
Dunlin activity[a]	309		349	46.66	3	<0.001	All interaction terms with $p > 0.05$			

For each parameter tested, we present the comparison between a full model including all main effects and their interactions, and the corresponding null model including only the main effects. In each case we present the AIC of the full and null models, the results of the analysis of deviance comparing the null and the full models, and the coefficients, standard errors, and p value estimates of the interaction terms. Estimates of the interactions are presented only when the full model was significantly superior to the null model, and when at least one interaction term was statistically significant ($p < 0.05$)

[a]The interaction tested was year × type of saltpan

[b]The interaction tested was (year × type of saltpan) × type of pond

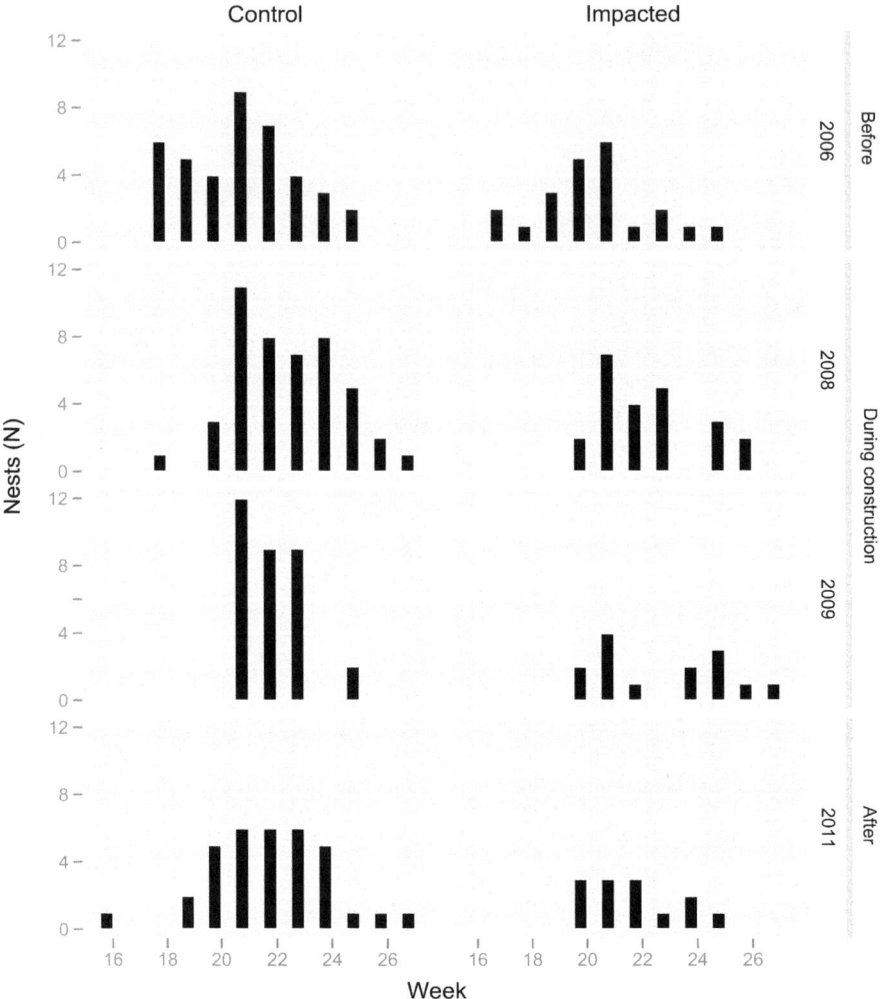

Fig. 12.3 Number of nests per week of Black-winged Stilt in impacted and control saltpans of Ria de Aveiro (Portugal), before (2006), during (2008–2009), and after (2011) the construction of a new railway in the study area

there were differences between the control and construction phases in impacted (D = 0.39, p < 0.05) and control (D = 0.32, p < 0.01) saltpans, with fewer early nests and more late season nests in the latter. We did not find significant differences in temporal nest distribution between the construction and the post-construction phase in either impacted (D = 0.19, p = 0.86) or control saltpans (D = 0.18, p = 0.40). This suggests an overall delay in nesting phenology, irrespective of proximity to the railway.

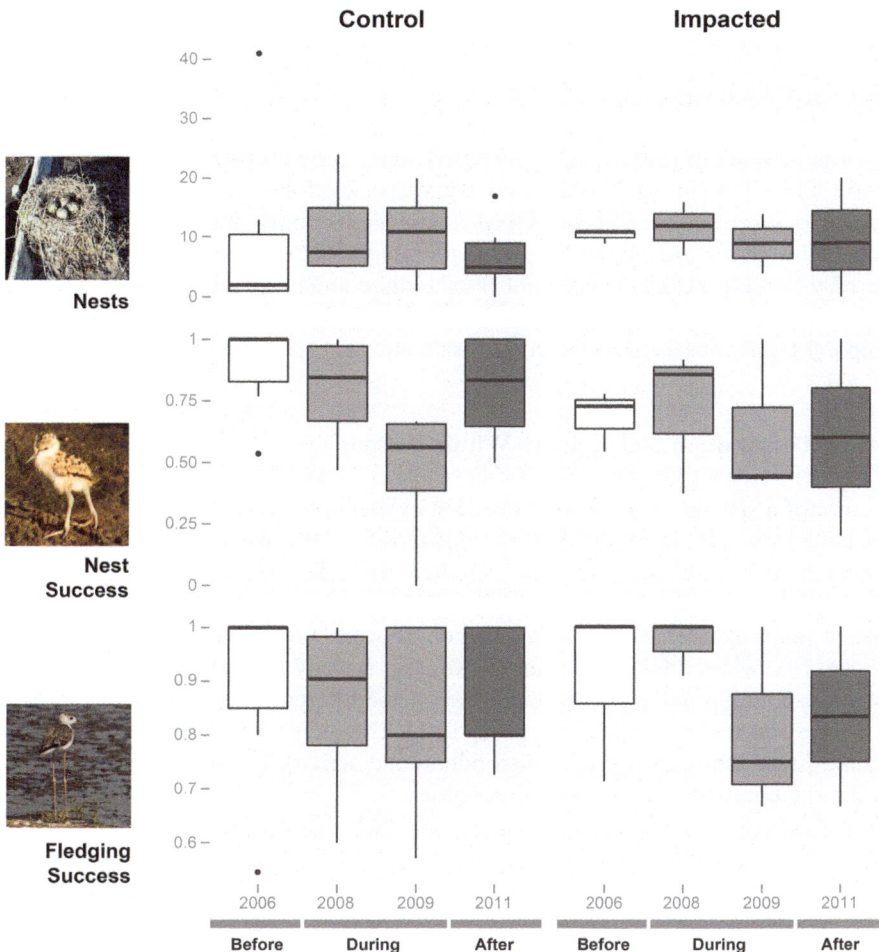

Fig. 12.4 Boxplots showing variation in the number of Black-winged Stilt nests, and nest and fledging success, in impacted and control saltpans of Ria de Aveiro (Portugal), before (2006), during (2008–2009), and after (2011) the construction of a new railway in the study area

The nest success (global = 62%, 2006 = 65%, 2008 = 68%, 2009 = 52%, 2011 = 64%) and fledging success (global = 52%, 2006 = 49%, 2008 = 62%, 2009 = 42%, 2011 = 51%) were fairly constant over time (Fig. 12.4). There was no evidence of impacts on breeding success, as the full GLMMs did not differ significantly from the corresponding null models without the interaction term (Table 12.1).

Effects on Wintering Shorebirds

Shorebird Abundances

The total number of shorebirds counted per winter varied between 9331 (2008) and 2166 (2009). Dunlin was the most abundant species, followed by the Black-tailed Godwit *Limosa limosa*. The full GLMM was significantly different from the null model without the interaction term, thus pointing out the presence of impacts from the railway (Table 12.1). The number of birds counted in the impacted saltpans during the operation phase (2011) was significantly lower than expected given the temporal trends observed at the control saltpans (Table 12.1).

Spatial Distribution and Activity Within Saltpans

The use of a specific type of pond (feeders, evaporators and crystallisers) in winter did not show significant differences with type of saltpan (impacted or control). Although the full GLMM was significantly different from a null model (Table 12.1), none of the variables presented significant interactions between the type of pond and the combined interaction between time of year and the type of saltpan. Likewise, although the full GLMM for activity differed significantly from the corresponding null model without the interaction term, none of the interaction terms was statistically significant (Table 12.1). This suggests that shorebirds did not change significantly their spatial distribution and activity within impacted saltpans during the construction and operation phases.

Discussion

During the breeding season, we detected significantly negative impact of the railway on the average abundance of Black-winged Stilt and Kentish plover. But, during the construction phase, it is likely that the interruption of the construction during the breeding phase in the whole area of the railway viaduct corridor, imposed by legal authorities, was crucial to reduce the impact of the construction phase and in the case of the Kentish plover (2008) and the Little tern (2008 and 2009) this had a positive impact on the average abundance in impacted saltpans. However, our data does not discriminate between birds that failed to breed from breeding individuals. These birds can be present at different saltpans in different sampling dates, and this increased the dispersion of the data and could have masked to some extent the real pattern.

In the case of the Black-winged Stilt, assumed to be the key indicator species to assess the impact of disturbance caused by the railway, we did not find significant evidence for any impact on the spatial distribution of nests, breeding phenology and breeding success. The results suggest that the distribution of Black-winged Stilt between types of pond (feeders, evaporators and crystallizers) was not particularly affected by the construction. The major disturbance took place in the crystallizers, where the pillars were erected. However, as the results above show, these were the areas less used by the species, even in the pre-construction phase, which may be the reason why no impact was observed during the entire construction period and afterwards. Concerning nest phenology, the interruption of construction probably prevented an impact on the temporal distribution of nests. Likewise, this may explain why we did not find any evidence that the railway affected hatching and survival of the fledglings that used impacted saltpans.

During the wintering season, the usage of the saltpans can be conditioned by their availability, particularly by unpredictable fluctuations in the water level. Thus, the birds may be more limited in their choice of the saltpans, causing a high variability in the counts. This effect is more restricted during the breeding season due to the management of saltpans for the extraction of salt. During the last phase of this study, two saltpans (one impacted) were transformed into aquacultures, conditioning the use of the saltpans even more. Additionally, many of the shorebird species do not show a random distribution since they group for foraging or roosting, even if there is enough area to spread (van de Kamp et al. 2004). Even taking into account these sources of variability, the data suggest that the wintering community, as a whole, did show a negative response to the disturbance caused by the railway, since significant differences were found between the two types of saltpans over the years, regarding the species' abundance. Nevertheless, although the use of a pooled number of individual birds allowed the analysis of data that is inherently over-dispersed and zero-inflated, it restricts the full interpretation of the data. On the other hand, the distribution of the main wintering species was not affected, while for the wintering Dunlin, the number of birds in each activity did not show a response to the disturbance caused by the railway. Disturbances can inhibit foraging activities in favor of vigilant behaviour, but since all saltpans are close to sources of human disturbance (e.g., roads), the effect of the railway may have been buffered.

Final Remarks

As an overall conclusion, the data collected suggest that the construction and the beginning of the operation of the new Aveiro railway showed mixed effects on both the breeding and wintering communities of shorebirds using the Aveiro saltpan complex. Additionally, it had indirect effects, as it probably accelerated the abandonment of salt extraction in one of the affected saltpans, and thus promoted its transformation into an aquaculture during the last phase of our study. The net consequence of this process was the loss of this saltpan for the breeding birds, and also for feeding of the smaller shorebird species. At this point and with the present data, however, it is impossible to evaluate the consequences of this loss of habitat, and whether it could be compensated by the use of the other saltpans or other places outside the study area.

We applied an experimental design that provides a spatial and temporal context to integrate the evaluation of potential impacts of railways and other line infrastructures on wetlands. Our approach surpasses the usual estimators of mortality due to the effect of barrier and evaluates the effects on the spatial and temporal distribution, survival and behaviour. This decision was made due to the nature of this railway and the mitigation measures that were decided *a priori* to decrease the probability of negative impacts over the aquatic birds that uses the area crossed by this infrastructure. Indeed, the construction of a viaduct over the most critical areas was a decision that decreased land reclamation on the saltpans. On the other hand, this railway was meant from the start to be used for commercial purposes, transporting containers from the harbor to the main national line. The expected traffic and speed of the trains was thus limited and conditioned to specific daily periods, thus decreasing the barrier effect of this infrastructure and, hopefully, lowered the expected mortality due to collisions with the trains. In fact, during the study period, only two casualties were reported: a common Sandpiper (*Actitis hypoleucos*) and an unidentified gull.

The main guidelines described in this book pertaining to future studies in railway ecology seem mostly designed to terrestrial environments (and species) and long stretches of continuous railway beds. This study dealt (albeit in a preliminary way) with some of them, namely by (1) studying species with different ecological traits (even if belonging to the same taxonomic group), (2) using an experimental analysis, which includes comparison between a control phase with the period of construction and after, (3) studying the effect of habitat fragmentation (in this case, how the birds responded to the loss/degradation of the impacted saltpans both in the breeding season and in winter, and (4) a preliminary study of the population persistence in the area after the railway construction.

In our view, the most interesting feature of the present study is the integration of several spatial scales, from the macroscale of other case studies (see Chaps. 7 and 11) to the microscale of the habitat. This option was imposed by the nature of the problem itself and allowed us to go beyond the usual evaluation of mortality of adult birds due to collisions (Chap. 7) or the general patterns of disturbance and habitat loss (Chap. 11), and changed the focus of analysis to the impact of disturbance (both on the breeding and wintering birds) and chick mortality at smaller scales (habitat), a feature that is seldom attempted. We have developed a conceptual methodology where the rationale is hypothesis-based and uses a Before-After-Control-Impact comparative approach (as advised in the guidelines of Chap. 19). We think that it can be easily extended to the same or other aquatic bird species and habitats affected by railways in the wetland areas, enabling a multi-scale habitat analysis that can be combined with the usual broad scope procedures. Although the method has the potential to be improved and refined, we feel that it represents an add-on to the arsenal of methods of analysis in future studies of railway ecology, as those proposed in Chap. 19.

Acknowledgements We would like to thank REFER (Rede Ferroviária Nacional E. P. E.), and especially Eng. João Sarmento for kindly making all data available for use in this publication. We also thank all the saltpan owners and workers for their support in the study area. Ricardo Jorge Lopes was supported by Grants SFRH/BPD/40786/2007 and SFRH/BPD/84141/2012, funded by FCT/MEC and POPH/QREN/FSE.

References

Bastos, M. R. (2009). On the track of salt: Adding value to the history of saltponds exploration in the coastal management scene of Aveiro lagoon. *Journal of Integrated Coastal Zone Management, 9,* 25–43.

Bates, D., Maechler, M., Bolker, B., & Walker, S. (2015). Fitting linear mixed-effects models using lme4. *Journal of Statistical Software, 67*(1), 1–48.

Cayford, J. T. (1993). Wader disturbance: A theoretical overview. *Wader Study Group Bulletin, 68,* 3–5.

Crawley, M. J. (2013). *The R book* (2nd ed.). West Sussex, UK: Wiley.

del Hoyo, J., Elliott, A., & Sargatal, J. (1996). *Handbook of the birds of the world* (Vol. 3) Hoatzin to Auks. Spain: Lynx Edicions, Barcelona

Dias, M. P., Lecoq, M., Moniz, F., & Rabaça, J. E. (2014). Can human-made saltpans represent an alternative habitat for shorebirds? Implications for a predictable loss of estuarine sediment flats. *Environmental Management, 53,* 163–171.

Fonseca, V., Grande, N., & Fonseca, L. C. (2005). Waterbird breeding on salinas in Ria Formosa, southern Portugal. *Wader Study Group Bulletin, 106,* 58–59.

Fournier, D. A., Skaug, H. J., Ancheta, J., Ianelli, J., Magnusson, A., Maunder, M. N., et al. (2012). AD model builder: Using automatic differentiation for statistical inference of highly parameterized complex nonlinear models. *Optimization Methods and Software, 27*(2), 233–249.

Griffin, L. R. (1999). Colonization patterns at Rook *Corvus frugilegus* colonies: Implications for survey strategies. *Bird Study, 46,* 170–173.

INE. (2016). *Estatísticas da Pesca 2015.* Lisboa: Instituto Nacional de Estatística.

International Wader Study Group. (2003). Waders are declining worldwide. *Wader Study Group Bulletin, 101,* 8–12.

Jardim, G. (1984). Recenseamento e distribuição de aves limícolas nidificantes no estuário do Tejo em 1983. *Cyanopica, 3,* 223–229.

Lenth, R. V. (2016). Least-squares means: The R package lsmeans. *Journal of Statistical Software, 69*(1), 1–33.

Lopes, R. J., Múrias, T., Cabral, J. C., & Marques, J. C. (2005). A ten year study of variation, trends and seasonality of shorebird community in the Mondego Estuary. *Portugal. Waterbids, 28*(1), 8–18.

Luís, A., Moreira, H., & Goss-Custard, J. (2002). The feeding strategy of the Dunlin (*Calidris alpina* L.) in artificial and non-artificial habitats at Ria de Aveiro, Portugal. *Hydrobiologia, 475–476*(1), 335–343.

Masero, J. A. (2003). Assessing alternative anthropogenic habitats for conserving waterbirds: Salinas as buffer areas against the impact of natural habitat loss for shorebirds. *Biodiversity and Conservation, 12*(6), 1157–1173.

McLusky, D. S., & Elliott, M. (2004). *The estuarine ecosystem.* Oxford: Oxford University Press.

Múrias, T., Cabral, J. A., Lopes, R. J., Marques, J. C., & Goss-Custard, J. D. (1997). Low-water use of the Mondego Estuary (West Portugal) by waders (Charadrii). *Ardeola, 44*(1), 79–91.

Múrias, T., Cabral, J. A., Lopes, R. J., Marques, J. C., & Goss-Custard, J. D. (2002). Use of traditional salines by waders in the Mondego estuary (Portugal): A conservation perspective. *Ardeola, 49,* 223–240.

Pearce-Higgins, J. W., Finney, S. K., & Yalden, D. W. (2007). Testing the effects of recreational disturbance on two upland breeding waders. *Ibis, 149*(s1), 45–55.

Prater, A. J. (1981). *Estuary birds of Britain and Ireland.* Calton, UK: T & A.D. Poyser.

R Core Team. (2015). *R: A language and environment for statistical computing.* Austria: R Foundation for Statistical Computing.

REFER. (2009). Ligação Ferroviária ao Porto de Aveiro. Rede Ferroviária Nacional E.P. (Direcção de Comunicação e Imagem), Lisboa, Portugal.

Robinson, J., & Pollitt, M. S. (2002). Sources and extent of human disturbance to waterbirds in the UK: An analysis of Wetland Bird Survey data, 1995/96 to 1998/99. *Bird Study, 49,* 205–211.

Rodrigues, C., Bio, A., Amat, F., & Vieira, N. (2011). Artisanal salt production in Aveiro/Portugal —An ecofriendly process. *Saline Systems, 7*(1), 3.

Rufino, R., & Araújo, A. (1987). Seasonal variation in wader numbers and distribution at the Ria de Faro. *Wader Study Group Bulletin, 51,* 48–53.

Rufino, R., Araújo, A., Pina, J. P., & Miranda, P. S. (1984). The use of salinas by waders in Algarve, South Portugal. *Wader Study Group Bulletin, 42,* 41–42.

Rufino, R., & Neves, R. (1992). The effects on wader populations of the conversion of salinas into fishfarms. In M. Finlayson, T. Hollis, & T. Davis (Eds.), *Managing Mediterranean Wetlands and Their Birds: Proceedings of an IWRB International Symposium* (pp. 177–182). Grado, Italy.

Sansom, A., Pearce-Higgins, J. W., & Douglas, D. J. T. (2016). Negative impact of wind energy development on a breeding shorebird assessed with a BACI study design. *Ibis, 158*(3), 541–555.

Tavares, P., Kelly, A., Maia, R., Lopes, R. J., Serrão Santos, R., Pereira, M. E., et al. (2008). Variation in the mobilization of mercury into Black-winged Stilt *Himantopus himantopus* chicks in coastal saltpans, as revealed by stable isotopes. *Estuarine, Coastal and Shelf Science, 77*(1), 65–76.

van de Kamp, J., Bruno, E., Piersma, T., & Zwarts, L. (2004). *Shorebirds: An illustrated behavioural ecology.* Utrecht, The Netherlands: KNNV Publishers.

Vieira, N., & Bio, A. (2014). Artisanal salina—Unique wetland habitats worth preserving. *Journal of Marine Science: Research & Development, 4*(1), e125.

Yasué, M., Dearden, P., & Moore, A. (2008). An approach to assess the potential impacts of human disturbance on wintering tropical shorebirds. *Oryx, 42*(3), 415–423.

Zuur, A. F., Ieno, E. N., Walker, N., Saveliev, A. A., & Smith, G. M. (2009). *Mixed effects models and extensions in ecology with R.* New York: Springer.

Chapter 13
Evaluating and Mitigating the Impact of a High-Speed Railway on Connectivity: A Case Study with an Amphibian Species in France

Céline Clauzel

Abstract The aim of this study is to evaluate and mitigate the impact of a high-speed railway (HSR) line on functional connectivity for the European tree frog (*Hyla arborea*), an amphibian species highly sensitive to habitat fragmentation. The method consists of modeling its ecological network using graph theory before and after the implementation of the infrastructure and of evaluating changes in connectivity. This diachronic analysis helps visualize the potential impact of the HSR line and to identify areas likely to be most affected by the infrastructure.

Keywords Impact assessment · Landscape connectivity · Ecological network · Graph theory · Amphibians

Introduction

Among world's vertebrates, amphibians have the highest proportion of threatened species, currently estimated at 32% by the IUCN Red List (IUCN 2016). This proportion could increase in the future because at least 42% of all amphibian species are declining in population (Stuart et al. 2008). Habitat loss, degradation and fragmentation are the greatest threats to amphibians. Fragmentation is a spatial process which affects habitat patches by decreasing their size or increasing their isolation (Fahrig 2003), making landscape less permeable to wildlife movements and gene flow (Cushman 2006; Forman and Alexander 1998). Amphibians are particularly affected because of the importance of movements during their life cycle

C. Clauzel (✉)
LADYSS, UMR 7533 CNRS, Sorbonne Paris Cité, University Paris-Diderot,
5 rue Thomas Mann, 75013 Paris, France
e-mail: celine.clauzel@univ-paris-diderot.fr

C. Clauzel
TheMA, CNRS, University Bourgogne Franche-Comté,
32 rue Mégevand, 25030 Besançon Cedex, France

© The Author(s) 2017
L. Borda-de-Água et al. (eds.), *Railway Ecology*,
DOI 10.1007/978-3-319-57496-7_13

(Joly et al. 2003). Most species occupy an aquatic habitat for breeding and during the larval period, and a terrestrial habitat after breeding and during hibernation. Daily movements and seasonal migrations across the landscape matrix connect these two types of habitat. Furthermore, many species are structured into metapopulations, in which several subpopulations occupy spatially distinct habitat patches separated by a more or less unfavorable matrix. Dispersal events allow individuals to colonize new ponds or to recolonize sites where the species is nearing extinction. A literature review about amphibian dispersal (Smith and Green 2005) showed that the median distance is less than 400 m but 7% of observed species may reach 10 km.

The major causes of landscape fragmentation are farming practices, urban development, and the construction of transportation infrastructures (Forman and Alexander 1998). Apart from direct loss of suitable habitats and road-kills, linear infrastructures cause the loss of landscape connectivity (Forman and Alexander 1998; Geneletti 2004), which is recognized as a key functional factor for the viability of species and their genetic diversity (Fahrig et al. 1995). Major infrastructures such as motorways or high-speed railway lines act as barriers to the movement of animals and isolate organisms in small subpopulations which become more sensitive to the risk of extinction (Forman and Alexander 1998). This is especially the case for populations of amphibians whose daily movements, seasonal migrations and dispersal events mean they regularly cross the landscape matrix (Allentoft and O'Brien 2010; Cushman 2006; Fahrig et al. 1995; Scherer et al. 2012).

Several case studies have contributed to identifying and quantifying the effects of linear infrastructures on species distribution in many regions of the world, using various methods. Authors have related data describing species (e.g. abundance, collisions) to proximity of infrastructures (Brotons and Herrando 2001; Fahrig et al. 1995; Huijser and Bergers 2000; Kaczensky et al. 2003; Li et al. 2010) and to the degree of habitat fragmentation (Fu et al. 2010; Serrano et al. 2002; Vos and Chardon 1998). These studies measure the real impact of the infrastructure using data on species collected after its construction. However, before the construction phase, an impact prediction stage is also necessary to compare alternative infrastructure routes (Fernandes 2000; Geneletti 2004; Vasas et al. 2009) or to guide the mitigation measures from the beginning of the project (Clauzel et al. 2015a, b; Girardet et al. 2016; Mörtberg et al. 2007; Noble et al. 2011).

Reviews by Geneletti (2006) and Gontier et al. (2006) show that the effects of landscape fragmentation are more difficult to predict than the direct loss of habitat. According to these authors, current assessment methods are often restricted to protected areas or to a narrow strip on either side of the infrastructure. However, landscape fragmentation may have consequences on a far broader scale (Forman 2000).

To assess the long-distance effects of linear infrastructures on species distributions, models must include connectivity metrics that take into account both structural (arrangement of habitat patches) and functional (behavior of the organisms) aspects. With this aim in mind, the development of methods based on graph theory in landscape modelling is promising (Dale and Fortin 2010; Urban et al. 2009). For our purposes, a graph is a set of habitat patches of a given species (called "nodes") potentially connected by functional relationships ("links"). It provides a simplified representation of ecological networks and is considered an interesting trade-off between information content and data requirements (Calabrese and Fagan 2004). Several metrics can be computed at global level to assess connectivity for the entire graph or at local level for individual nodes or links (Galpern et al. 2011). Graph-based studies have been used to identify the key elements for connectivity (Bodin and Saura 2010; Saura and Pascual-Hortal 2007), to evaluate the importance of connectivity for species distribution (Foltête et al. 2012a; Galpern and Manseau 2013; Lookingbill et al. 2010), to assess the impact assessment of a given development on connectivity (Clauzel et al. 2013; Fu et al. 2010; Girardet et al. 2013; Gurrutxaga et al. 2011; Liu et al. 2014) or to propose mitigation measures for improving connectivity (Clauzel et al. 2015a, b; Girardet et al. 2016; Mimet et al. 2016).

In this case study, we propose to assess and to mitigate the potential impact of a HSR line on connectivity for the European tree frog (*Hyla arborea*) in eastern France. Populations of this species are declining in western Europe due to several causes: climate change, increased UV-radiation (Alford and Richards 1999), predation or competition, pollution and eutrophication of ponds (Borgula 1993), road-kill (Elzanowski et al. 2008), and above all the destruction and fragmentation of its habitat (Andersen et al. 2004; Cushman 2006; Vos and Stumpel 1996). In the region under study, the development of the HSR line and consecutive changes in connectivity may therefore impact the viability of tree frog populations.

The analysis consists of modeling the ecological network of the European tree frog with and without the HSR line in the landscape map. The presence of the HSR line probably entails a loss of connectivity among suitable habitats thus restricting the potential movements of the species. Comparison of the connectivity values between the two graphs provides an assessment of the changes in connectivity induced by the HSR line. The analysis is supplemented by a search for the best locations for amphibian crossing structures in order to improve connectivity. This study is based on a predictive modelling approach that estimates a potential impact but does not measure a real impact. This approach is therefore put in place before the construction of an infrastructure and allows areas potentially affected by isolation to be mapped.

Materials and Methods

Data Preparation

Study Area

Our study was carried out in a zone of 4600 km^2 in the Burgundy-Franche-Comté region of eastern France (Fig. 13.1). In this zone altitude ranges from 184 to 768 m, the landscape mosaic is dominated by forests (42% of total area), intensive agriculture (27%) and grasslands (20%). This area is strategic for environmental conservation because it contains many threatened species of birds (such as the little

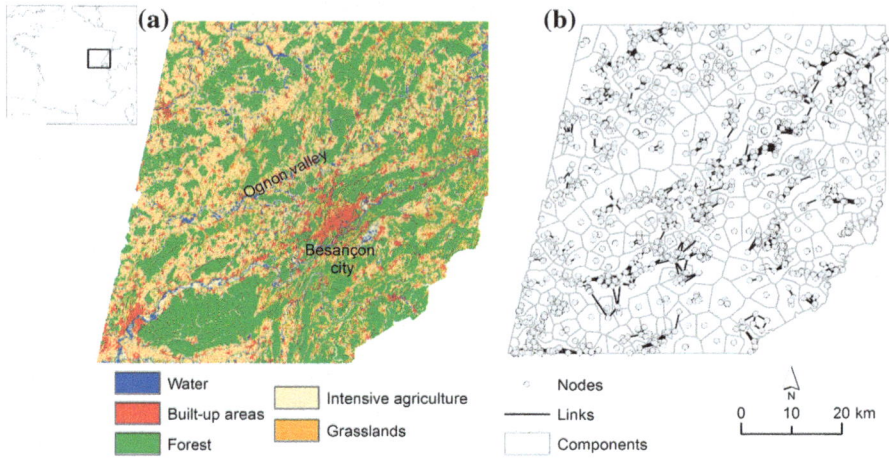

Fig. 13.1 Landscape **(a)** and ecological network of the tree frog **(b)** in the Burgundy-Franche-Comté region

bustard, *Tetrax tetrax*), mammals (lesser horseshoe bat, *Rhinolophus hipposideros*), reptiles (European pond turtle, *Emys orbicularis*) and amphibians (European tree frog, *Hyla arborea*) (Paul 2011).

In December 2011, a high-speed railway line came into service in the region after 4 years of construction. It is part of a larger project that improves connections for eastern France with both Paris and the south of France. This infrastructure is 140 km long and 30 m wide on average and crosses the study area from west to east following the Ognon valley. The line includes a total 1300 m of viaducts and 2000 m of tunnels. In this study, this infrastructure is considered as impassable either because it forms a physical barrier or because traffic noise can disrupt the animals' behaviors (Eigenbrod et al. 2009; Lengagne 2008) even when the infrastructure is on a viaduct. This simplified representation of reality is used to better predict the potential impact of the infrastructure on landscape connectivity and by repercussion on tree frog populations.

Study Species

The tree frog is widely distributed in Europe from Spain to western Russia, but its populations have declined in north-western Europe over the last 50 years (Corbett 1989). The species is classified as endangered and is on the IUCN Regional Red List of Threatened Species for Burgundy-Franche-Comté where it is mainly present in the Ognon valley (Fig. 13.1) (Pinston et al. 2000).

The tree frog has a two-phase life cycle with aquatic and terrestrial stages. The breeding pond consists of shallow and sunny ponds, marshlands, gravel-pits or river pools. Pond size does not appear decisive and ranges from 1 to 4000 m^2 (Grosse and Nöllert 1993). In fact, the presence of the tree frog does not seem to be related to pond size but rather to the amount of terrestrial habitat surrounding the pond (Vos and Stumpel 1996). Although the aquatic habitat is essential for reproduction, the species spends most of its time in terrestrial habitats. In agricultural environments, the species often prefers edge habitats like ponds located on river banks, ditches, fields and forest edges (Pellet et al. 2004).

The seasonal migrations between this terrestrial habitat and the breeding pond range from 250 to 1000 m (Stumpel 1993). These dispersal events are very important in the tree frog's life cycle. Juveniles disperse from the breeding pond into the surrounding landscape to access other ponds. Despite the fidelity of the tree frog to its breeding ground, some adults may disperse and change ponds (Fog 1993). Observed dispersal distances are generally less than 2000 m but may reach up to 4000 m (Vos and Stumpel 1996).

Landscape Data

The study required the creation of two landscape maps, the first describing the initial state before construction of the HSR line, the second including the linear infrastructure. Except for the HSR line, the land cover was identical on both maps so only the impact of the infrastructure was estimated. Different data sources were combined using ArcGis 10. The 1 m-accuracy vectorial landscape databases provided by French cartographic services (DREAL Franche-Comté, BD Topo IGN) were used to represent wetlands, hedgerows, forests, buildings, roads, railways and rivers. In agricultural areas, the French Record of Agricultural Plots were used to separate grassland and bare ground. A morphological spatial pattern analysis was also applied to the forest category to identify forest edges. All these data elements were combined into a single raster layer with a resolution of 10 m. In this raster layer, the HSR line was represented by a 3-pixel wide line to reflect its actual size. It

Table 13.1 Landscape categories and cost values

Landscape categories	Function	Cost values
Ponds	Aquatic habitat	1
Hedgerows	Terrestrial habitat	1
Forest edges	Terrestrial habitat	1
Rivers	Suitable	10
Wetlands	Suitable	10
Wooded grasslands	Suitable	10
Grasslands	Unfavorable	100
Roads	Unfavorable	100
Railways	Unfavorable	100
Bare ground	Highly unfavorable	1000
Forests	Highly unfavorable	1000
Buildings	Highly unfavorable	1000
Motorway (A36)	Highly unfavorable	1000

also avoided potential discontinuities induced by the conversion of linear elements in a raster format (Adriaensen et al. 2003). Altogether, thirteen landscape categories were obtained (Table 13.1).

Landscape Graph Construction

The nodes of the graphs corresponding to the habitat patches were defined as the aquatic habitat units (breeding ponds) adjacent to an area of potential terrestrial habitat. The quality of the habitat patches, i.e. capacity, was defined as the amount of terrestrial habitat and suitable elements around ponds. The links between nodes were defined in cost distance because the ability of the tree frog to disperse depends greatly on the surrounding matrix. The ecological literature and expert opinions were thus used to classify each landscape category according to its resistance to movement (Table 13.1). Radio tracking experiments (Pellet et al. 2004; Vos and Stumpel 1996) have concluded that wooded grasslands and linear elements like hedgerows or forest edges facilitate movement and often provided the species' terrestrial habitat. Rivers and wetlands are less favorable because their permeability depends on the density of vegetation. Conversely, grasslands, roads and railways tend to constrain movement. Finally, the cores of forest patches, bare ground, buildings and motorways are considered highly impassable and are mostly avoided by the tree frog.

Clauzel et al. (2013) performed several tests by varying the cost values and the number of classes to find the model that best explained the occurrence of the tree frog in the Franche-Comté region. The results showed that highly contrasting values between favorable and unfavorable landscape categories were the most relevant. Consequently, in the present study (Table 13.1), aquatic and terrestrial habitats were assigned a cost of 1. Suitable elements such as wetlands or wooded grasslands were assigned a low cost (10), unfavorable landscape elements a cost of 100, and highly unfavorable elements a high cost (1000). In the second landscape map including the HSR line, a cost of 10,000 was assigned to the infrastructure so as to remove any links crossing it, i.e. considering it as an absolute barrier. From the two landscape maps, without and with the HSR line, two graphs were built and thresholded at a distance of 2500 m corresponding approximately to the dispersal distance for the tree frog. This distance was selected in line with the results by Clauzel et al. (2013), where several maximum distances were tested. The model using the distance of 2500 m proved the most relevant. Consequently, only links shorter than this distance were kept.

Landscape Graph Analysis

For each graph, connectivity metrics were calculated, and then compared over time by computing their rates of variation. The magnitude of the difference between metric values provided an assessment of the impact of the HSR line. Two levels of analysis were investigated. The regional-scale analysis provided an assessment of changes on the overall connectivity throughout the study area. The local-scale analysis provided a finer assessment by identifying the most severely affected patches and corridors, i.e., those that experienced the largest changes in local connectivity or that removed by the infrastructure.

The identification of the best locations for potential wildlife crossings was based on a cumulative method developed by Foltête et al. (2014) and Girardet et al. (2016). The method consisted of testing each graph link crossed by the HSR line and to validate the one maximizing the global connectivity of the tree frog network. In the first step, all links cut by the HSR line were removed and the global connectivity was calculated. Then, an iterative process added each link and computed again connectivity. After testing all links individually, the one that produced the biggest increase in the connectivity was validated. The process was repeated until the desired number of new crossings was reached by integrating changes in the graph topology induced by the addition of previous crossings.

For all analysis, connectivity was quantifying by the Probability of Connectivity (PC) developed by Saura and Pascual-Hortal (2007). The PC index is a global metric given by the expression:

$$PC = \left(\sum_{i=1}^{n} \sum_{j=1}^{n} a_i a_j p_{ij}^* \right) \Big/ A^2$$

where a_i and a_j are the capacities of the patches i and j, p_{ij}^* is the maximum probability of all potential paths between patches i and j, and A is the total area under study. The maximum p_{ij}^* is obtained from p_{ij} which is determined by an exponential function such that:

$$p_{ij} = \exp(-\alpha d_{ij})$$

where d_{ij} is the least-cost distance between these patches and α $(0 < \alpha < 1)$ expresses the intensity of the decrease of the dispersal probabilities resulting from this exponential function (Foltête et al. 2012a).

From the global metric PC, a patch-based metric was derived, the PC_{flux} (Foltête et al. 2014), which is the contribution of each patch to the global PC index. For a given patch j, $PC_{flux(j)}$ is given by:

$$PC_{flux(j)} = \left(\sum_{i=1}^{n} a_i a_j p_{ij}^* \right) \Big/ A^2$$

where a_i and a_j are the capacities of the patches i and j, p_{ij}^* is the maximum probability of all potential paths between patches i and j, and A is the total area under study.

Graphab 2.0 software (Foltête et al. 2012b) was used to construct landscape graphs and perform all analysis (Software available at http://thema.univ-fcomte.fr/productions/graphab/en-home.html).

Results

The graph modeling the ecological network of the tree frog contains 1464 nodes ranging in size from 0.01 to 86 ha (mean 0.6 ha) and 2624 links (Fig. 13.1). The network is highly fragmented with 264 components, i.e. the subparts of the graph within which the species may move from one patch to another at the dispersal scale. The largest components are located in the Ognon valley, where landscape is dominated by wetlands and grasslands. This area has the largest number of sites of this species in the region.

The diachronic analysis of the global PC value without and with the HSR line shows that the infrastructure has a low effect (−1.36%) on connectivity at the regional scale. This is explained by the already high level of fragmentation. However, as the HSR line runs through the Ognon valley, a much stronger local impact can be expected. Indeed, the implementation of the HSR line removes 61 links and let to a decline in local connectivity in 339 habitat patches (23%) mainly located in the Ognon valley (Fig. 13.2). The decrease in the PC_{flux} values on these impacted patches is, on average, −71%, with a maximum of −99%. Some patches located more than 12 km north also experience a decline in their connectivity (about −77%).

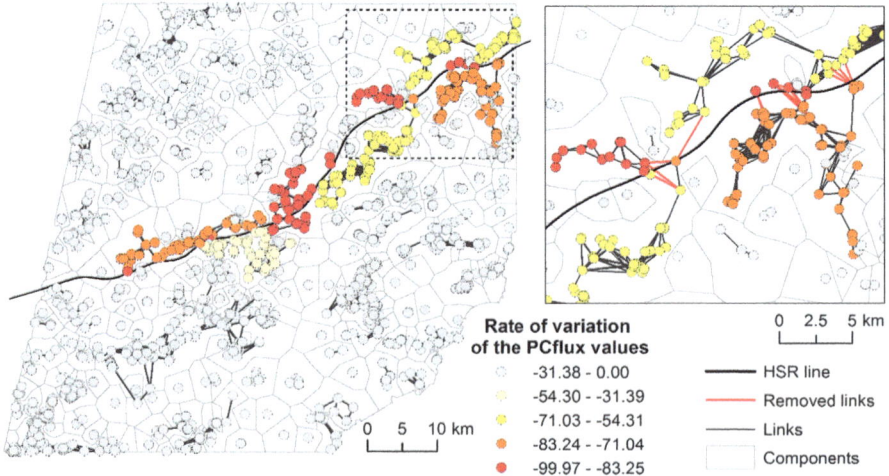

Fig. 13.2 Rate of variation of the PC_{flux} values (%) due to the implementation of the HSR line

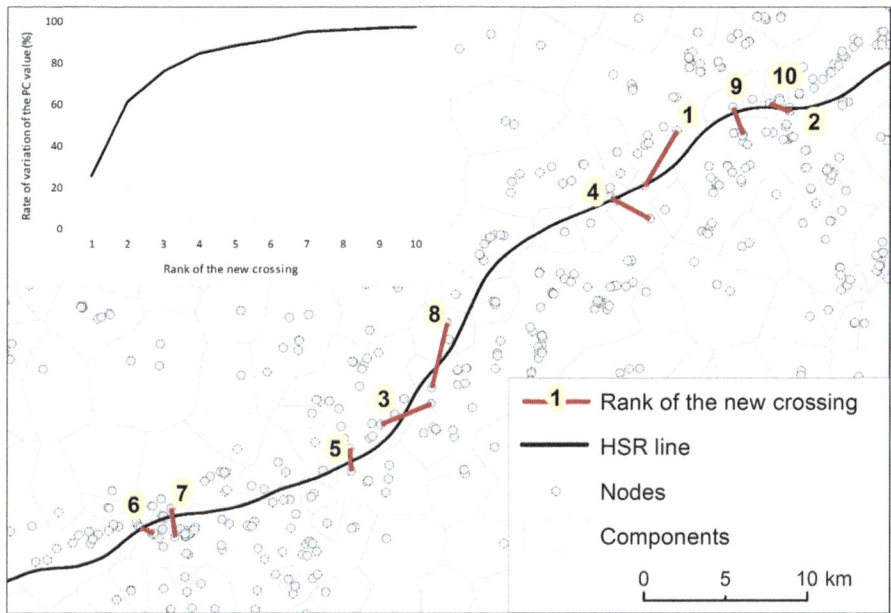

Fig. 13.3 Location of ten new wildlife crossing structures maximizing connectivity. The *top left inset* shows the curve of the increase in connectivity provided by new amphibian passes. The *numbers 1–10* refer to the rank of crossings according to the gain in connectivity they provide

The search algorithm tests the 61 links removed by the HSR line to select the 10 best locations for new amphibian crossing structures, i.e. those maximizing the global connectivity. The curve in Fig. 13.3 shows that the PC value increases greatly with the first two new crossing structures (+60%) and tends to gradually stabilize from the seventh one. The construction of only 4 amphibian passes is theoretically sufficient to restore 85% of the initial connectivity. These 4 first crossing structures reconnect several components to the main one along the Ognon valley. The combination of these four amphibian passes provides the greatest increase in connectivity because it concerns the largest components of the network. The other crossing structures reconnect smallest components or reinforce connections inside the main component.

Discussion

This study proposes an integrative approach to assess the potential impact of a high-speed railway line on connectivity for the European tree frog in the eastern France. The analysis can improve knowledge in the fields of environmental impact forecasting and species conservation.

The diachronic analysis of the local connectivity values shows that the impact of the HSR line ranges from a few meters to several kilometers. The impact is often located near the infrastructure but, in some sections, it may occur up to 12 km from the line. This variability is related to the landscape configuration and the initial state of the connectivity of the habitat patches. Indeed, all impacted habitat patches are into components fragmented by the HSR line. The extent of disturbance therefore depends on the size of these components, with a large component increasing the distance of the impact as in the north-east of the study area. This highlights the value of a regional-scale analysis for taking into account the long-distance effect of an infrastructure on connectivity. From this perspective, graph-based methods are interesting because they include both structural and functional aspects of connectivity. Our results are consistent with several previous studies that highlighted the importance of integrating the barrier effect in addition to the direct loss of habitat in environmental impact assessment (Clauzel et al. 2013; Forman and Alexander 1998; Fu et al. 2010; Girardet et al. 2013; Liu et al. 2014).

This graph-based approach provides an approximation of the potential impact but not a hard and fast measurement of the true impact. In order to validate our approach, these findings could be confirmed by field observations to test whether the real impact of the infrastructure is similar to that predicted by the model. In June 2011, a specific field survey was conducted in the Ognon valley to observe the tree frog presence after the construction of the HSR line. A total of 227 sites was visited with 42 presences and 185 absences. The results from this survey were compared with the connectivity changes predicted by the model. All presence points were located where there was no impact according to the model. The absence points were located in more or less affected areas, with a potential connectivity change between 0 and −90%. These survey results should be considered carefully because the Spring of 2011 was very unfavorable for the tree frog due to early drying up of water bodies. These climatic conditions could therefore explain the many points of absence of the species. Furthermore, the time delay between the construction of the HSR line and the surveys was not sufficient to assess the real impact of the infrastructure. Other field surveys should be conducted in the coming years to assess precisely the conservation status of the tree frog populations and of their habitats in the region. These surveys will also identify the causes of extinction of breeding populations, as several environmental factors may lie behind the extinction process, and be compounded to the long-distance effects of the infrastructure.

The method used to identify the best locations for new amphibian crossings goes beyond the prioritization of candidate sites developed by García-Feced et al. (2011). It is cumulative and so includes changes made to graph topology by adding previous links before searching for the next one. Graph modelling is used to include broad-scale connectivity as a criterion to be maximized, which is a key factor for the ecological sustainability of landscapes and for the viability of metapopulations (Opdam et al. 2006). In this study, the tested locations corresponded to the links, i.e. corridors potentially used by the tree frog, cut by the HSR line. Relying on the initial network of the species increases the likelihood of functional crossings because these links already connected habitat patches before the implementation of the infrastructure. In addition to cartographic results, a curve of the increased connectivity provided by new crossings is generated. This graphical tool indicates the number of potential crossings to be created in order to reach a given level of connectivity or to detect levels above which crossing creation fails to increase connectivity sufficiently. The method used here can also be applied to other species with different ecological requirements, or to other perspectives such as habitat restoration as in the study of Clauzel et al. (2015a) for the tree frog conservation.

Conclusions

The methodological approach used appears to be a handy tool for planners in forecasting the impact of linear infrastructures at different spatial scales, including the regional level, which is recognized as a gap in current methods (Fernandes 2000; Geneletti 2006; Mörtberg et al. 2007). The map of connectivity changes can help optimize the location of new protected areas or mitigation measures by identifying the areas most affected by the infrastructure. The results also provide information about the maximum distance of the impact, which is often difficult to assess. In the case of the HSR line in the Burgundy-Franche-Comté region, the environmental assessment studies focused only on a strip of 800 m on either side of the HSR line, which allowed the creation of new ponds to replace those destroyed by the construction of the infrastructure.

Acknowledgements The research has been funded by the French Ministry of Ecology, Energy, Sustainable Development and the Sea (ITTECOP Program). The graph analysis was conducted as part of the Graphab project managed by the USR 3124 MSHE 744 Ledoux. Computations were performed on the supercomputer facilities of the MSHE Ledoux.

References

Adriaensen, F., Chardon, J. P., De Blust, G., Swinnen, E., Villalba, S., Gulinck, H., et al. (2003). The application of "least-cost" modelling as a functional landscape model. *Landscape and Urban Planning, 64,* 233–247.
Alford, R. A., & Richards, S. J. (1999). Global amphibian declines: A problem in applied ecology. *Annual Review of Ecology and Systematics, 30,* 133–165.
Allentoft, M. E., & O'Brien, J. (2010). Global amphibian declines, loss of genetic diversity and fitness: A review. *Diversity, 2,* 47–71.
Andersen, L. W., Fog, K., & Damgaard, C. (2004). Habitat fragmentation causes bottlenecks and inbreeding in the European tree frog *(Hyla arborea)*. *Proceedings of the Royal Society B: Biological Sciences, 271,* 1293–1302.
Bodin, Ö., & Saura, S. (2010). Ranking individual habitat patches as connectivity providers: Integrating network analysis and patch removal experiments. *Ecological Modelling, 221,* 2393–2405.
Borgula, A. (1993). Causes of the decline in *Hyla arborea*. In A. H. P. Stumpel & U. Tester (Eds.), *Ecology and conservation of the European tree frog. Proceedings of the 1st International*

Workshop on Hyla arborea (pp. 71–80). Wageningen: Institute for Forestry and Nature Research.

Brotons, L., & Herrando, S. (2001). Reduced bird occurrence in pine forest fragments associated with road proximity in a Mediterranean agricultural area. *Landscape and Urban Planning, 57,* 77–89.

Calabrese, J. M., & Fagan, W. F. (2004). A comparison-shopper's guide to connectivity metrics. *Frontiers in Ecology and the Environment, 2,* 529–536.

Clauzel, C., Bannwarth, C., & Foltête, J.-C. (2015a). Integrating regional-scale connectivity in habitat restoration: An application for amphibian conservation in eastern France. *Journal for Nature Conservation, 23,* 98–107.

Clauzel, C., Girardet, X., & Foltête, J.-C. (2013). Impact assessment of a high-speed railway line on species distribution: Application to the European tree frog (*Hyla arborea*) in Franche-Comté. *Journal of Environmental Management, 127,* 125–134.

Clauzel, C., Xiqing, D., Gongsheng, W., Giraudoux, P., & Li, L. (2015b). Assessing the impact of road developments on connectivity across multiple scales: Application to Yunnan snub-nosed monkey conservation. *Biological Conservation, 192,* 207–217.

Corbett, K. (1989). *Conservation of European reptiles and amphibians*. Bromley, Kent: Helm.

Cushman, S. A. (2006). Effects of habitat loss and fragmentation on amphibians: A review and prospectus. *Biological Conservation, 128,* 231–240.

Dale, M. R. T., & Fortin, M.-J. (2010). From graphs to spatial graphs. *Annual Review of Ecology Evolution and Systematics, 41,* 21–38.

Eigenbrod, F., Hecnar, S. J., & Fahrig, L. (2009). Quantifying the road effect zone: Threshold effects of a motorway on anuran populations in Ontario, Canada. *Ecology and Society, 14.* http://www.ecologyandsociety.org/vol14/iss1/art24/

Elzanowski, A., Ciesiołkiewicz, J., Kaczor, M., Radwańska, J., & Urban, R. (2008). Amphibian road mortality in Europe: A meta-analysis with new data from Poland. *European Journal of Wildlife Research, 55,* 33–43.

Fahrig, L. (2003). Effects of habitat fragmentation on biodiversity. *Annual Review of Ecology Evolution and Systematics, 34,* 487–515.

Fahrig, L., Pedlar, J. H., Pope, S. E., Taylor, P. D., & Wegner, J. F. (1995). Effect of road traffic on amphibian density. *Biological Conservation, 73,* 177–182.

Fernandes, J. P. (2000). Landscape ecology and conservation management—Evaluation of alternatives in a highway EIA process. *Environmental Impact Assessment Review, 20,* 665–680.

Fog, K. (1993). Migration in the tree frog *Hyla arborea*. In A. H. P Stumpel & U. Tester (Eds.), *Ecology and conservation of the European tree frog. Proceedings of the 1st International Workshop on Hyla arborea* (pp. 55–64). Wageningen: Institute for Forestry and Nature Research.

Foltête, J.-C., Clauzel, C., & Vuidel, G. (2012a). A software tool dedicated to the modelling of landscape networks. *Environmental Modelling and Software, 38,* 16–32.

Foltête, J.-C., Clauzel, C., Vuidel, G., & Tournant, P. (2012b). Integrating graph-based connectivity metrics into species distribution models. *Landscape Ecology, 27,* 1–13.

Foltête, J.-C., Girardet, X., & Clauzel, C. (2014). A methodological framework for the use of landscape graphs in land-use planning. *Landscape and Urban Planning, 124,* 240–250.

Forman, R. T. T. (2000). Estimate of the area affected ecologically by the road system in the United States. *Conservation Biology, 14,* 31–35.

Forman, R. T. T., & Alexander, L. E. (1998). Roads and their major ecological effects. *Annual Review of Ecology and Systematics, 29,* 207–231.

Fu, W., Liu, S., Degloria, S. D., Dong, S., & Beazley, R. (2010). Characterizing the "fragmentation–barrier" effect of road networks on landscape connectivity: A case study in Xishuangbanna, Southwest China. *Landscape and Urban Planning, 95,* 122–129.

Galpern, P., & Manseau, M. (2013). Modelling the influence of landscape connectivity on animal distribution: A functional grain approach. *Ecography, 36,* 1004–1016.

Galpern, P., Manseau, M., & Fall, A. (2011). Patch-based graphs of landscape connectivity: A guide to construction, analysis and application for conservation. *Biological Conservation, 144,* 44–55.

García-Feced, C., Saura, S., & Elena-Rosselló, R. (2011). Improving landscape connectivity in forest districts: A two-stage process for prioritizing agricultural patches for reforestation. *Forest Ecology and Management, 261,* 154–161.

Geneletti, D. (2004). Using spatial indicators and value functions to assess ecosystem fragmentation caused by linear infrastructures. *International Journal of Applied Earth Observation and Geoinformation, 5,* 1–15.

Geneletti, D. (2006). Some common shortcomings in the treatment of impacts of linear infrastructures on natural habitat. *Environmental Impact Assessment Review, 26,* 257–267.

Girardet, X., Foltête, J. C., & Clauzel, C. (2013). Designing a graph-based approach in the landscape ecological assessment of linear infrastructures. *Environmental Impact Assessment Review, 42,* 10–17.

Girardet, X., Foltête, J.-C., Clauzel, C., & Vuidel, G. (2016). Restauration de la connectivité écologique: proposition méthodologique pour une localisation optimisée des passages à faune. VertigO - la revue électronique en sciences de l'environnement. http://vertigo.revues.org/17337

Gontier, M., Balfors, B., & Mörtberg, U. (2006). Biodiversity in environmental assessment— Current practice and tools for prediction. *Environmental Impact Assessment Review, 26,* 268–286.

Grosse, W. R., & Nöllert, A. K. (1993). Migration in the tree frog *Hyla arborea.* In A. H. P. Stumpel & U. Tester (Eds.), *Ecology and conservation of the European tree frog. Proceedings of the 1st international workshop on Hyla arborea* (pp. 37–45). Wageningen: Institute for Forestry and Nature Research.

Gurrutxaga, M., Rubio, L., & Saura, S. (2011). Key connectors in protected forest area networks and the impact of highways: A transnational case study from the Cantabrian range to the Western Alps (SW Europe). *Landscape and Urban Planning, 101,* 310–320.

Huijser, M. P., & Bergers, P. J. (2000). The effect of roads and traffic on hedgehog (*Erinaceus europaeus*) populations. *Biological Conservation, 95,* 111–116.

IUCN. (2016). *Amphibians on the IUCN Red List.* http://www.iucnredlist.org/initiatives/amphibians

Joly, P., Morand, C., & Cohas, A. (2003). Habitat fragmentation and amphibian conservation: Building a tool for assessing landscape matrix connectivity. *Comptes Rendus Biologies, 326,* 132–139.

Kaczensky, P., Knauer, F., Krze, B., Jonozovic, M., Adamic, M., & Gossow, H. (2003). The impact of high speed, high volume traffic axes on brown bears in Slovenia. *Biological Conservation, 111,* 191–204.

Lengagne, T. (2008). Traffic noise affects communication behaviour in a breeding anuran, *Hyla arborea. Biological Conservation, 141,* 2023–2031.

Li, Z., Ge, C., Li, J., Li, Y., Xu, A., Zhou, K., et al. (2010). Ground-dwelling birds near the Qinghai-Tibet highway and railway. *Transportation Research Part D: Transport and Environment, 15,* 525–528.

Liu, S., Dong, Y., Deng, L., Liu, Q., Zhao, H., & Dong, S. (2014). Forest fragmentation and landscape connectivity change associated with road network extension and city expansion: A case study in the Lancang River Valley. *Ecological Indicators, 36,* 160–168.

Lookingbill, T. R., Gardner, R. H., Ferrari, J. R., & Keller, C. E. (2010). Combining a dispersal model with network theory to assess habitat connectivity. *Ecological Applications, 20,* 427–441.

Mimet, A., Clauzel, C., & Foltête, J.-C. (2016). Locating wildlife crossings for multispecies connectivity across linear infrastructures. *Landscape Ecology, 32,* 1955–1973.

Mörtberg, U. M., Balfors, B., & Knol, W. C. (2007). Landscape ecological assessment: A tool for integrating biodiversity issues in strategic environmental assessment and planning. *Journal of Environmental Management, 82,* 457–470.

Noble, B., Hill, M., & Nielsen, J. (2011). Environmental assessment framework for identifying and mitigating the effects of linear development to wetlands. *Landscape and Urban Planning, 99*, 133–140.

Opdam, P., Steingröver, E., & Rooij, S. (2006). Ecological networks: A spatial concept for multi-actor planning of sustainable landscapes. *Landscape and Urban Planning, 75*, 322–332.

Paul, J.-P. (2011). Liste rouge des vertébrés terrestres de Franche-Comté. Groupe Naturaliste de Franche-Comté et Conseil Régional de Franche-Comté (Eds.), Besançon.

Pellet, J., Hoehn, S., & Perrin, N. (2004). Multiscale determinants of tree frog (*Hyla arborea* L.) calling ponds in western Switzerland. *Biodiversity and Conservation, 13*, 2227–2235.

Pinston H., Craney E., Pépin D., Montadert M., & Duquet M. (2000). Amphibiens et Reptiles de Franche-Comté. Atlas commenté de répartition. Groupe Naturaliste de Franche-Comté et Conseil Régional de Franche-Comté (Eds.), Besançon.

Saura, S., & Pascual-Hortal, L. (2007). A new habitat availability index to integrate connectivity in landscape conservation planning: Comparison with existing indices and application to a case study. *Landscape and Urban Planning, 83*, 91–103.

Scherer, R. D., Muths, E., & Noon, B. R. (2012). The importance of local and landscape-scale processes to the occupancy of wetlands by pond-breeding amphibians. *Population Ecology, 54*, 487–498.

Serrano, M., Sanz, L., Puig, J., & Pons, J. (2002). Landscape fragmentation caused by the transport network in Navarra (Spain). *Landscape and Urban Planning, 58*, 113–123.

Smith, A., & Green, D. (2005). Dispersal and the metapopulation paradigm in amphibian ecology and conservation: Are all amphibian populations metapopulations? *Ecography, 28*, 110–128.

Stuart, S. N., Hoffmann, M., Chanson, J. S., Cox, N. A., Berridge, R. J., Ramani, P., et al. (2008). Threatened amphibians of the world. Lynx Edicions, Barcelona, Spain; IUCN, Gland, Switzerland; and Conservation International, Arlington, VA.

Stumpel, A. H. P. (1993). The terrestrial habitat of *Hyla arborea*. In A. H. P. Stumpel & U. Tester (Eds.), *Ecology and conservation of the European tree frog. Proceedings of the 1st international workshop on Hyla arborea* (pp. 47–54). Wageningen: Institute for Forestry and Nature Research.

Urban, D. L., Minor, E. S., Treml, E. A., & Schick, R. S. (2009). Graph models of habitat mosaics. *Ecology Letters, 12*, 260–273.

Vasas, V., Magura, T., Jordán, F., & Tóthmérész, B. (2009). Graph theory in action: Evaluating planned highway tracks based on connectivity measures. *Landscape Ecology, 24*, 581–586.

Vos, C. C., & Chardon, J. P. (1998). Effects of habitat fragmentation and road density on the distribution pattern of the moor frog *Rana arvalis*. *Journal of Applied Ecology, 35*, 44–56.

Vos, C. C., & Stumpel, A. H. P. (1996). Comparison of habitat-isolation parameters in relation to fragmented distribution patterns in the tree frog (*Hyla arborea*). *Landscape Ecology, 11*, 203–214.

Chapter 14
Habitat Fragmentation by Railways as a Barrier to Great Migrations of Ungulates in Mongolia

Takehiko Y. Ito, Badamjav Lhagvasuren, Atsushi Tsunekawa and Masato Shinoda

Abstract Mongolia's Gobi-Steppe Ecosystem is the largest grassland in the world and the habitat of long-distance movement ungulates, such as the Mongolian gazelle (*Procapra gutturosa*) and the Asiatic wild ass (*Equus hemionus*). The international railway between Russia and China bisects this habitat, and there has been concern that it may impede the movements of wild ungulates. We tracked ungulate movements on both sides of the Ulaanbaatar–Beijing Railway, and found that most of the tracked animals never crossed the railway. The construction of additional railways to permit mining projects in the area is therefore a further threat to maintaining the great migrations of ungulates across Mongolia.

Keywords Asiatic wild ass · Dryland · Grassland · Desert · Long-distance movement · Mongolian gazelle · NDVI · Remote sensing · Satellite tracking · Terrestrial mammal

Introduction

Mongolia occupies an area of 1,564,100 km^2 in central Asia. Mainly due to its low human population density (2.0 ind./km^2 in 2015 for all of Mongolia; National Statistical Office of Mongolia 2016), it harbors the largest grassland in the world. Several species of ungulates inhabit this grassland, including the Mongolian gazelle (*Procapra gutturosa*), the goitered gazelle (*Gazella subgutturosa*), and the Asiatic wild ass (*Equus hemionus*), with some of them, such as the Mongolian gazelle, forming large herds (Fig. 14.1).

T.Y. Ito (✉) · A. Tsunekawa
Arid Land Research Center, Tottori University, Tottori, Japan
e-mail: ito@alrc.tottori-u.ac.jp

B. Lhagvasuren
Institute of General and Experimental Biology, Mongolian Academy of Sciences, Ulaanbaatar, Mongolia

M. Shinoda
Graduate School of Environmental Studies, Nagoya University, Nagoya, Japan

L. Borda-de-Água et al. (eds.), *Railway Ecology*,
DOI 10.1007/978-3-319-57496-7_14

Fig. 14.1 A herd of Mongolian gazelles moving along the Ulaanbaatar–Beijing Railway

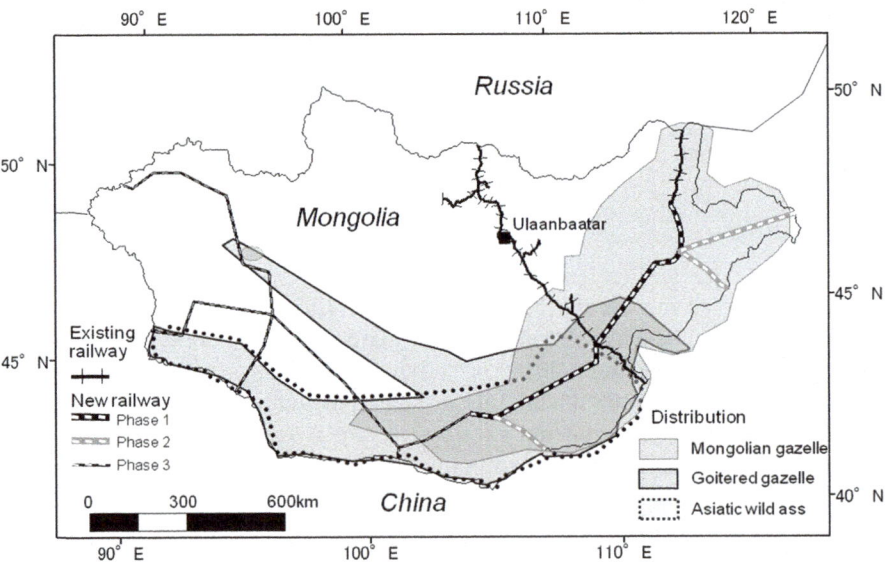

Fig. 14.2 Existing and planned railways in Mongolia and the distributions of the Mongolian gazelle, goitered gazelle, and Asiatic wild ass. Data of the wildlife distribution were downloaded from the IUCN red list website (http://www.iucnredlist.org)

This ecosystem is under threat by human activities, including the presence of the Ulaanbaatar–Beijing Railway and its further development. The name of the railway is deceptive, because it actually connects Russia with China, going through Mongolia's capital, Ulaanbaatar. In doing so, it bisects the country (Fig. 14.2), forming a barrier to the movements, including migrations, of several animals, and particularly the Mongolian gazelle. Since 2002, we have been studying the impact of the railway on the Mongolian gazelle by using satellite tracking methods to follow individuals of this species. Our results show that the railway is a barrier to migrations of the gazelle, contributing directly and indirectly to its mortality. Harsh winters in Mongolia force the Mongolian gazelle and other ungulates to perform long-distance movements to find food and the presence of the railway acts a barrier preventing animals from accessing certain food sources.

Here, we report the results of our studies on the impact of the Ulaanbaatar–Beijing Railway on the movements of the Mongolian gazelle and the spatial distribution of gazelle carcasses that results from the presence of this barrier. Moreover, we assess the impact of further development of the railway network in Mongolia and the need to implement mitigation measures if we want to preserve the great migrations of ungulates across the Mongolian grasslands.

Environments of Mongolia's Gobi-Steppe Ecosystem

Mongolia's Gobi-Steppe Ecosystem is the world's largest area of intact grassland (827,000 km^2), and it is much larger than other globally famous grasslands as wildlife habitat, such as the Greater Yellowstone Ecosystem (108,000 km^2) in the western United States and the Serengeti-Mara Ecosystem (25,000 km^2) in East Africa (Batsaikhan et al. 2014). The Gobi-Steppe Ecosystem is a wide and intact steppe and semi-desert ecosystem in northern China, Mongolia, and southern Russia where great migrations of wild ungulates still occur. Nomadic pastoralism has been the main lifestyle of the Mongolian people for several thousand years, and the grassland ecosystem has been maintained in relatively good condition compared to that in other regions thanks to low human and livestock densities, low grazing pressure, and movements of people with their livestock between grazing sites to take advantage of better grazing conditions in different seasons and years.

Mongolia has one of the smallest human populations in the world and, although the population is growing, its density in the countryside outside of the capital city was only 1.1 ind./km^2 in 2015 (National Statistical Office of Mongolia 2016), mainly because half of the country's population is concentrated in the capital city, Ulaanbaatar. The number of livestock, especially goats, has increased since the capitalist system replaced the communist system in 1991, and grassland degradation by overgrazing has been reported in some areas (e.g., Sasaki et al. 2008), although relatively fine and continuous habitats for wildlife remain.

The Gobi-Steppe Ecosystem dominates most of Mongolia, except in mountainous areas and in the northern forests, and is characterized by relatively flat topography. Annual average precipitation in this area is less than about 350 mm, but its interannual variation is large (Vandandorj et al. 2015; Yu et al. 2004), which is typical for drylands. Because precipitation and snowfall increase and temperature decreases from south to north (Morinaga et al. 2003; Nandintsetseg and Shinoda 2011), vegetation correspondingly changes from desert to drylands and forest steppe ecosystems. Droughts and severe winters are a cause of high mortality among live-stock (Fernandez-Gimenez et al. 2012, 2015; Tachiiri and Shinoda 2012) and wild animals (Kaczensky et al. 2011a).

Railways and Vulnerable Wild Animals in Mongolia

The Ulaanbaatar–Beijing Railway is an international railway that connects China with Russia thorough Ulaanbaatar (Fig. 14.2). It was built in the 1950s and runs from northwest to southeast; the section that is south of Ulaanbaatar crosses the Gobi-Steppe Ecosystem. Barbed wire fences have been built on both sides of the railway, mainly to prevent accidents with livestock. The influence of the railway on wild animal movements and concomitant population declines has been a cause of concern. Animals can suffer mortality directly when crossing the railway or indirectly because of barrier effects that prevent them from reaching suitable sites for food, water, and reproduction. These concerns led to animal tracking studies (e.g., Ito et al. 2005, 2013a). Mammals inhabiting the grassland ecosystems of Mongolia and showing long-distance movements that are potentially affected by the railway are the Mongolian gazelle, the goitered gazelle, and the Asiatic wild ass (Fig. 14.2; Mallon and Jiang 2009). Current train frequency on the railway is not high (about 1 train/h in daytime based on our rough observations), and freight trains are more frequent than passenger trains.

Both historically and today, the Mongolian gazelle has had one of the longest migrations among terrestrial animals (Berger 2004; Teitelbaum et al. 2015). Until the 1930s, this species had a distribution that occupied most of the grasslands in northern China, Mongolia, and southern Russia, but since then its distribution has been reduced to the eastern half of Mongolia and to areas close to the border between Mongolia, China, and Russia (Fig. 14.2; Jiang et al. 1998; Lhagvasuren and Milner-Gulland 1997; Mallon 2008b). In the latest IUCN Red List, the Mongolian gazelle is ranked as least concern (LC) due to population estimates over the last 10 years ranging from 400,000 to 2,700,000 and because its range is expanding toward the northwest (Mallon 2008b).

Mongolia has the largest population of goitered gazelles (40–50% of the global population; Mallon 2008a). This species has a body size and morphology similar to those of the Mongolian gazelle, but it does not form large herds, and details of its ecology are still unknown. The Asiatic wild ass has lost 70% of its global range since the 19th century, and at present, more than 75% of the population (about 55,000 animals) lives in Mongolia (Kaczensky et al. 2011a, 2015; Reading et al. 2001).

Movement Ecology of Mongolian Gazelles

Among ungulates, the Mongolian gazelle is the most well studied species in terms of the impact of the Ulaanbaatar–Beijing Railway on its movements. Although the species' behavioral ecology has not been fully elucidated, Mongolian gazelles inhabit grassland and semi-desert areas and sometimes form large herds of several thousand animals. Under severe climate conditions, the estimated lifespan in the wild is 7–8 years (Batsaikhan et al. 2010). The rutting period is in winter, and females over 2 years old usually give birth to one calf in late June or early July (Lhagvasuren and Milner-Gulland 1997). The long-distance movements of this species for migration or nomadism were already understood before scientific tracking started (Jiang et al. 1998; Lhagvasuren and Milner-Gulland 1997). Since then, analyses of gazelle movements in relation to habitat selection and environmental factors, including the presence of a railway, have brought many new findings on the ecology of this species and related conservation issues.

Using modern technology, we have been able to prove the capability of the Mongolian gazelle to travel long distances. For instance, Argos systems and GPS with satellite communication systems have been used to track wild ungulates in Mongolia (Kaczensky et al. 2010). We showed that Mongolian gazelles moved distances greater than 300 km (the maximum linear distance between two locations traveled by one individual gazelle in a year) and changed their range seasonally (Ito et al. 2006, 2013b). The gazelles moved more than 100 km per week during some periods of the year, whereas the distances moved were short in other periods. Interannual differences in the seasonal range locations among the same individuals were also observed, which in some cases were larger than 300 km in winter, suggesting nomadic movements rather than typical seasonal migrations between specific locations (Ito et al. 2013b; Olson et al. 2010).

Understanding why and how animals move long distances is important both for purely scientific purposes and for conservation. Studies on this topic have shown that environmental factors play a pivotal role, and the normalized-difference vegetation index (NDVI) has mainly been used as an index of the amount of live plants in studies of the Mongolian gazelle. For instance, Leimgruber et al. (2001) showed that the winter and the calving grounds in the eastern steppes of Mongolia (identified based on expert knowledge of scientists and pastoralists, but not tracking data), had the highest NDVI scores during periods when gazelles used these areas. In a study comparing gazelle distribution and NDVI values in different seasons in the eastern steppes, Mueller et al. (2008) showed that gazelles preferred areas with intermediate NDVI values in the spring and autumn. Similarly, during a drought period in September 2005, Olson et al. (2009a) reported a mega-herd of more than 200,000 gazelles in areas with a high probability of gazelle occurrence predicted by a NDVI-based model. In the southeastern Gobi, the shifts in NDVI values between the summer and winter ranges explained the gazelles' seasonal movements (Ito et al. 2006), and interannual differences in the spatial distribution of NDVI explained the interannual differences in the seasonal range of the tracked gazelles (Ito et al. 2013b). The interannual differences in locations were much larger in winter than in summer, likely because of the large differences in the spatial distribution of snow cover. Avoidance of areas with deep snow cover by Mongolian gazelles was also reported in Inner Mongolia, China (Luo et al. 2014). In addition, regional differences in the amount of vegetation across the species' spatial distribution also led to intraspecific variations of their movement patterns (Imai et al. 2017).

The spatial distribution and seasonal change of food plants are considered to be important factors determining the movement patterns of ungulates (Mueller and Fagan 2008). Therefore, environmental unpredictability at the landscape level affects movement patterns of animals through seasonal and interannual changes in food availability. Mongolian gazelles in the eastern steppe of Mongolia have nomadic movements that are more irregular than those of other ungulate species, such as caribou (*Rangifer tarandus granti*) in Alaska, which exhibit regular seasonal migration, or the guanaco (*Lama guanicoe*) in Argentina and moose (*Alces alces*) in the northeastern United States, both of which move shorter distances with more predictable movements (Mueller et al. 2011). Therefore, in order to conserve nomadic animals, like Mongolian gazelles, that live in unpredictable environments, the maintenance of good environmental conditions and access to vast areas are essential, especially during those periods when conditions become unsuitable in much of their ranges.

Effects of the Ulaanbaatar–Beijing Railway on Wild Ungulates

Barrier Effect of the Existing Railway

The barrier effect of the Ulaanbaatar–Beijing Railway on the movements of wild ungulates in Mongolia has been apparent since the beginning of satellite tracking of these animals. We first conducted satellite tracking of Mongolian gazelles in 2002 (Ito et al. 2005, 2006). In October 2002, two female gazelles were captured on the southwestern side of the railway and collared with a satellite transmitter. The distance between the site of capture and the railway was about 7 km. Although both gazelles were captured at exactly the same place, their movements after release were different (Fig. 14.3). This is a small sample size, but it is representative of the movements of more individuals because at the time when the animals were captured, the gazelles were part of a herd with several hundred animals. The different movements of the two gazelles also indicate loose relations among gazelle individuals. Sometimes Mongolian gazelles form large herds, whereas other times they separate into relatively small groups. Such uncoordinated movement patterns were also reported in a comparative study of the Mongolian gazelle, caribou, guanaco, and moose (Mueller et al. 2011).

The two gazelles captured near the railway moved along it for a year, but never crossed it (Ito et al. 2005). They moved closer to the railway during winter (October–March) than during summer (April–September), and the nearest distances of the two gazelles to the railway were 0.3 and 1.8 km, respectively (Fig. 14.3). These results suggest that the railway is a barrier for gazelle movements mainly because of the fences along the rails. Some wildlife crossing locations exist along the railway (Fig. 14.4), but our data suggest that these crossings are ineffective, probably due to their limited number and their lack of suitable natural features that could reduce the natural wariness of the gazelles.

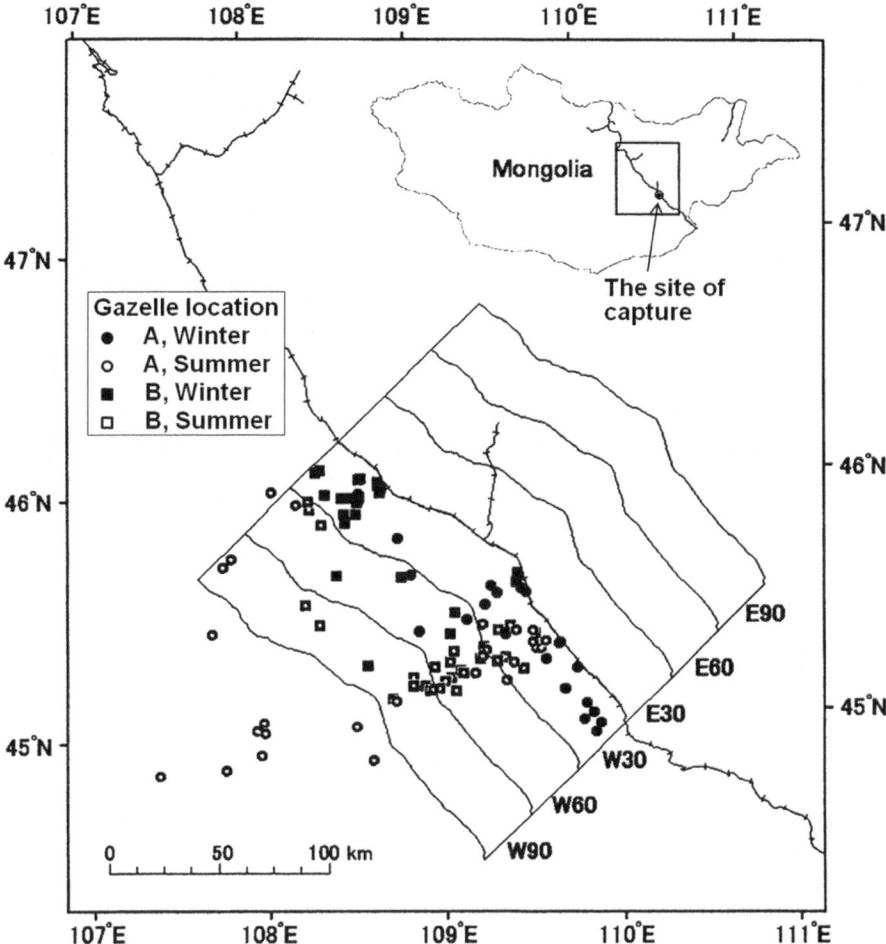

Fig. 14.3 Locations of two tracked Mongolian gazelles from 2002 to 2003 in Mongolia, and the zones in which the normalized-difference vegetation index (NDVI) was analyzed. Zones W30, W60, and W90 are 0–30, 30–60, and 60–90 km northwest of the railway, respectively, and zones E30, E60, and E90 are 0–30, 30–60, and 60–90 km southeast of the railway, respectively. The *hatched line* represents the railway (from Ito et al. 2005)

To understand why the gazelles used areas near the railway during the winter and whether there were any disadvantages for gazelles caused by being restricted to only one side of the railway, we compared NDVI values on both sides of the railway. We established three 30-km zones on each side of the railway, parallel to the rails (Fig. 14.3). These zones spanned most northern and southern gazelle locations observed during the winter. We detected a significant gradient of NDVI values from 17 November to 2 December 2002, the period when the gazelles moved the greatest distance, from southwest to northeast (Ito et al. 2005). The highest average NDVI value was in the 30- to 60-km northeastern zone (Fig. 14.5), suggesting that the

Fig. 14.4 Underpasses on the Ulaanbaatar–Beijing Railway

tracked gazelles, (and thus, many gazelles living on the southwestern side of the railway) could not use better sites existing on the opposite side due to the barrier caused by the railway.

The barrier effect of the railway became more evident as more animal tracking data were gathered. An Asiatic wild ass tracked from July 2005 to February 2006 also moved along the southwestern side of the railway but did not cross it (Kaczensky et al. 2006, 2011b). The railway now likely forms the eastern edge of

Fig. 14.5 Average normalized-difference vegetation index (NDVI) in each zone adjacent to the railway from 17 November to 2 December 2002. *Error bars* represent 95% confidence intervals. See Fig. 14.3 for explanation of zone names (from Ito et al. 2005)

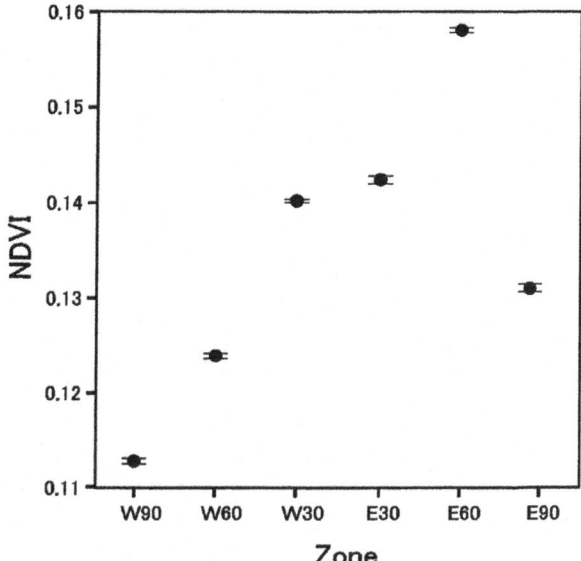

the current Asiatic wild ass distribution because no wild ass has been observed on the eastern side for many years, despite the existence of suitable habitat on the northeastern side of the railway (Kaczensky et al. 2011b, 2015). Further data on tracked Mongolian gazelles have confirmed that the railway impeded 24 gazelles tracked between 2002 and 2012 from crossing to both sides (Fig. 14.6; Ito et al. 2013a), except for one individual (Olson 2012).

International border fences further exacerbate the fragmentation of the landscape and have a similar barrier effect on wild ungulates. Almost none of the tracked ungulates in Mongolia crossed the borders (Fig. 14.6; Ito et al. 2013a; Kaczensky et al. 2006, 2008, 2011a; Olson 2012; Olson et al. 2009b), except for one reported case of an Asiatic wild ass in western Mongolia (Kaczensky et al. 2011a). These results indicate that the railway and the border fences are likely causes of habitat fragmentation and impediments to long-distance movements of ungulates.

Influences of the Barrier Effect on Wild Ungulates

What are the problems caused by railway barrier effects on wild ungulates? Although the railway bisects Mongolia's Gobi-Steppe Ecosystem, each part of the ecosystem is still vast. The seasonality of the ungulates' location is a key to evaluate how the railway influences the animals restricted to one side of this barrier. Tracked Mongolian gazelles and Asiatic wild asses used areas far from the barriers during summer but used areas close to the railway and the borders during winter (Fig. 14.7; Ito et al. 2005, 2013a). The only case of a tracked wild ass crossing the border to China and returning to Mongolia was observed during the harsh winter

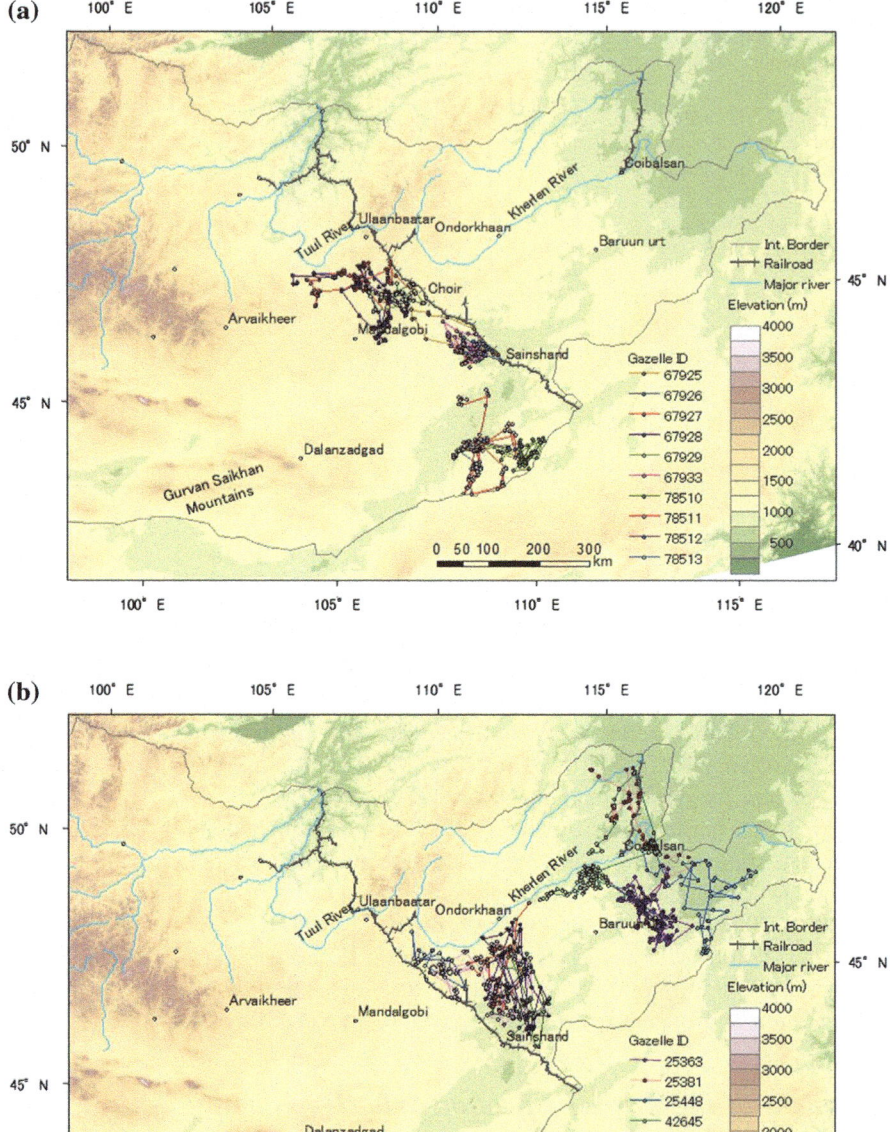

Fig. 14.6 Movements of the tracked Mongolian gazelles: **a** gazelles captured on the southwestern side of the Ulaanbaatar–Beijing Railway in 2007, and **b** gazelles captured on the northeastern side from 2003. The tracking continued until 2012 (from Ito et al. 2013a)

(*dzud*) of 2009/2010 in western Mongolia (Kaczensky et al. 2011a), but no other tracked animals ever crossed it. The reason why animals tend to be closer to the railway in the winter is likely to be related to food availability. During the plant growing period (summer in Mongolia), suitable areas for herbivores are widespread. In contrast, the period from winter to early spring is the severest period for wild and livestock animals in Mongolia because of low temperatures and poor food availability caused by plant withering, plant consumption by continuous grazing, and snow cover on the vegetation. Herbivorous animals, therefore, have to move in order to find suitable sites with enough vegetation to survive during this harsh time, and many animals consequently reach areas close to the railway and international borders, which often have fences.

The limitation of movements and habitat use by ungulates may lead to higher mortality. If suitable sites for ungulates are mainly located on the opposite side of the barriers during winter, the animals may not get enough food to survive. Therefore we predicted a higher density of carcasses of Mongolian gazelles along the Ulaanbaatar–Beijing Railway than in remote areas away from the railway. Indeed, many carcasses were actually found near the railway, although we have not compared this with the carcass density in remote areas. We conducted a carcass survey along a 630-km section of the railway in June 2005 and found a total of 241 Mongolian gazelle carcasses. Carcass densities differed not only between regions but also between the southwestern and northeastern sides of the railway (Fig. 14.8; Ito et al. 2008). A similar carcass survey was conducted in summer 2011, and 393

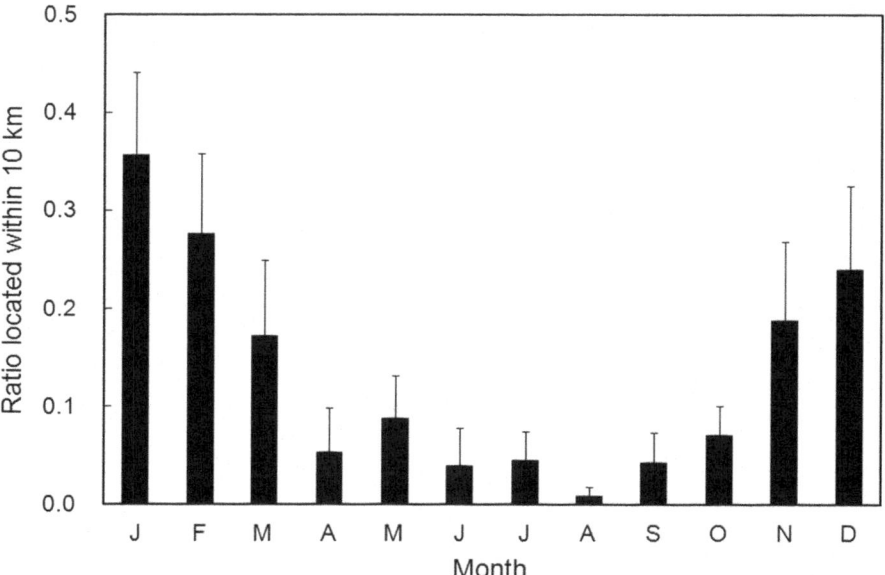

Fig. 14.7 Monthly ratio (mean + SE) of location data within 10 km of the anthropogenic barriers (the railway and the international border fence) to all monthly location data for tracked Mongolian gazelles that used areas within 10 km of the barriers at least once during the tracking periods (*n* = 16; from Ito et al. 2013a)

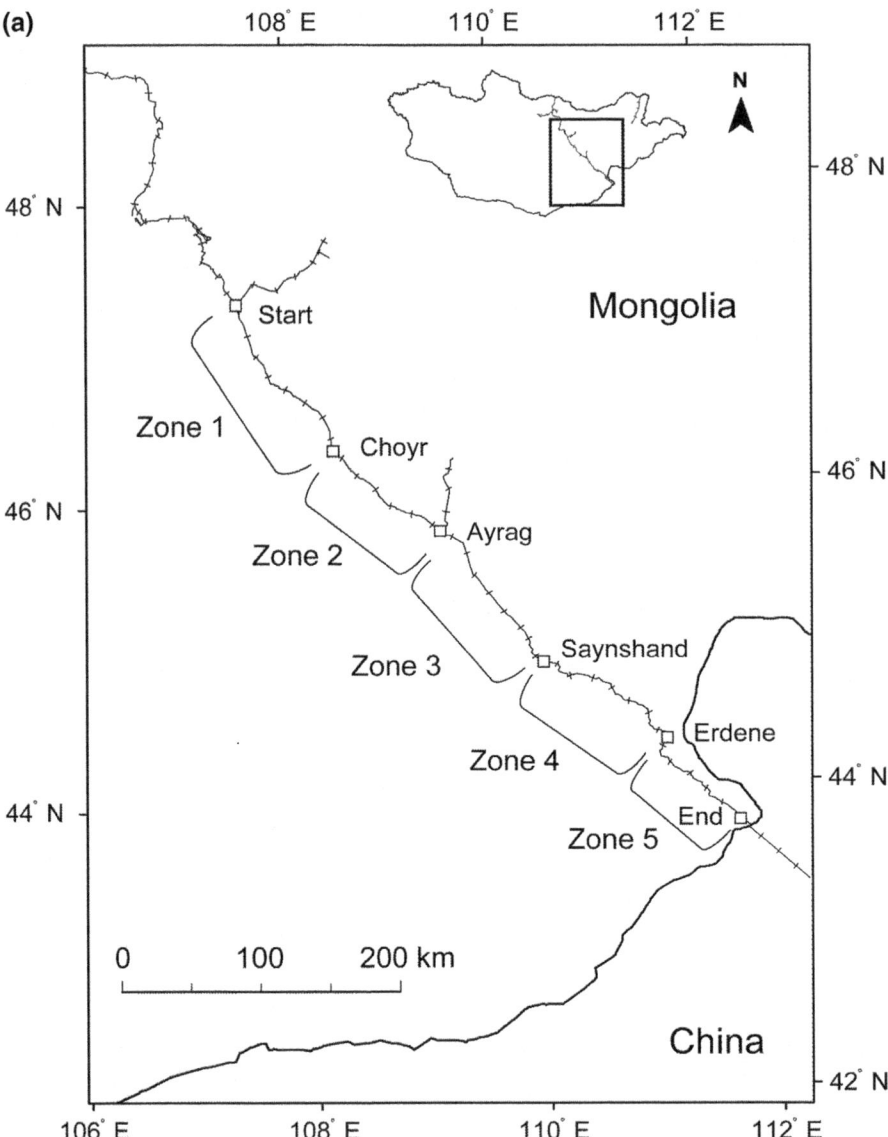

Fig. 14.8 a Study area of a Mongolian gazelle carcass census conducted along the Ulaanbaatar–Beijing Railway in June 2005 and ranges of each zone. *Open squares* are locations of major towns and the start- and end-points of the carcass census. The *hatched line* represents the railway. **b** Carcass numbers of Mongolian gazelles on the southwestern and northeastern sides of the Ulaanbaatar–Beijing Railway in each zone. We categorized carcasses according to whether we found them outside or inside railway fences (from Ito et al. 2008)

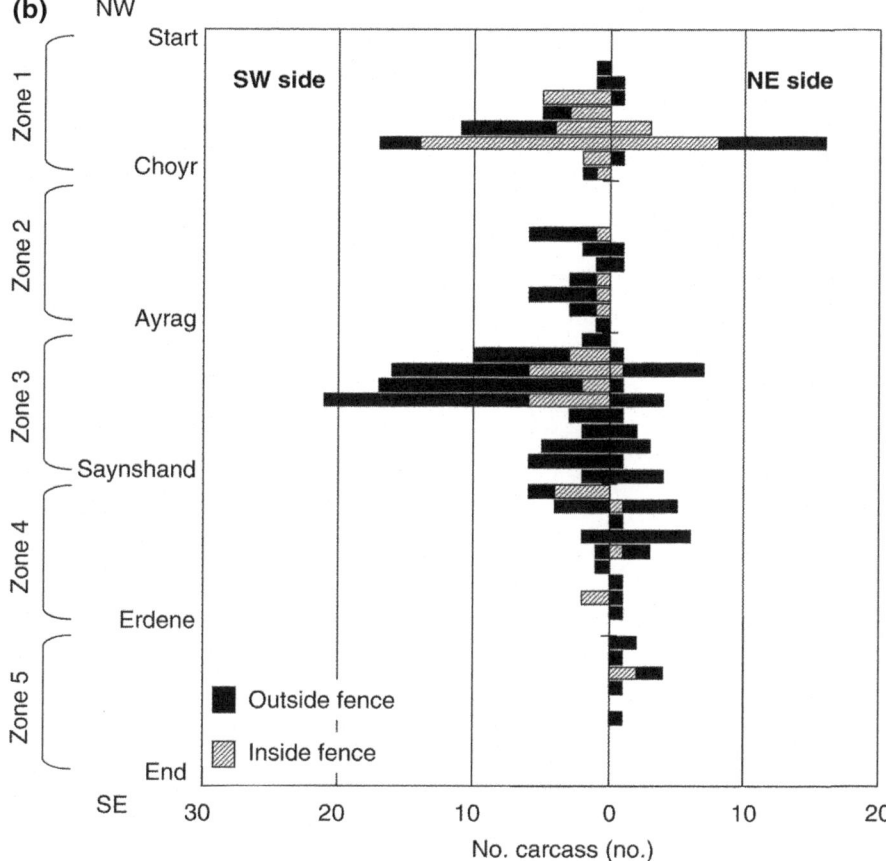

Fig. 14.8 (continued)

carcasses were recorded (Olson 2012). Although the carcass distribution patterns in different years were similar, there were also some differences. An area of high carcass density existed to the southeast of Ayrag Town in the 2005 survey (Fig. 14.8), but a high density was found to the northwest of the town in the 2011 survey. We attribute this interannual difference among areas of high carcass density to the high environmental unpredictability, which leads to changes in the locations of suitable sites for wild ungulates. Thus, the areas with high carcass density in one year do not necessarily indicate that these are the best sites to implement wildlife crossings. However, because Mongolian gazelles likely have a learning capacity, the sites are still good candidates given that gazelles had previously used these areas. For example, some tracked gazelles came back from several hundred kilometers to the same location in the following summer (Ito et al. 2013b).

Carcasses were also found inside the fenced area (Fig. 14.8), suggesting that the fences did not form a total barrier to wild ungulates. The existence of carcasses inside the fences might be explained by the fact that plant biomass is more abundant

Fig. 14.9 Carcass of a Mongolian gazelle entangled in a railway fence of the Ulaanbaatar–Beijing Railway

inside the fence at the beginning of winter, because these areas are not usually grazed by livestock or wild ungulates during summer. Therefore, the area inside the fences may be attractive for herbivores in winter, leading to some ungulates trying to enter it. Once inside the fenced area, the animal may not be able to find suitable locations to jump or crawl out of the fence again or may become too weak to do so. Note, however, that we found some carcasses entangled in the fence's barbed wires (Fig. 14.9), suggesting that the fence itself is a cause of mortality.

Several questions remain concerning the interpretation of areas of high carcass density. For example, was animal density simply higher in these areas? Did a large number of animals try to cross there because the area looked easy to pass? Did the area have any intrinsic factors that cause higher mortality? For example, do the fence structures easily entangle animals, making it difficult to escape once inside?

Finally, the genetic structure of the gazelle populations sampled in the 2005 survey was not different between the two sides of the railway (Okada et al. 2012, 2015). This can be explained by animals occasionally crossing the railway via underpasses and areas with broken fences, and other permeable areas. In addition, the survey was conducted just about 50 years after the railway's construction, which is not enough time for genetic differentiation given the relatively long lifespan of the species. Therefore, the genetic structure of wild ungulate populations may differentiate in the future if the railway barrier effects persist.

Railway Development as a Threat to Wildlife

Mongolia's Gobi-Steppe Ecosystem is facing the threats of new railway development (Batsaikhan et al. 2014). Large mining projects are being developed in southern Mongolia, and there are plans to develop a railway network to transport mining products and connect cities. In fact, construction has already started in some regions. The new railways run through the central distribution ranges of the Mongolian gazelle, the goitered gazelle, and the Asiatic wild ass (Fig. 14.2). If the new railways have barrier effects similar to those of the existing railway, then the ungulates' habitat will be further fragmented into smaller areas. The railways also cross the distribution ranges of the critically endangered wild Bactrian camel (*Camelus ferus*), the reintroduced Przewalski's horse (*Equus ferus przewalskii*), and the saiga antelope (*Saiga tatarica*). Przewalski's horse was once extinct in the wild, but it is currently only ranked as an endangered species by the IUCN Red List because its populations have successfully increased after its reintroduction (King et al. 2016). However, future recovery of the distributions and population numbers of these endangered ungulates may be threatened by the new railways.

Railways also attract human activities to their vicinities that have potential negative impacts on wildlife. Olson et al. (2011) reported that in eastern Mongolia, the Mongolian gazelle density was lower in areas with households than in areas with no households. One detrimental impact of an increased human population is the increase in livestock. Mongolian gazelles have food habits similar to those of domestic sheep and goats (Campos-Arceiz et al. 2004; Yoshihara et al. 2008), and the food habits of the Asiatic wild ass are similar to those of the domestic horse (Taro Sugimoto, personal communication). Thus, food competition between wild ungulates and livestock would likely become more severe if the livestock density increases. Hunting and poaching by humans may also decrease populations of wild ungulates.

Habitat fragmentation has caused regional extinctions of many animal species in various regions of the world (see Chap. 4). Even if the ungulate populations in Mongolia's Gobi-Steppe Ecosystem persist in the smaller habitat fragments produced by railways, their great migrations might disappear in the future. These spectacular migrations need both continuous vast areas and large animal populations. To maintain large wild ungulate populations in Mongolia, it is essential that the accessibility to wide ranges is maintained, due to the high environmental unpredictability in this region.

To maintain accessibility after the construction of new railways, it will be necessary to construct suitable wildlife crossings at regular intervals. Sections without fences in areas with low livestock densities will be also effective. In addition, maintaining the ecosystem in good condition is also necessary, for example, by avoiding land-use changes to farmland, human residential areas, or road networks and by avoiding land degradation due to overgrazing by livestock. Maintaining good conditions of the ecosystem across wide ranges will be effective for the conservation of both wild ungulates and biodiversity in the ecosystem.

Acknowledgements This study was funded by the Japan Ministry of Education, Culture, Sports, Science, and Technology's Grants-in-Aid for Scientific Research 14405039, 18255002, 20255001, 24510326, 25220201, 15K06931.

References

Batsaikhan, N., Buuveibaatar, B., Chimed, B., Enkhtuya, O., Galbrakh, D., Ganbaatar, O., et al. (2014). Conserving the world's finest grassland amidst ambitious national development. *Conservation Biology, 28,* 1736–1739.

Batsaikhan, N., Samiya, R., Shar, S., & King, S. R. B. (2010). *A field guide to the mammals of Mongolia.* London: Zoological Society of London.

Berger, J. (2004). The last mile: How to sustain long-distance migration in mammals. *Conservation Biology, 18,* 320–331.

Campos-Arceiz, A., Takatsuki, S., & Lhagvasuren, B. (2004). Food overlap between Mongolian gazelles and livestock in Omnogobi, southern Mongolia. *Ecological Research, 19,* 455–460.

Fernandez-Gimenez, M. E., Batkhishig, B., & Batbuyan, B. (2012). Cross-boundary and cross-level dynamics increase vulnerability to severe winter disasters (*dzud*) in Mongolia. *Global Environmental Change-Human and Policy Dimensions, 22,* 836–851.

Fernandez-Gimenez, M. E., Batkhishig, B., Batbuyan, B., & Ulambayar, T. (2015). Lessons from the *dzud*: Community-based rangeland management increases the adaptive capacity of Mongolian herders to winter disasters. *World Development, 68,* 48–65.

Imai, S., Ito, T. Y., Kinugasa, T., Shinoda, M., Tsunekawa, A., & Lhagvasuren, B. (2017). Effects of spatiotemporal heterogeneity of forage availability on annual range size of Mongolian gazelles. *Journal of Zoology, 301,* 133–140.

Ito, T. Y., Miura, N., Lhagvasuren, B., Enkhbileg, D., Takatsuki, S., Tsunekawa, A., et al. (2005). Preliminary evidence of a barrier effect of a railroad on the migration of Mongolian gazelles. *Conservation Biology, 19,* 945–948.

Ito, T. Y., Miura, N., Lhagvasuren, B., Enkhbileg, D., Takatsuki, S., Tsunekawa, A., et al. (2006). Satellite tracking of Mongolian gazelles (*Procapra gutturosa*) and habitat shifts in their seasonal ranges. *Journal of Zoology, 269,* 291–298.

Ito, T. Y., Okada, A., Buuveibaatar, B., Lhagvasuren, B., Takatsuki, S., & Tsunekawa, A. (2008). One-sided barrier impact of an international railroad on Mongolian gazelles. *Journal of Wildlife Management, 72,* 940–943.

Ito, T. Y., Lhagvasuren, B., Tsunekawa, A., Shinoda, M., Takatsuki, S., Buuveibaatar, B., et al. (2013a). Fragmentation of the habitat of wild ungulates by anthropogenic barriers in Mongolia. *PLoS ONE, 8,* e56995.

Ito, T. Y., Tsuge, M., Lhagvasuren, B., Buuveibaatar, B., Chimeddorj, B., Takatsuki, S., et al. (2013b). Effects of interannual variations in environmental conditions on seasonal range selection by Mongolian gazelles. *Journal of Arid Environments, 91,* 61–68.

Jiang, Z., Takatsuki, S., Gao, Z., & Jin, K. (1998). The present status, ecology and conservation of the Mongolian gazelle, *Procapra gutturosa*: A review. *Mammal Study, 23,* 63–78.

Kaczensky, P., Sheehy, D. P., Johnson, D. E., Walzer, C., Lhkagvasuren, D., & Sheehy, C. M. (2006). *Room to roam? The threat to khulan (wild ass) from human intrusion.* Washington, D. C.: World Bank.

Kaczensky, P., Ganbaatar, O., von Wehrden, H., & Walzer, C. (2008). Resource selection by sympatric wild equids in the Mongolian Gobi. *Journal of Applied Ecology, 45,* 1762–1769.

Kaczensky, P., Ito, T. Y., & Walzer, C. (2010). Satellite telemetry of large mammals in Mongolia: What expectations should we have for collar function? *Wildlife Biology in Practice, 6,* 108–126.

Kaczensky, P., Ganbataar, O., Altansukh, N., Enkhsaikhan, N., Stauffer, C., & Walzer, C. (2011a). The danger of having all your eggs in one basket: Winter crash of the re-introduced Przewalski's horses in the Mongolian Gobi. *PLoS ONE, 6,* e28057.

Kaczensky, P., Kuehn, R., Lhagvasuren, B., Pietsch, S., Yang, W. K., & Walzer, C. (2011b). Connectivity of the Asiatic wild ass population in the Mongolian Gobi. *Biological Conservation, 144,* 920–929.

Kaczensky, P., Lkhagvasuren, B., Pereladova, O., Hemami, M., & Bouskila, A. (2015). *Equus hemionus.* The IUCN Red List of Threatened Species 2015, e.T7951A45171204. doi:10.2305/IUCN.UK.2015-4.RLTS.T7951A45171204.en

King, S. R. B., Boyd, L., Zimmerman, W., & Kendall, B. E. (2016). *Equus ferus.* The IUCN Red List of Threatened Species 2016, e.T41763A97204950.

Leimgruber, P., McShea, W. J., Brookes, C. J., Bolor-Erdene, L., Wemmer, C., & Larson, C. (2001). Spatial patterns in relative primary productivity and gazelle migration in the eastern steppes of Mongolia. *Biological Conservation, 102,* 205–212.

Lhagvasuren, B., & Milner-Gulland, E. J. (1997). The status and management of the Mongolian gazelle *Procapra gutturosa* population. *Oryx, 31,* 127–134.

Luo, Z., Liu, B., Liu, S., Jiang, Z., & Halbrook, R. S. (2014). Influences of human and livestock density on winter habitat selection of Mongolian gazelle (*Procapra gutturosa*). *Zoological Science, 31,* 20–30.

Mallon, D. P. (2008a). *Gazelle subgutturosa.* The IUCN Red List of Threatened Species 2008, e.T8976A12945246. doi:10.2305/IUCN.UK.2008.RLTS.T8976A12945246.en/

Mallon, D. P. (2008b). *Procapra gutturosa.* The IUCN Red List of Threatened Species 2008, e.T18232A7858611. doi:10.2305/IUCN.UK.2008.RLTS.T18232A7858611.en

Mallon, D. P., & Jiang, Z. (2009). Grazers on the plains: Challenges and prospects for large herbivores in central Asia. *Journal of Applied Ecology, 46,* 516–519.

Morinaga, Y., Tian, S. F., & Shinoda, M. (2003). Winter snow anomaly and atmospheric circulation in Mongolia. *International Journal of Climatology, 23,* 1627–1636.

Mueller, T., & Fagan, W. F. (2008). Search and navigation in dynamic environments: From individual behaviors to population distributions. *Oikos, 117,* 654–664.

Mueller, T., Olson, K. A., Dressler, G., Leimgruber, P., Fuller, T. K., Nicolson, C., et al. (2011). How landscape dynamics link individual- to population-level movement patterns: A multispecies comparison of ungulate relocation data. *Global Ecology and Biogeography, 20,* 683–694.

Mueller, T., Olson, K. A., Fuller, T. K., Schaller, G. B., Murray, M. G., & Leimgruber, P. (2008). In search of forage: Predicting dynamic habitats of Mongolian gazelles using satellite-based estimates of vegetation productivity. *Journal of Applied Ecology, 45,* 649–658.

Nandintsetseg, B., & Shinoda, M. (2011). Seasonal change of soil moisture in Mongolia: Its climatology and modelling. *International Journal of Climatology, 31,* 1143–1152.

National Statistical Office of Mongolia. (2016). Mongolian statistical information service. http://www.1212.mn/en/contents/stats/contents_stat_fld_tree_html.jsp

Okada, A., Ito, T. Y., Buuveibaatar, B., Lhagvasuren, B., & Tsunekawa, A. (2012). Genetic structure of Mongolian gazelle (*Procapra gutturosa*): the effect of railroad and demographic change. *Mongolian Journal of Biological Sciences, 10,* 59–66.

Okada, A., Ito, T. Y., Buuveibaatar, B., Lhagvasuren, B., & Tsunekawa, A. (2015). Genetic structure in Mongolian gazelles based on mitochondrial and microsatellite markers. *Mammalian Biology, 80,* 303–311.

Olson, K. (2012). *Wildlife crossing options along existing and planned Mongolian railway corridors.* Washington D.C.: World Bank.

Olson, K. A., Fuller, T. K., Mueller, T., Murray, M. G., Nicolson, C., Odonkhuu, D., et al. (2010). Annual movements of Mongolian gazelles: Nomads in the eastern steppe. *Journal of Arid Environments, 74,* 1435–1442.

Olson, K. A., Mueller, T., Kerby, J. T., Bolortsetseg, S., Leimgruber, P., Nicolson, C. R., et al. (2011). Death by a thousand huts? Effects of household presence on density and distribution of Mongolian gazelles. *Conservation Letters, 4,* 304–312.

Olson, K. A., Mueller, T., Bolortsetseg, S., Leimgruber, P., Fagan, W. F., & Fuller, T. K. (2009a). A mega-herd of more than 200,000 Mongolian gazelles *Procapra gutturosa*: a consequence of habitat quality. *Oryx, 43*, 149.

Olson, K. A., Mueller, T., Leimgruber, P., Nicolson, C., Fuller, T. K., Bolortsetseg, S., et al. (2009b). Fences impede long-distance Mongolian gazelle (*Procapra gutturosa*) movements in drought-stricken landscapes. *Mongolian Journal of Biological Sciences, 7*, 45–50.

Reading, R. P., Mix, H. M., Lhagvasuren, B., Feh, C., Kane, D. P., Dulamtseren, S., et al. (2001). Status and distribution of khulan (*Equus hemionus*) in Mongolia. *Journal of Zoology, 254*, 381–389.

Sasaki, T., Okayasu, T., Jamsran, U., & Takeuchi, K. (2008). Threshold changes in vegetation along a grazing gradient in Mongolian rangelands. *Journal of Ecology, 96*, 145–154.

Tachiiri, K., & Shinoda, M. (2012). Quantitative risk assessment for future meteorological disasters: Reduced livestock mortality in Mongolia. *Climatic Change, 113*, 867–882.

Teitelbaum, C. S., Fagan, W. F., Fleming, C. H., Dressler, G., Calabrese, J. M., Leimgruber, P., et al. (2015). How far to go? Determinants of migration distance in land mammals. *Ecology Letters, 18*, 545–552.

Vandandorj, S., Gantsetseg, B., & Boldgiv, B. (2015). Spatial and temporal variability in vegetation cover of Mongolia and its implications. *Journal of Arid Land, 7*, 450–461.

Yoshihara, Y., Ito, T. Y., Lhagvasuren, B., & Takatsuki, S. (2008). A comparison of food resources used by Mongolian gazelles and sympatric livestock in three areas in Mongolia. *Journal of Arid Environments, 72*, 48–55.

Yu, F., Price, K. P., Ellis, J., Feddema, J. J., & Shi, P. (2004). Interannual variations of the grassland boundaries bordering the eastern edges of the Gobi Desert in central Asia. *International Journal of Remote Sensing, 25*, 327–346.

Chapter 15
Railway Ecology—Experiences and Examples in the Czech Republic

Z. Keken and T. Kušta

Abstract The range of direct and indirect effects of railway transport on animals, plants, ecological processes and the actual ecosystems vary considerably. Railway transport operations and infrastructure building lead to environmental pollution, loss or conversion of habitats, landscape fragmentation and, last but not least, to animal mortality caused by collisions with passing trains. The impact of railways is determined by the nature of railway infrastructure, which is not as significant in the Czech Republic as road infrastructure, yet it is one of the densest in Europe. An important feature is relatively low electrification (about 33% of the lines) and the length of multi-track lines (about 20%). In the coming years, we can expect massive investments in revitalization, optimization and modernization of the railways in the Czech Republic, and eventually their electrification. To connect the crucial trans-European lines and all regions it will be necessary to complete the basic network of high-speed railways. Based on these facts we can say that the significance of railway ecology in the Czech Republic will grow with the amount of investment activities implemented in the railway network. In the past, similar development took place with road infrastructure, and therefore there is an opportunity to learn from it. To mitigate the direct effects of railways on wildlife, on the basis of previous experience in the Czech Republic we recommend working primarily with management measures. These are both in terms of wildlife management and the management of habitats in the area of transport infrastructure.

Keywords Habitat fragmentation · Train accident · Traffic mortality · Wildlife-train collision · Wildlife-vehicle collision

Z. Keken (✉)
Department of Applied Ecology, Faculty of Environmental Sciences, Czech University of Life Sciences Prague, Kamýcká 129, 165 21 Praha 6, Suchdol, Czech Republic
e-mail: keken@knc.czu.cz

T. Kušta
Department of Game Management and Wildlife Biology, Faculty of Forestry and Wood Sciences, Czech University of Life Sciences Prague, Kamýcká 129, 165 21 Praha 6, Suchdol, Czech Republic

Introduction

Railway transport is more environmentally sustainable than road transport (Operational programme Transport 2014) and therefore more preferable for investment from EU structural funds (Partnership Agreement with the Czech Republic 2014). Therefore, we can expect that major construction activities, such as revitalization of train tracks, optimization, and the actual expansion of railway infrastructure, will continue on railway networks across the EU member states in future decades, also in connection with fulfilling objectives in combating climate change by developing low-carbon transport systems (European Commission 2011).

One of the major problems of nature and landscape protection at the turn of the 21st century is the fragmentation of natural habitats by settlements and transport infrastructure (Kušta et al. 2017). Railway transport is involved in this phenomenon as well (Anděl et al. 2010a). Railways can be a significant barrier to the migration of large mammals (Kušta et al. 2014b), especially high-speed railways, which in Europe are usually enclosed, thereby preventing large mammals from crossing them (Groot and Hazebroek 1996). Noise barriers along railway corridors also significantly contribute to the fragmentation of the landscape, creating an impenetrable barrier. Other factors negatively affecting migration near railway infrastructure include noise and light pollution, but also a higher level of human activities, for example maintenance (Dussault et al. 2006; Jaren et al. 1991) or development of residential and commercial activities.

To show the degree of landscape fragmentation, the method of expressing Unfragmented Area by Traffic (UAT) is very often used (Illmann et al. 2000; Gawlak 2001; Binot-Hafke et al. 2002; Anděl et al. 2005; Anděl et al. 2010b). UAT is the part of the landscape which simultaneously fulfils two conditions: (1) it is bounded either by roads with an annual average daily traffic volume higher than 1000 vehicles/day, or by multi-track railways; and (2) it has an area greater than or equal to 100 km^2 (Anděl et al. 2010a). In the context of railway transport, the problem when expressing UAT is mainly seen with multi-track railways; however, research carried out in recent years suggests that even single-track railways with lower traffic intensity can be problematic in terms of railway ecology (Kušta et al. 2014b), and therefore more attention should be paid to them.

The key task is the integration of landscape fragmentation issues and their resulting effects into decision-making processes at all levels, from international and national concepts up to actual investment projects (Fig. 15.1).

Railway Network in the Czech Republic

Railway transport in the Czech Republic has its origins in the first third of the 19th century. The Czechoslovak Republic took over the network at its inception in 1918, after the collapse of the Austro-Hungarian Empire. The dominant owner, builder

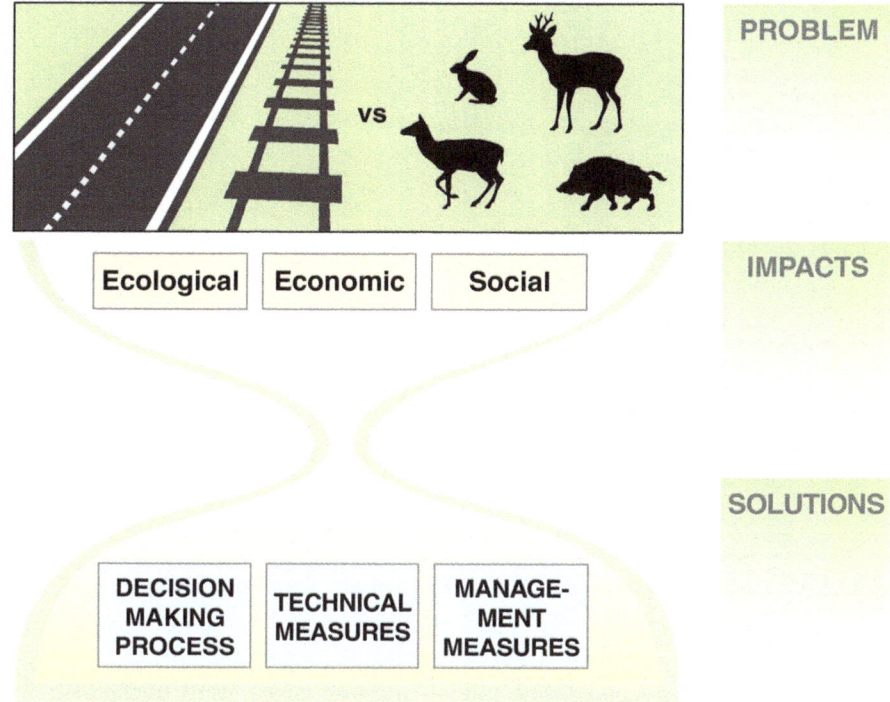

Fig. 15.1 Road and railway and wildlife, overview of implication and general way of solution

and operator of railway lines in the Czech Republic throughout history has been the state. Currently, the state owns the majority of railway lines in the Czech Republic, represented by the state organization Railway Infrastructure Administration. Czech Railways are the largest national carrier.

The railway network in the Czech Republic is relatively inefficient by European standards, being characterized by its low technical level (insufficient ground speed and frequent drops in speed, low capacity, insufficient interoperability and insufficient standards for freight). The reasons for this include the fact that much of the infrastructure is obsolete, with previously neglected maintenance and high operational costs.

The total length of railway lines in the Czech Republic is 9566 km (CSO 2015). About 33% (3153 km) of the length of constructed lines are electrified, and 20% (1858 km) of lines are two tracks. Only 0.38% (36 km) of lines are three and multi-track. In terms of maximum speed, about 38% of tracks are in the category <60 km/h, 26% of tracks in the category 60–80 km/h, 22% of tracks in the category 80–120 km/h and 14% of tracks in the category 120–160 km/h (Fig. 15.2).

In 2014, 176.05 million passengers were transported by railway in the Czech Republic. The average transport distance was approximately 44.3 km. The average transport distance is increasing: in 1995 it was only 35.2 km. In terms of the

Fig. 15.2 Categorization of railways in the Czech Republic (location in Europe in *inset*) by maximal speed

amount of transported goods, 91.6 million tons of goods were transported by railway with an average transport distance of 159.2 km in 2014 (CSO 2015).

In 2014, 104 serious accidents happened on the railway in the Czech Republic, resulting in 29 fatalities and 60 people seriously injured. Property damage was 151.7 million CZK (CSO 2015).

Railway Versus Road Infrastructure in the Czech Republic

The problem of animal mortality, often discussed in relation to road transport, is only marginally known in the case of railways. There are only a few studies that quantify the impact of railways on the mortality of wildlife (e.g. Andersen et al. 1991; Child 1983; Kušta et al. 2011; Kušta et al. 2014a, b; Paquet and Callaghan 1996; Wells et al. 1999).

Although railway infrastructure in the Czech Republic is significantly less extensive compared to road infrastructure (Fig. 15.3), it is one of the densest in Europe. The overall density of railways is 0.12 km/km^2, while road density is approximately six times higher (0.7 km/km^2) (for comparison, the density of railways in Poland is 0.071 km/km^2, in Italy 0.065 km/km^2, in Sweden 0.025 km/km^2, and the average in the EU is 0.053 km/km^2). In the Czech Republic in many cases, the main railway corridors are parallel with the main roads.

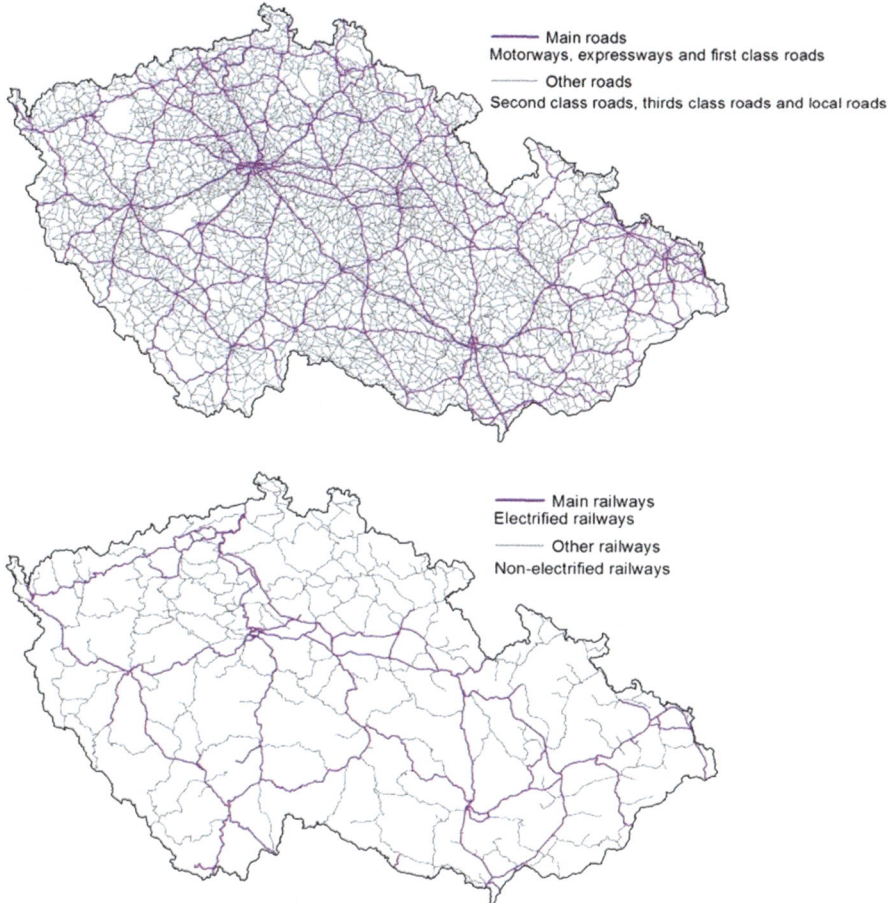

Main roads
Motorways, expressways and first class roads

Other roads
Second class roads, thirds class roads and local roads

Main railways
Electrified railways

Other railways
Non-electrified railways

Fig. 15.3 Road versus railway infrastructure in the Czech Republic

Future Perspectives for Railway Infrastructure Development

To define good practice within Railway Ecology, as well as Road Ecology, it is necessary to put the acquired knowledge and experience into context with the planned strategic development of transport infrastructure. This is because a summary measure of impact affecting both the environment and public health will be generated from the nature of future development directions and extension as well.

Based on White Paper on transport (2011), The European Commission adopted a roadmap of 40 concrete initiatives for the next decade to build a competitive transport system that will increase mobility, remove major barriers in key areas and fuel growth and employment. At the same time, the proposals will dramatically reduce Europe's dependence on imported oil and cut carbon emissions in transport by 60% by 2050. According to the general requirements of this roadmap, planned railway development in the EU member states can be described as follows:

- Implement structural changes that would enable railways to compete effectively and take a significantly higher proportion of freight and passengers over medium and long distances.
- By 2030, 30% of road freight transport over 300 km should be transferred to other means such as railway or ship transportation, and by 2050 it should be more than 50%.
- Finish the European high-speed railway network by 2050. Triple the length of the existing high-speed railway networks by 2030 and maintain a dense railway network in all EU member states. By 2050, most passengers over medium distance should travel by railway.
- By 2050, link all core network airports (T-ENT) with the railway network, preferably high-speed; ensure that all major seaports are connected to the railway freight transport and, where appropriate, to inland waterways.

Meeting these goals will require massive development of the railway infrastructure across EU member states, both in towns and their immediate surroundings, as well as in the open countryside. In the case of the Czech Republic, various kinds of investments will be needed, from revitalization, optimization and modernization of railways to retrofitting the existing railways with modern security systems, and eventually their electrification. Next, we need to take into account the completion of a basic network of railway lines that will provide connection to the crucial trans-European lines, linking all regions, connection of Prague international airport to the long-distance railway network, and solutions for major urban and suburban linkages (Operational programme Transport 2014).

Railway Ecology Experience in the Czech Republic

Obtaining data on train accidents with wild animals is very difficult. One reason is probably that these accidents very rarely involve human injuries or property damage (Steiner et al. 2014). It is common knowledge that many animals are killed due to railway transport (Kušta et al. 2014b).

In the Czech Republic there is no database where accidents between trains and animals are recorded. This contrasts with road accidents, where the reported accidents are registered by the Police of the Czech Republic. In case of railways some data is obtained from employees of Czech Railways, and it is often the only way to get more detailed information (in particular the exact time and location of the accident). A site visit will only give data on the location of the accident, not time or other information. Also, it is not easy to find the carcasses near the track because they can be thrown into the surrounding vegetation. Most of the studies addressing this topic struggled with the issue of obtaining input on wildlife-train collision

(WTC) (e.g. Becker and Grauvogel 1991; Jaren et al. 1991; Muzzi and Bisset 1990).

The first known study in the Czech Republic dealing with the influence of railways on animals was the study by Havlín (1987), which focused on animal mortality caused by railways in the agrarian landscape. A 314-km long railway line in the South Moravia Region was studied between 1981 and 1986. In that period, 91 birds (19 species) and 149 mammals (11 species) were killed. The most common species were European hare (*Lepus europaeus*) and common pheasant (*Phasianus colchicus*).

Another well-known study in the Czech Republic was carried out on the track between Trhový Štěpánov and Benešov u Prahy (Jankovský and Čech 2001). The track is 33 km long and it crosses many very different habitats and therefore enables a more comprehensive view of the impacts of railways. The first research on this track was carried out in the winter of 1999–2000. The research was based on field trips and analysis of skeletal findings of animals killed by train. The results show that Leporidae were hit in 32% of cases, Even-toed ungulates in 22% [in the vast majority of cases, roe deer (*Capreolus capreolus*)], Carnivora in 18%, Birds in 10%, Insectivora in 4% and Reptiles in 2%. The findings of physical remnants were tied to those sections where the track does not form a significant barrier height, whether by an embankment or a cutting. At areas significantly banked and often vegetated by shrubs, many pheasant carcasses were found. In May 2006, repeated research was carried out on the track between Trhový Štěpánov and Benešov u Prahy which, among other findings, found an increase in the mortality of roe deer (Jankovský and Čech 2008).

Kušta et al. (2014b) published the results of the monitoring of roe deer deaths in 2009 on four train tracks (regional as well as local): (1) Plzeň–Horažďovice; (2) Bělčice–Závišín; (3) Obrataň–Jindřichův Hradec; (4) Dobrá Voda u Pelhřimova–Hříběcí. The research showed that collisions of trains with roe deer occur most frequently in winter. The results of this study show that when evaluating animal mortality the specific context should be taken into account (location, climatic conditions, animal abundance in the location, its biology and ethology, railway transport intensity, etc.). For example, on the monitored track Dobrá Voda u Pelhřimova–Hříběcí, collisions with roe deer occurred most often in summer months, while on the track Jedlová–Chřibská in winter months. Both tracks lie at approximately the same altitude (650 m a.s.l.) and about the same number of trains use them weekly (132 and 197).

Interestingly, railway accidents most often occur in winter, while accidents on roads occur less frequently in winter (Hothorn et al. 2012; Kušta et al. 2014a; Kušta et al. 2017; Pokorný 2006). Therefore railway transport can influence animal behaviour in a different way than road transport. It is necessary to look for the reasons why accidents happen so often in this season. In the Czech Republic it may be due to the attractive food availability, which is in the immediate vicinity of the tracks in the winter (wild raspberry bushes, etc.). This explanation is also supported in other studies (Bowman et al. 2010; Rea et al. 2010; Marchand 1996). In the winter months the cleared track can also act as a migration corridor. On sunny days, it can serve as a resting place for animals (Seiler 2005).

Kušta et al. (2014b) found a connection between the number of accidents and the frequency of passing trains, a pattern also reported in other studies (Danks and Porter 2010; Hussain et al. 2007; Seiler 2004). On the other hand, there was no correlation between accidents and the abundance of wildlife around the monitored tracks. The study also found that there were more accidents on tracks with more meadows and fields around them than on tracks with prevailing forest. However, significant differences in accidents in different types of habitats (meadows, forests and scrubland) were not statistically confirmed. According to the information from employees of Czech Railways, the most frequent collisions with animals occur in open flat sections of the track, especially at dawn, dusk or at night. This is probably caused by grazing ungulates in these locations and their greater physical activity at dusk and dawn (Kušta et al. 2017), which can correlate with temporal patterns of wildlife-vehicle collision (WVC) on roads (Kušta et al. 2017).

Good Practice and Recommendation for Future Action

Decision Making Process

For individual investment plans, the permissible level of environmental impact is already pre-defined within a strategic level by SEA procedure (Strategic Environmental Assessment) and at a project level by EIA procedure (Environmental Impact Assessment). In the future, the results of the evaluation need to be considered, especially through the implementation of EIA Follow-up (Environmental Impact Assessment post-project analysis). To ensure more accurate predictions and more effective mitigation measures, the existing railways need to compare the predicted impacts with the ones that occurred in reality and define recommendations for future authorization procedures based on these comparisons.

Technical Measures

Technical measures against accidents with animals on railway tracks are not very frequent. For example, measures for reducing drive-through speed of trains or warning devices are placed (Becker and Grauvogel 1991; Muzzi and Bisset 1990), noise warning devices are installed (Boscagli 1985; Child 1983), the ground speed of trains is reduced (Becker and Grauvogel 1991; Child 1983), snow along the railway is cleaned up, enabling animals to escape from the corridor when snow is deep (Child 1983; Child et al. 1991), carcasses are removed from around the tracks (Gibeau and Heuer 1996), spilled cereal grains and other food sources are cleaned from around the tracks and occurrence of this spillage is prevented (Gibeau and Heuer 1996; Wells et al. 1999), or fences are built along the tracks (Boscagli 1985).

In the Czech Republic, the above mentioned technical measures are used quite rarely. The most common technical solution is the application of odour repellents near railway tracks. However, their effectiveness is questionable. Their effectiveness depends on many factors. Among them is, for example, seasonal food availability and its nutritional value (Van Beest et al. 2010), weather, frequency of product application and its concentration (Diaz-Varela et al. 2011; Knapp et al. 2004; Wagner and Nolte 2001). The reason for the ineffectiveness of repellents can often be the excessive disturbance of animals (e.g. by hunting), which then changes their behaviour (Benhaiem et al. 2008) making them less responsive to the repellents. The most probable reason for their frequent ineffectiveness is the fact that the environment around tracks is highly food-attractive for the animals and the applied odours are not able to discourage the animals from being near the railway. Other studies (Elmeros et al. 2011; Schlageter and Haag-Wackernagel 2012a, b) came to similar conclusions, i.e. ineffectiveness of odour repellents in places with attractive food for animals. In contrast, in the vicinity of roads, odour repellents appear to be an efficient tool to reduce the number of animals killed in collisions with cars (e.g. Kušta et al. 2015).

Management Measures

On the other hand, we recommend paying close attention to management measures, whose implementation is often the cheapest and can be most effective option of solving collisions.

We should study in detail the ecological and behavioural needs of animals that are threatened by railway transport. It is very important to identify critical areas where collisions occur most frequently and establish databases on animal mortality on railways with the information on location, time and external collision factors.

Fencing is an option for sites with high collision rates, and also in the case of main corridors with significantly high capacity, but it requires maintenance and has a major impact on the fragmentation of habitats and populations. For the wider countryside, the most important management measures include the management of wildlife in such numbers that match the capacity of the environment (applies to the species that can be managed by hunting). Animals must not be overpopulated in a given area in any case. Their high numbers do not only cause frequent damage to agricultural land, crops and forests, but also cause significantly frequent WTC as well as WVC.

Furthermore, hunting must not interfere with the appropriate social structure (i.e. age and sex) of the population and a correct sex ratio must be retained. In this context, the heads of the herd and their young need to be protected. This prevents excessive migration of animals due to excessive stress and the need to find new locations.

In the Czech countryside it is common practice to give supplementary feed to game in the winter season. Knowledge of this feeding can be used to reduce migration and therefore even collisions (Kušta et al. 2017). It is important to place feeding devices near animal resting areas so that they do not have to migrate in search of food in winter, or even have to search for it intensively. Feeding devices should never be placed near railways, which would enable animals to concentrate near them.

Also, around railway tracks where the animals occur in higher numbers, very intense hunting should be practiced within the hunting plan in the given territory. On the contrary, hunting should be reduced in areas where there is no danger of WTC or wildlife-vehicle collision (WVC). In game management this corresponds to establishing so-called quiet zones: there is no hunting, peace and quiet is ensured and food is supplied intensively. Suitable conditions for animals should also be created in these locations (e.g. planting appropriate woody plants, planting feeding field, etc.). Animals are then focused into them and protected. They can thus be detained in a certain place, which will prevent their migration in search of food across railway tracks. The results of the study Kušta et al. (2017) show that when animals have suitable resting, cover and feeding conditions, they limit their physical activity to a minimum.

In the vicinity of transport infrastructure (road and railway) there are often hiking trails, country lanes, or the area is used for recreation (walking, dog walking, etc.). Thus, resting animals are often disturbed and forced to cross a road or railway in their escape, which may result in a collision with a passing vehicle or train. Peace and quiet must be guaranteed, especially in winter, which is extremely energy-intensive for animals. For this reason, it is undesirable to let dogs move freely when walking in the countryside in this period, especially around roads and railways; in the Czech Republic there are very often successional areas with sparse vegetation or woody plants growing outside the forest, and the animals rest there (Keken et al. 2016).

A practical tool for reducing collisions with animals also appears to be removing vegetation from the immediate vicinity of the tracks (deciduous trees, herbs, brambles) and trimming browsing trees that are very attractive to animals from the food point of view (Jaren et al. 1991) and also growing unattractive plant species along railway tracks (Jaren et al. 1991; Gundersen and Andreassen 1998; Lavsund and Sandegren 1991). Some studies demonstrated a reduction of animal mortality due to a decrease in food supply along the tracks, for example a 56% reduction in collisions on a railway studied in Norway (Jaren et al. 1991). The measure consisted of the removal of vegetation 20–30 m along the train track. However, Sielecki (2000) states that this method is very costly.

Conclusion

Taking into account the worldwide research on this subject, it is obvious that this issue is very topical and requires further thorough investigation. The planned future growth of railway transport associated with an increase in the probability of collisions with animals must be treated as a nationwide issue with the involvement of the largest possible number of affected organizations, including government administration, businesses farming on adjacent land, hunters, conservationists and others.

Although it is very difficult and expensive to prevent animal mortality due to collisions with road and railway transport, it is important to find solutions which will at least minimize these accidents. Above all, we propose: evaluating the risks associated with the environment more efficiently and accurately in the planned projects within the decision-making process; providing feedback to individual licensing procedures; paying close attention to management measures; and focussing on technical measures in the riskiest sections.

Acknowledgements This chapter was supported by the Ministry of Agriculture of the Czech Republic Grant No. QJ1220314. Further, authors would like to thank the Faculty of Environmental Sciences and the Faculty of Forestry and Wood Sciences at the Czech University of Life Sciences Prague for support during our research career.

References

Anděl, P., Gorčicová, I., Hlaváč, V., Miko, L., & Andělová, H. (2005). *Assessment of landscape fragmentation caused by traffic*. Liberec: Evernia.

Anděl, P., Miná̌riková, T., & Andreas, M. (Eds.). (2010a). *Protection of landscape connectivity for large mammals*. Liberec: Evernia.

Anděl, P., Petržílka, L., & Gorčicová, I. (2010b). *Indikátory fragmentace krajiny*. Liberec: Evernia.

Andersen, R., Wiseth, B., Pedersen, P. H., & Jaren, J. (1991). Moose–train collisions: Effects of environmental conditions. *Alces, 27*, 79–84.

Becker, E. F., & Grauvogel, C. A. (1991). Relationship of reduced train speed on moose-train collisions in Alaska. *Alces, 27*, 161–168.

Benhaiem, S., Delon, M., Lourtet, B., Cargnelutti, B., Aulagnier, S., Hewison, A. J. M., et al. (2008). Hunting increases vigilance levels in roe deer and modifies feeding site selection. *Animal Behaviour, 76*, 611–618.

Binot-Hafke, M., Illmann, J., Schäfer, H. J., & Wolf, D. (Eds.). (2002). *Nature data 2002*. Bonn: Bundesamt für Natruschutz.

Boscagli, G. (1985). Wolves, bears and highways in Italy: a short communication. In J. M. Bernard (Eds.), *Proceedings highways and wildlife relationships* (pp. 237–239). Ministsre de løÐquipement, du Logement, de løAmTnagement du Territoire et des Transports, Colmar, France.

Bowman, J., Ray, J. C., Magoun, A. J., Johnson, D. S., & Dawson, F. N. (2010). Roads, logging, and the large-mammal community of an eastern Canadian boreal forest. *Canadian Journal of Zoology, 88*, 454–467.

Child, K. N. (1983). Railways and moose in the central interior of BC: A recurrent management problem. *Alces, 19,* 118–135.

Child, K. N., Barry, S. P., & Aitken, D. A. (1991). Moose mortality on highways and railways in British Columbia. *Alces, 27,* 41–49.

Czech Statistical Office. (2015). *Statistical databases and registers of Czech Republic.* Available in internet. https://www.czso.cz/csu/czso/doprava_a_spoje (cited August 15, 2016).

Danks, Z. D., & Porter, F. P. (2010). Temporal, spatial, and landscape habitat characteristics of moose-vehicle collisions in western Maine. *The Journal of Wildlife Management, 74,* 1229–1241.

Diaz-Varela, E. R., Vazquez-Gonzalez, I., Marey-Perez, M. F., & Alverez-Lopez, C. J. (2011). B57 assessing methods of mitigating wildlife-vehicle collisions by accident characterization and spatial analysis. *Transport Research Part D, 16,* 281–287.

Dussault, C., Poulin, M., Courtois, R., & Ouellet, J. P. (2006). Temporal and spatial distribution of moose-vehicle accidents in the Laurentides Wildlife Reserve, Quebec, Canada. *Wildlife Biology, 12,* 415–425.

Elmeros, M., Winbladh, J. K., Andersen, P. N., Madsen, A. B., & Christensen, J. T. (2011). Effectiveness of odour repellents on red deer (*Cervus elaphus*) and roe deer (*Capreolus capreolus*): a field test. *European Journal of Wildlife Research, 57,* 1223–1226.

European Commission. (2011). *Europe 2020: A European strategy for smart, sustainable, and inclusive growth.* Brussels.

Gawlak, C. (2001). Unzerschnittene verkehrsarme Räume in Deutschland 1999. *Natur und Landschaft, 76,* 481–484.

Gibeau, M. L., & Heuer, K. (1996). Effects of transportation corridors on large carnivores in the Bow River Valley, Alberta. In G. L. Evink, P. Garrett, D. Zeigler, & J. Berry (Eds.), *Proceedings of the transportation related wildlife mortality seminar* (pp. 58–96). Tallahassee, Florida: Florida Department of Transportation.

Groot, B., & Hazebroek, E. (1996). Ungulate traffic collisions in Europe. *Conservation Biology, 10,* 1059–1067.

Gundersen, H., & Andreassen, H. P. (1998). The risk of moose (*Alces alces*) collision: A predictive logistic model for moose-train accidents. *Wildlife Biology, 4,* 103–110.

Havlín, J. (1987). On the importance of railway lines for the life of avifauna in agrocoenoses. *Folia Zoologica, 36,* 345–358.

Hothorn, T., Brandl, R., & Müller, J. (2012). Large-scale model-based assessment of deer–vehicle collision risk. *PLoS ONE, 7,* e29510.

Hussain, A., Armstrong, J. B., Brown, D. B., & Hogland, J. (2007). Land-use pattern, urbanization, and deer-vehicle collisions in Alabama. *Human-Wildlife Conflicts, 1,* 89–96.

Illmann, J., Lehrke, S., & Schäfer, H. J. (Eds.). (2000). *Nature data 1999* (266 p.). Bonn: Bundesamt für Naturschutz.

Jankovský, M., & Čech, M. (2001). Railway tracks as the place for collisions with animals. *Živa, 1,* 39–40.

Jankovský, M., & Čech, M. (2008). Železniční doprava a fauna v okolí tratě. *Živa, 3,* 136.

Jaren, V. R., Anderson, R., Ulleberg, M., Pederson, P., & Wiseth, B. (1991). Moose–train collisions: The effects of vegetation removal with a cost-benefit analysis. *Alces, 27,* 93–99.

Keken, Z., Kušta, T., Langer, P., & Skaloš, J. (2016). Landscape structural changes between 1950 and 2012 and their role in wildlife–vehicle collisions in the Czech Republic. *Land Use Policy, 59,* 543–556.

Knapp, K. K., Yi, X., Oakasa, T., Thimm, W., Hudson, E., & Rathmann, C. (2004). *Deer–Vehicle crash countermeasure toolbox: A decision and choice resource* (263 p.). Wisconsin: Midwest Regional University Transportation Center.

Kušta, T., Holá, M., Keken, Z., Ježek, M., Zíka, T., & Hart, V. (2014a). Deer on the railway lines: Spatiotemporal trends in mortality patterns of roe deer. *Turkish Journal of Zoology, 38,* 479–485.

Kušta, T., Ježek, M., & Keken, Z. (2011). Mortality of large mammals on railway tracks. *Scientia Agriculturae Bohemica, 42,* 12–18.

Kušta, T., Keken, Z., Barták, V., Holá, M., Ježek, M., Hart, V., et al. (2014b). The mortality patterns of wildlife-vehicle collisions in the Czech Republic. *North-Western Journal of Zoology, 10,* 393–399.

Kušta, T., Keken, Z., Ježek, M., Holá, M., & Šmíd, P. (2017). The effect of traffic intensity and animal activity on probability of ungulate-vehicle collisions in the Czech Republic. *Safety Science, 91,* 105–113.

Kušta, T., Keken, Z., Ježek, M., & Kůta, Z. (2015). Effectiveness and costs of odor repellents in wildlife–vehicle collisions: A case study in Central Bohemia, Czech Republic. *Transport Research Part D, 38,* 1–5.

Lavsund, S., & Sandegren, F. (1991). Moose-vehicle relations in Sweden: A review. *Alces, 27,* 118–126.

Marchand, P. J. (1996). *Life in the cold: An introduction to winter ecology.* Hanover, NH: University Press of New England.

Muzzi, P. D., & Bisset, A. R. (1990). Effectiveness of ultrasonic wildlife warning devices to reduce moose fatalities along railway corridors. *Alces, 26,* 37–43.

Operational programme Transport. (2014). *Operational programme transport for the programming period 2014–2020.* Ministry of Transport of the Czech Republic, Prague.

Paquet, P., Callaghan, C. (1996). Effects of linear developments on winter movements of gray wolves in the Bow River Valley of Banff National Park, Alberta. In G. L. Evink, P. Garrett, D. Ziegler, & J. Berry (Eds.), *Proceedings of the transportation related wildlife mortality seminar* (113 pp.). Florida: Florida Department of Transportation, Tallahassee.

Partnership Agreement. (2014). *Partnership Agreement for the programming period 2014–2020.* Ministry of Regional Development of the Czech Republic, Prague.

Pokorný, B. (2006). Roe deer-vehicle collisions in Slovenia: Situation, mitigation strategy and countermeasures. *Veterinary Archives, 76,* 177–187.

Rea, R. V., Child, K. N., Spata, D. P., & MacDonald, D. (2010). Road and rail side vegetation management implications of habitat use by moose relative to brush cutting season. *Environmental Management, 46,* 101–109.

Schlageter, A., & Haag-Wackernagel, D. (2012a). Evaluation of an odor repellent for protecting crops from wild boar damage. *Journal of Pest Science, 85,* 209–215.

Schlageter, A., & Haag-Wackernagel, D. (2012b). A gustatory repellent for protection of agricultural land from wild boar damage: An investigation on effectiveness. *Journal of Agricultural Science, 4,* 61–68.

Seiler, A. (2004). Trends and spatial patterns in ungulate-vehicle collisions in Sweden. *Wildlife Biology, 10,* 301–313.

Seiler, A. (2005). Predicting locations of moose–vehicle collisions in Sweden. *Journal of Applied Ecology, 42,* 371–382.

Sielecki, L. E. (2000). *WARS wildlife accident reporting system 1999 annual report (1995–1999 synopsis).* Victoria, BC: Environmental Management Section, Ministry of Transportation and Highways.

Steiner, W., Leisch, F., & Hackländer, K. (2014). A review on the temporal pattern of deer-vehicle accidents: Impact of seasonal, diurnal and lunar effects in cervids. *Accident Analysis and Prevention, 66,* 168–181.

Van Beest, F. M., Mysterud, A., Loe, L. E., & Milner, J. M. (2010). Forage quantity, quality and depletion as scale-dependent mechanism driving habitat selection of a large browsing herbivore. *Journal of Animal Ecology, 79,* 910–922.

Wagner, K. K., & Nolte, D. L. (2001). Comparison of active ingredients and delivery systems in deer repellents. *Wildlife Society Bulletin, 29,* 322–330.

Wells, P., Woods, J. G., Bridegewater, G., Morrison, H. (1999). Wildlife mortalities on railways: Monitoring, methods and mitigation strategies. In G. L. Evink, P. Garrett, & D. Ziegler (Eds.), *Proceedings of the Third International Conference on Wildlife Ecology and Transportation* (pp. 85–88). FL-ER-73-99, Florida Department of Transportation, Florida.

White Paper on Transport. (2011). *White paper—Roadmap to a single European transport area—Towards a competitive and resource efficient transport system.* Brussels.

Chapter 16
Ecological Roles of Railway Verges in Anthropogenic Landscapes: A Synthesis of Five Case Studies in Northern France

J.-C. Vandevelde and C. Penone

Abstract This chapter presents the results of several studies realised between 2010 and 2014 on the French railway network. We assessed several potential ecological roles for this network on various taxa: (1) habitat, (2) corridor/longitudinal connectivity along railways and (3) barrier/transversal connectivity across railways. Our results show that railway verges, contrary to common belief, can have positive effects in anthropogenic landscapes for several taxa and communities. They can be habitats for semi-natural grassland plants, bats and orthopteran. They provide functional connectivity for some plants, but do not seem to increase it for highly mobile invasive species in urban landscapes. Railways also seem to be weaker barriers than roads for one butterfly species. We conclude by proposing a change in management practices, stressing that extensive management coupled with small-scale revegetation processes, either artificial or natural, may help increase the positive effects of railway verges and counteract the negative large-scale effects of urbanisation.

Keywords Habitat · Corridor · Barrier · Agricultural landscape · Urban landscape · Bat · Orthoptera · Plant · Butterfly · Railway verges management

Introduction

Changes in land-use are important drivers of biodiversity loss (MEA 2005). Linear infrastructures such as roads and railways play a particular role in this process by altering, artificializing and fragmenting landscapes (see previous chapters).

J.-C. Vandevelde (✉)
UMR 7204 (MNHN-CNRS-UPMC), Centre d'Ecologie et de Sciences
de la Conservation (CESCO), 43 rue Buffon, 75005 Paris, France
e-mail: vandeveldejeanchristophe@gmail.com

C. Penone
Institute of Plant Sciences, University of Bern,
Altenbergrain 21, 3013 Bern, Switzerland

© The Author(s) 2017   261
L. Borda-de-Água et al. (eds.), *Railway Ecology*,
DOI 10.1007/978-3-319-57496-7_16

Nonetheless, linear infrastructure verges can also have positive roles for a large number of taxa (Hodkinson and Thompson 1997; Merriam and Lanoue 1990). They can be substitution habitats for grassland plants and insects (Saarinen et al. 2005; Wehling and Diekmann 2009), hence contribute to the conservation of indigenous flora (O'Farrell and Milton 2006) and fauna (Ries et al. 2001). They can also ensure structural and functional connectivity when they penetrate artificial areas such as dense urban areas or agricultural-intensive landscapes (Tikka et al. 2001).

Among studies examining the effects of linear infrastructure verges, very few focused on railway verges, the majority concerning roadside verges (Forman et al. 2002). Indeed, compared to roadside verges, railway verges occupy smaller surfaces. Nevertheless, railway verges have at least two peculiarities worth considering when compared to roadside verges. First, their potential positive effects on biodiversity may be significant in human-dominated areas such as intensive agricultural or urban landscapes due to their greater width margins and lower traffic intensity. Second, railway lines of many European countries are managed by one unique manager, allowing realistic biodiversity-friendly management (see discussion).

This chapter presents the results of five studies realised between 2009 and 2014 on the French railway network. We assessed three potential ecological roles for this network: (1) habitat, (2) corridor/longitudinal connectivity along the railway and (3) barrier/transversal connectivity across the railway.

General Context and Methodology

We conducted five different studies in the Paris region, France (Fig. 16.1). Although densely populated, with 20% of the national population living in just 2% of the nation's land area, and with a spreading urbanisation, the Paris region is still predominantly rural. Intensive farming covers 50% of the regional territory, woods and natural land more than 25%, and urban areas and transport infrastructures cover about 25%.

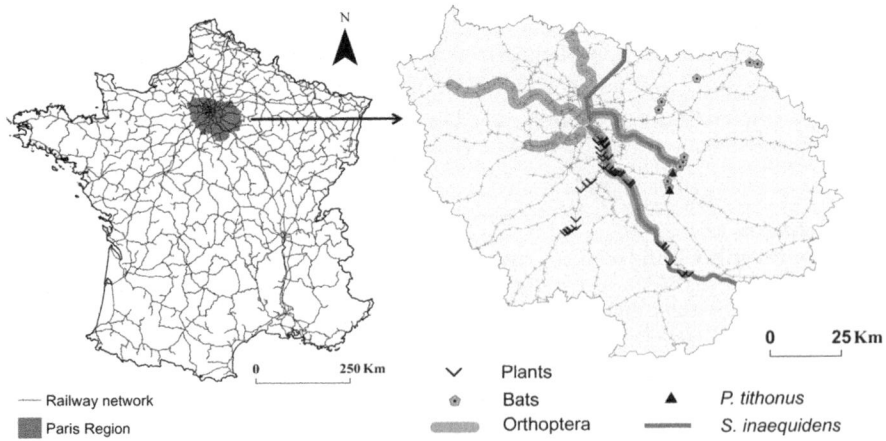

Fig. 16.1 Location of study sites in France (*left panel*) and in the Paris region (*right panel*)

Vegetation of railway verges in France

The studies described in this chapter were conducted along railway verges. Railway verges are all the green areas that are beside the railway tracks. The total area occupied by railway verges in France is estimated by the railway owner, the *SNCF Réseau* company, at 50 000 hectares. The vegetation type on these green spaces is dictated by security reasons. Management is thus necessary but different in different parts of the railway concerned:

- Tracks and lanes: the goal is the absence of vegetation to avoid platform/ infrastructure deterioration. Total weed control is recommended once a year but depends on the age of the infrastructure. For instance, a recent track, better drained, is less sensitive to plant colonization and is usually not treated for 10 year;

  The verges can be separated in two main zones:

- The first zone is adjacent to the tracks and has variable width depending on the slope and surrounding landscape. Here the objective is to have an herbaceous vegetation to allow visibility and limit fire problems. These semi-natural grasslands are ideally managed by annual or multi-annual mowing depending on fire risks and on potential recolonisation by woody species;
- In the second zone, the desired vegetation can be higher depending on their configuration: if the verge is lower compared to the tracks (fill section), the trees and shrubs are allowed to grow higher compared to a configuration where the verges are higher than the tracks (cut section). Management here does not need to be intensive and can be done every 3–5 years depending on the slope and the vegetation in the surrounding landscape.

We examined two types of railways: recently built high-speed railway lines (1994 and 2007) and long-established railway lines (built in the nineteenth century). In the high-speed lines, we studied verges from 10 to 20 m-wide, composed of herbaceous species and shrubs. No vegetation management plan existed for these verges. Clearcutting of trees and mowing were applied on a case-by-case basis when presenting threats for train security, e.g. risk of tree falling on overhead lines. The long-established lines traverse a landscape structured by different degrees of urbanisation, as they cross numerous cities and towns from the centre of Paris towards the limits of the region. The vegetation found along the borders of the train tracks is mainly spontaneous as the verges have not been planted or sown since their construction and is principally interrupted by railway stations and overpasses.

Roles of Railway Verges: Habitat, Corridor, Barrier

Railways as Habitats

Linear infrastructures can be substitution habitats for various taxa (O'Farrell and Milton 2006; Ries et al. 2001), however, hardly any studies focused properly on railways. We thus decided to examine the role of railway verges as a habitat for two groups: common bats and Tettigoniidae (Orthoptera) (Vandevelde et al. 2014; Penone et al. 2013).

Bats Along Railway Verges in an Intensive Agricultural Landscape

Studies have shown that bat activity and species richness decrease when approaching a motorway (Berthinussen and Altringham 2012). Some studies stressed particular negative effects, such as road casualties (Lesinski et al. 2011). Others found that roads formed strong barriers to the movements of bats within the landscape (Bach et al. 2004; Kerth and Melber 2009; Abbott et al. 2012).

Yet, very few studies have focused on the potential positive effects of linear infrastructures such as railway verges on bats. Some bat species fly along such features when commuting from roosts to foraging areas because these linear elements could constitute commuting paths away from predators and wind (Limpens and Kapteyn 1991). In addition, some species forage regularly along linear elements (Verboom and Huitema 1997). Indeed, foraging activity is facilitated close to these features because of greater abundance of some preys (Verboom and Spoelstra 1999). These contrasting behaviours among species may be linked to their specific foraging ecology (Kerth and Melber 2009).

We hypothesized that railway verges impacted negatively some species, particularly gleaner species generally linked with forest habitat (such as species from the *Myotis* genus). These species could perceive railway verges as an inadequate habitat and in addition could be impacted by the fragmentation effect of these verges. Conversely, aerial hawking species that generally forage in more open habitats, such as species of the genus *Pipistrellus*, *Nyctalus* or *Eptesicus*, could benefit from the edge effect of railway verges, using them as a foraging/commuting habitat (Vandevelde et al. 2014).

To test this hypothesis, we examined the potential use by bats of railway verges crossing woodland patches within an agricultural matrix as foraging/commuting habitats. We tested whether (i) at a large scale (national level), railways lines were globally a preferred foraging/commuting habitat for these different common bats species, and (ii) at a local scale (landscape level), woodland-railway verges had an effect on bat activity compared to other habitat types like woodland-field verges, woodland habitats and field habitats. At local scale, we also tested the influence of landscape composition on bat activity over habitat types (Vandevelde et al. 2014).

We identified ten similar sites via aerial photographs and field visits: each site consisted of a railway portion of 600 m to 2 km long fragmenting a woodland patch within a farmland matrix. We used two distinct sets of data for bat calls. 'Local-scale' data from railway sites were sampled in the summer of 2010 following a similar protocol to the one designed for the French Bat Monitoring Programme (named Vigie-Chiro), from which we used 'large-scale' (national) data collected from 2006 to 2011. For both local- and large-scale data sets, bat calls were detected. Each point was monitored twice: once in the period between June 15 and

July 31, during which females give birth and feed their offspring; second, in the period between August 15 and September 31, during which young are flying and individuals are expected to be less dependent on their reproductive roost. Habitat characteristics were analysed in a radius of 100 m around the sampled point, using a detailed habitat classification (Kerbiriou et al. 2010). Species calls were identified through spectrogram analyses.

At large scale, we found that activity of common bats in railway verges was of the same order of magnitude as in other habitats, except for aquatic habitats, which are known as key habitats for numerous European bat species (Russo and Jones 2003; Nicholls and Racey 2006). Furthermore, we found that activity was even greater in railway verges than in some other habitats for two aerial species: *N. leislerii* and *P. pipistrellus*. For *P. pipistrellus*, a very generalist species (Russ and Montgomery 2002), activity in railway verges was greater than in seven other habitats. Overall, bat activity among railway verges was not inferior to the activity in highly modified habitats such as continuous artificial surfaces, discontinuous artificial surfaces and arable land (confirmed for 4 of the 5 taxa studied).

At local scale, we found a negative effect of railway areas on the *Myotis* spp. for the two sample periods and a positive effect on *Nyctalus* spp. for the post-reproductive period. When focusing the analysis on the site type, we detected few obvious significant differences in bat activity between the verges of railways and the three other site types (field verges, plain field and plain wood). The significant difference for *Nyctalus* spp. during the post-reproductive period (lower activity in woods than on railway verges) may seem contradictory with the positive effect of surrounding areas of wood for these species at the same period. The foraging ecology of these aerial hawking bats and their habitat requirements could explain these results: *Nyctalus* spp. are species that forage mainly along edges, thus more in forest verges than in the heart of forests, but otherwise inhabit the interior of the forests, thus the amount of woodland habitat at the landscape level would have also an overall positive effect.

It must be noted that except for the *Myotis* spp. group in the reproductive period, we detected no difference between the verges of railways and those of fields. This result seems to indicate a similar functioning role for railway verges on bats as that of other linear verges, particularly on common aerial hawking bats (Verboom and Huitema 1997).

Our results suggest that the presence of railway verges does not influence significantly the foraging/commuting activity of common bats, except for species like those from the *Myotis* group, a group including mostly gleaner species, less generalist species (Dietz et al. 2007) and some threatened species (Temple and Terry 2009).

In several cases (for *P. pipistrellus* at global scale and for *Nyctalus* spp. at local scale), railway verges even seem to provide a significant habitat in intensive agricultural landscapes where semi-natural elements, in particular linear elements like hedgerows, tend to disappear. Railways, along with other artificial linear infrastructures, may thus contribute to maintaining common bat populations in such landscapes.

Orthoptera Along Railway Verges in an Urban Landscape

Railway verges could potentially provide a habitat for Orthoptera in an urban landscape because the majority of these insects in Europe are grassland species and some occur at roadsides (Theuerkauf and Rouys 2006). We thus studied Tettigonidae along railway verges in urban landscapes using an original method. Because Orthoptera produce mating calls that are species-specific, it is possible to collect large standardised datasets using sound recording devices. To sample a large number of landscapes, we thus recorded Orthoptera sounds from running trains along five long-established railway lines in the Paris region (Penone et al. 2013). We did not sample the entire community of singing Orthoptera but rather focused on insects that produced powerful stridulations, which were not masked by the noise produced by trains.

   We detected 2003 individuals of 10 species of bush-crickets from the Tettigoniidae family along 209 km of railway verges (Penone et al. 2013). This represented 59% of Tettigoniidae species known to exist in the region (Voisin 2003) and included specialist species (e.g. *Conocephalus fuscus*). These results suggest that railway verges play a habitat role for these arthropods even for specialist species that are more heavily affected by anthropogenic changes. This was further strengthened by the fact that Orthoptera abundance was significantly lower when the trains crossed areas with paved (non vegetated) railway verges (Penone et al. 2013). We also detected a significant negative effect of urbanisation on species richness, abundance and community specialization, over a broad set of spatial scales (Penone et al. 2013). This effect was partly mitigated when railway verges were vegetated, suggesting that these zones are not only habitats but could even act as refugia for Orthoptera in an urban landscape. Therefore, if extensively managed (Marini et al. 2008), railway verges could play a role in the conservation of biodiversity in urban areas. More generally, we believe that acoustic surveys are a simple and cost-effective method, allowing long-term and large-scale surveys that can help advance railway ecology studies.

Railways as Corridors in an Urban Landscape (Longitudinal Connectivity)

Habitat fragmentation is a major perturbation that affects biodiversity at local, regional and global scales (Fahrig 2003; Krauss et al. 2010). Fragmentation has effects not only at the population level but also on taxonomic and functional composition of communities and on species interactions (Fahrig 2003; Fischer and Lindenmayer 2007; Krauss et al. 2010). Fragmented landscapes, including urban ones, are characterised by small habitat patches isolated by a matrix that filters species according to their mobility and dispersal abilities (Ewers and Didham 2006; Ricketts 2001). In this context, functional connectivity is crucial to maintain populations, communities and potentially, ecosystem services (Taylor et al. 1993). As shown before in this book, railways are mainly considered to be barriers for wildlife or corridors for invasive species. However, railways are linear and almost continuous patches of vegetation that penetrate into dense urban areas, ensuring a structural connectivity. Despite their potential as corridors, railway verges functional connectivity has been poorly studied (but see Tikka et al. 2001), and most studies have focussed on invasive species (Hansen and Clevenger 2005).

As mentioned previously, semi-natural grasslands are one of the vegetation types in railway verges, especially in urban contexts, where the space is precious and the verges have a reduced width. In general, semi-natural grasslands are important habitats for conservation because they support plants, invertebrates and vertebrates, including threatened ones, and provide ecosystem services (Ridding et al. 2015). However, surfaces are decreasing all around Europe, therefore it is crucial to conserve them and ensure connectivity between the patches to enhance gene and species flows. In our study (Penone et al. 2012), we examined whether these areas could provide functional connectivity for semi-natural grassland plants in urban landscapes.

Using Structural Breaks to Detect Functional Connectivity Along Railways

If linear structures provide functional connectivity an interruption of structure should result in an interruption of the function. The structural connectivity of railway verges is regularly interrupted by spatial breaks, such as overpasses and stations. We hypothesised that if railway verges favour connectivity, spatially connected communities within railway verges should be more similar than disconnected communities (H1). We tested this assumption by comparing floristic dissimilarities between sites located along railway verges that were either connected or separated by a railway break (Penone et al. 2012). In order to take into account differences among plant species, we considered species traits linked to dispersal abilities. These analyses were conducted on 71 plots located on two different railway lines going from Paris to the south of the Paris region (Fig. 16.1).

We found a significant effect of railway breaks on most species and dispersal traits consistent to our hypothesis (H1): connected communities were more similar than disconnected ones, suggesting a potential corridor role of railway verges (Penone et al. 2012). However, this effect was influenced by species mobility (summarized in Fig. 16.2). Highly mobile species (e.g. wind pollinated) were not affected by railway breaks, likely because they can disperse over long distances in any direction. Similarly, we did not find a significant effect of railway breaks on poorly mobile species (e.g. gravity-dispersed seeds). Their dispersal is more related to very local factors (e.g. slope) and the effect of breaks might be only visible over multiple generations. The most interesting result was found for moderately mobile species (e.g. wind-dispersed seeds), for which we found a significant effect of railway breaks. For these species railways seemed to provide functional

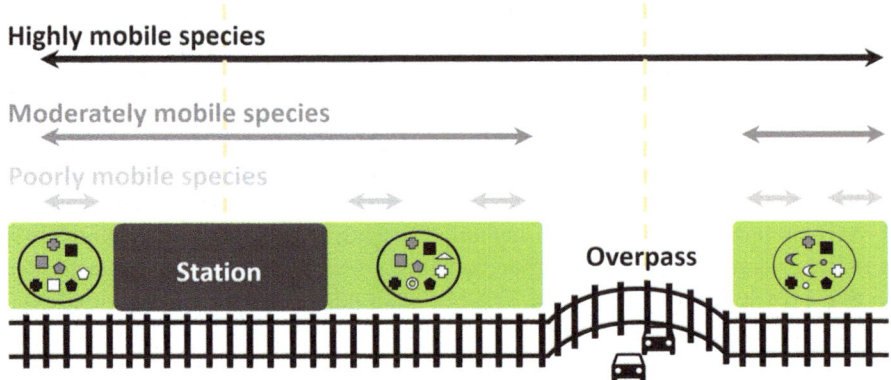

Fig. 16.2 Functional connectivity of railway verges for semi-natural plant species. *Green patches* represent railway verges. *Arrows* represent propagules movement. *Circles with symbols* represent plant communities with highly (*black*), moderately (*dark grey*) and poorly (*light grey*) mobile species

connectivity, at least between two railway breaks, which can represent large distances (e.g. more than 5 km in our study) (Penone et al. 2012).

Interestingly, we detected an effect of overpasses but not of stations on community dissimilarities (Penone et al. 2012 and Fig. 16.2). The slipstream of trains, which possibly carry wind-dispersed seeds (as shown for cars, Von der Lippe and Kowarik 2007), can be interrupted by air turbulence or crosswind when the trains traverse overpasses (Ernst 1998). Since those have very narrow paved verges, the seeds are more likely to be lost. The airflow may also be interrupted when trains slow down or stop at stations. But at stations, seeds may then be picked up again by air turbulence or wind. More generally, using structural connectivity to detect functional connectivity allows assessing corridor efficiency for several species groups. The precision is lower compared to genetic methods, which are the best suited methods for this purpose, but our methods are cost and time effective. Therefore, we believe that is an interesting and valuable cost-effective approach that can be widely applied to railway research to assess corridor efficiency in the future.

Using Urbanisation Effects to Detect Functional Connectivity Along Railways

Plant community composition is the result of stochastic effects as well as different biotic and abiotic processes, or filters, that select species with suitable functional traits from a species pool through dispersal and survival (Roy and de Blois 2006; Weiher and Keddy 1999; Williams et al. 2009). The effects of these processes are particularly strong in urban landscapes (Stenhouse 2004). The abiotic constraints in cities include higher temperatures and nitrogen deposition (McDonnell et al. 1997; Pellissier et al. 2008). Fragmentation and isolation in urban landscapes also have an effect on communities through habitat size reduction and dispersal limitation. All these effects should be reflected in communities' trait composition (Weiher and Keddy 1999). While the first abiotic effects (temperature, nitrogen), should not be influenced by connectivity between patches within urban landscapes, the seconds (fragmentation effects) should be attenuated if communities are connected. We used this assumption as a supplementary and indirect approach to assess functional connectivity in railway verges. We hypothesised that if railway verges favour connectivity this should attenuate the effect of fragmentation on plant communities (H2). We thus examined the effects of urbanisation on species richness, diversity and trait composition in railway verges and compared our results with the patterns described in literature. This study was conducted using the same data as the previous analysis, i.e. 71 plots on two lines.

We did not detect urbanisation effects on plant communities due to fragmentation (Penone et al. 2012). Indeed, we did not find significant relationships between urban cover and plant richness, diversity or dispersal traits, while most studies detected positive or negative effects, depending on urbanisation intensity (McKinney 2008). In contrast we found an effect of abiotic conditions linked to the urban landscape, which were consistent with the literature. For instance, we detected a positive effect of urbanisation on the abundance of species adapted to higher temperatures and nitrogen levels (Penone et al. 2012). These results matched our hypothesis (H2) and thus further suggested that railway verges provide functional connectivity in urban landscapes for semi-natural grassland species.

What About Exotic and Invasive Species?

Our study showed that railway verges can play a positive role for semi-natural grassland plants in urban landscapes by connecting communities and potentially improving gene and species flow between communities (Penone et al. 2012). However, as showed in Chap. 5, railway verges can also be corridors for invasive species. In order to explore this topic in an urban context, we performed the same analyses described above (railway breaks and urbanisation effects) separately for exotic and invasive species (Penone et al. 2012). We also focussed on one particular invasive species, *Senecio inaequidens*, which is one of Europe's fastest plant invaders and is suspected to spread along railways (Lachmuth et al. 2010).

We detected a significant effect of railway breaks on community similarity in terms of exotic but not invasive species (the latter being a subset of the exotic). As for native plants, this effect was only observed for overpasses and not railway stations (Penone et al. 2012). In addition, we measured the similarity between plots in the two different railway lines, accounting for urbanisation effects. We also found a difference between the two lines for exotic but not for invasive species (Penone et al. 2012). According to our hypothesis (H1: if railways verges provide functional connectivity, structurally connected communities should be more similar than disconnected ones) this result suggests that railway verges might play a connectivity role for exotic but not for invasive species in urban landscapes.

We also analysed the effects of urbanisation on the frequency of exotic and invasive plants to test our hypothesis H2 (urbanisation effects should be attenuated if communities are connected). We found the well-known positive urbanisation effect on exotic and invasive species, i.e., increased species frequency in highly urbanised areas (Penone et al. 2012). This result possibly suggests that the urbanisation effect was not attenuated by railways connectivity and thus railways did not provide additional functional connectivity for exotic and invasive species in urban landscapes.

To further confirm these results we performed a genetic analysis on *S. inaequidens* in the same region (Blanchet et al. 2014). This perennial herb presents a self-incompatible reproductive system and is pollinated by insects; the effective fruit production is high, with 75% viable achenes, which are well dispersed by wind and animals (Lopez-Garcia and Maillet 2005). This species was introduced in the south of France (Mazamet) in 1936 and then observed in Gare d'Austerlitz (a Parisian railway station) in 1990 (Henry and Maurin 1999). Consequently, we hypothesised that if railway verges play a corridor role for *S. inaequidens*, genetic diversity of this species should be structured along railways. We collected 450 individuals from 15 populations regularly on 3 different railway lines connecting the centre of Paris to its periphery. Given that the species was first seen in Parisian railway stations and assuming that railways act as corridors for *S. inaequidens* in the Parisian region, we expected a significant genetic differentiation among the main departure stations within Paris centre and among the three railway lines. We also expected significant isolation by distance (i.e., positive correlation between genetic distance and geographical distance) if the main central stations act as sources for other populations in the surroundings.

Our results did not confirm these scenarios: despite a clear differentiation among the three departure stations of the centre of Paris and a genetic cline along one of the lines, the differentiation among lines was low. Similarly, we did not detect significant isolation by distance among populations within lines, supporting the absence of population structure along railways (Blanchet et al. 2014). One explanation could lie in multiple sources of introduction of *S. inaequidens* from other European regions (e.g. Belgium or Germany). This species has high dispersal ability and is also present in other urban habitats (e.g. wastelands), which could also contribute to the high gene flow detected. In sum, the impact of railways as corridors on the genetic structure of *S. inaequidens* was not confirmed in our study, especially outside the centre of Paris where the deep interconnection with other corridors such as motorways in the suburban matrix may have a significant impact by reducing genetic differentiation among railway lines.

All the results that we put together converge and tend to suggest that in a highly urbanized and anthropogenic region, railway verges provide functional connectivity for exotic species but not for invasive species. This is inconsistent with many other studies on other transport networks (roadsides, highways), and should then been taken with care (but see Kalwij et al. 2008 for a similar result). Although railway verges did not appear as clear corridors for invasive species, trains might constitute an efficient transportation vector for seeds and contribute to their spread outside cities. Furthermore, railway verges are habitats and thus potential sources for these species for other habitats. The positive urbanisation effect that we found on these species, suggests that their presence is more linked to the urban context (sources, abiotic conditions) than railways connectivity. More studies are needed to assess railways connectivity in other landscapes and for a greater number of species.

Railways as Barriers in an Agricultural Landscape (Transversal Connectivity)

Linear infrastructures such as railways can be barriers to the movements of individuals and, hence, may have strong impacts on populations (see Chap. 4). We tested if this effect existed or if there was transversal connectivity between the two sides of a high-speed railway for the common butterfly *Pyronia tithonus* (Vandevelde et al. 2012). We hypothesised that *P. tithonus* individuals, if captured on the railway verge on one side of the railway and translocated to the other side in a similar habitat, will not be able to return to their original site because of the alleged barrier effect of the railroad. This hypothesis is based on empirical evidence that *P. tithonus* shows homing behaviour when displaced: when given a choice between a familiar and an unfamiliar habitat, these butterflies prefer to return to their familiar habitat (Conradt et al. 2001).

Pyronia tithonus is a European Satyrinae (Nymphalidae) occurring in several grassland habitats, usually in association with scrub or hedges. Mean life-span of adults ranges from 3.5 to 8 days (Brakefield 1987). *P. tithonus* produces only one generation per year and uses several grasses as host plants (Merckx and Van Dyck 2002).

We captured, marked and displaced 152 individuals in two different locations. One-third of the butterflies were released at a capture plot, one-third on the other side of the railway (in a similar habitat) and one-third on the same side but 100 m away from the capture plot. Many (31%) butterflies crossed the railway, showing homing behaviour. There were even a slightly greater number of exchanges between the two release plots across the railway (12) than between the two release plots along the railway (8); the opposite of what we would expect if railways acted as a barrier to butterfly movements.

Even if based on a small sample, our results tend to show that contrary to wide busy roads (Munguira and Thomas 1992), high-speed railways, which are narrower and have infrequent traffic, are not barriers to Gatekeeper movements. These results also suggest that in a predominantly agricultural-intensive context, railways with large grassy verges could be substitution habitats for grassland butterflies, with no effect of the 20-m-wide-ballast-tracks. Furthermore, it suggests that passing trains do not have an impact on *P. tithonus* movements. However, the mortality risk from train traffic still has to be properly assessed, as well as the possible effects of the railway structure (tracks on cuttings vs. those on embankments; wide vs. narrow verges) on movements of individuals.

Conclusion

In a context of rapid biodiversity decline, our results (see synthesis Table 16.1) suggest that railway verges should be considered by managers and engineers not only as a side aspect of the railroad, but also as elements having a potential role in maintaining common biodiversity, especially in human-dominated landscapes such as urban areas or agricultural systems. We found that railway verges could play habitat and corridor roles and that railways were not barriers for butterfly species. The total length of the railway network worldwide is more than one million kilometres (CIA 2008), and its verges represent important green areas (e.g. nearly 0.1% of the French surface). Therefore, the latter could represent an interesting topic for biodiversity conservation in human-dominated landscapes, as already suggested (Jarošík et al. 2011; Le Viol et al. 2008). More precisely, railway verges may contribute, along with private gardens and green spaces (Vergnes et al. 2012), to improve biodiversity in urban environments and could be potentially included in urban green network planning. Moreover, railway verges plants are mainly herbs and grasses that can be of interest for semi-natural grassland conservation, as already suggested for road verges (Cousins and Lindborg 2008).

In order to contribute to the maintenance of biodiversity, the management of these verges is thus crucial, as shown by previous studies (e.g. Noordijk et al. 2010). The following management rules should be considered: (1) during the construction phase, the revegetation process should try to maximise plant diversity to increase interactions with higher trophic levels as well as ecosystem functioning, (2) the relationship between the network of these linear features and the overall landscape connectivity should be considered (Karim and Mallik 2008; Boughey et al. 2011), (3) finally, during the exploitation phase of the railroad, management decisions could take biodiversity into account through a move from clearcutting to more extensive management (e.g. selected cutting and late-mowing), enabling the maintenance of some linear structures and allowing the reproduction of several plant and insect species.

Table 16.1 The three roles of railway verges studied on selected taxa

	Habitat	Longitudinal connectivity	Transversal connectivity
Bats	Generalist species: yes Specialist species: no	nt	nt
Orthoptera	Yes	nt	nt
Plants	Yes	Yes	nt
Invasive species (*S. inaequidens*)	Yes	Not detected	Possible
Common butterfly (*P. tithonus*)	nt	nt	Yes

nt not tested

Overall, we suggest that extensive management coupled with small-scale revegetation processes, either artificial or natural, may help counteract the negative large-scale effects of urbanisation and of the homogenisation of agricultural landscapes.

Acknowledgements We deeply thank the editors for inviting us to participate to this stimulating book on railway ecology. We would also like to thank in particular L. Borda de Água as well as I. Takehito for their constructive and helpful comments on the submitted version of this chapter.

References

Abbott, I. M., Butler, F., & Harrison, S. (2012). When flyways meet highways—The relative permeability of different motorway crossing sites to functionally diverse bat species. *Landscape and Urban Planning, 106,* 293–302.

Bach, L., Burkhardt, P., & Limpens, H. J. (2004). Tunnels as a possibility to connect bat habitats. *Mammalia, 68,* 411–420.

Berthinussen, A., & Altringham, J. (2012). The effect of a major road on bat activity and diversity. *Journal of Applied Ecology, 49,* 82–89.

Blanchet, É., Penone, C., Maurel, N., Billot, C., Rivallan, R., Risterucci, A.-M., et al. (2014). Multivariate analysis of polyploid data reveals the role of railways in the spread of the invasive South African Ragwort (*Senecio inaequidens*). *Conservation Genetics, 16,* 523–533.

Boughey, K. L., Lake, I. R., Haysom, K. A., & Dolman, P. M. (2011). Improving the biodiversity benefits of hedgerows: How physical characteristics and the proximity of foraging habitat affect the use of linear features by bats. *Biological Conservation, 144,* 1790–1798.

Brakefield, P. M. (1987). Geographical variability in, and temperature effects on, the phenology of *Maniola jurtina* and *Pyronia tithonus* (Lepidoptera, Satyrinae) in England and Wales. *Ecological Entomology, 12,* 139–148.

Central Intelligence Agency. (2008). Transportation. In *The world factbook.*

Conradt, L., Roper, T. J., & Thomas, C. D. (2001). Dispersal behaviour of individuals in metapopulations of two British butterflies. *Oikos, 95,* 416–424.

Cousins, S. A. O., & Lindborg, R. (2008). Remnant grassland habitats as source communities for plant diversification in agricultural landscapes. *Biological Conservation, 141,* 233–240.

Dietz, C. Von, Helversen, O., & Nill, D. (2007). *Handbuch der Fledermäuse Europas und Nordwestafrikas.* Stuttgart: Franckh-Kosmos Verlag.

Ernst, W. H. O. (1998). Invasion, dispersal and ecology of the South African neophyte *Senecio inaequidens* in The Netherlands: From wool alien to railway and road alien. *Acta Botanica Neerlandica, 47,* 131–151.

Ewers, R. M., & Didham, R. K. (2006). Confounding factors in the detection of species responses to habitat fragmentation. *Biological Reviews, 81,* 117–142.

Fahrig, L. (2003). Effects of habitat fragmentation on biodiversity. *Annual Review of Ecology Evolution and Systematics, 34,* 487–515.

Fischer, J., & Lindenmayer, D. B. (2007). Landscape modification and habitat fragmentation: A synthesis. *Global Ecology and Biogeography, 16,* 265–280.

Forman, R. T., Sperling, D., Bissonette, J. A., Clevenger, A. P., Cutshall, C. D., Dale, V. H., et al. (2002). *Road ecology: Science and solutions.* Washington, DC: Island Press.

Hansen, M. J., & Clevenger, A. P. (2005). The influence of disturbance and habitat on the presence of non-native plant species along transport corridors. *Biological Conservation, 125,* 249–259.

Henry, H.-P., & Maurin, H. (1999). Les inventaires du patrimoine naturel en milieu urbain [The inventories of the natural heritage in urban environments]. In B. Lizet, A.-E. Wolf, & J. Celecia (Eds.), *Sauvages dans la ville. De l'inventaire naturaliste à l'écologie urbaine [Wild in the city.*

From naturalist inventories to urban ecology] (pp. 333–355). Paris: Publications scientifiques du Muséum.

Hodkinson, D. J., & Thompson, K. (1997). Plant dispersal: The role of man. *Journal of Applied Ecology, 34,* 1484–1496.

Jarošík, J., Konvička, M., Pyšek, P., Kadlec, T., & Beneš, J. (2011). Conservation in a city: Do the same principles apply to different taxa? *Biological Conservation, 144,* 490–499.

Kalwij, J. M., Milton, S. J., & McGeoch, M. A. (2008). Road verges as invasion corridors? A spatial hierarchical test in an arid ecosystem. *Landscape Ecology, 23,* 439–451.

Karim, M. N., & Mallik, A. U. (2008). Roadside revegetation by native plants: I. Roadside microhabitats, floristic zonation and species traits. *Ecological Engineering, 32,* 222–237.

Kerbiriou, C., Bas, Y., Dufrêne, L., Robert, A., & Julien, J. F. (2010). *Long term trends monitoring of bats, from biodiversity indicator production to species specialization assessment.* Paper presented at the 24th Annual Meeting of the Society for Conservation Biology, Edmonton, Canada.

Kerth, G., & Melber, M. (2009). Species-specific barrier effects of a motorway on the habitat use of two threatened forest-living bat species. *Biological Conservation, 142,* 270–279.

Krauss, J., Bommarco, R., Guardiola, M., Heikkinen, R. K., Helm, A., Kuussaari, M., et al. (2010). Habitat fragmentation causes immediate and time-delayed biodiversity loss at different trophic levels. *Ecology Letters, 13,* 597–605.

Lachmuth, S., Durka, W., & Schurr, F. M. (2010). The making of a rapid plant invader: Genetic diversity and differentiation in the native and invaded range of *Senecio inaequidens. Molecular Ecology, 19,* 3952–3967.

Lesinski, G., Olszewski, A., & Popczyk, B. (2011). Forest roads used by commuting and foraging bats in edge and interior zones. *Polish Journal of Ecology, 59,* 611–616.

Le Viol, I., Julliard, R., de Kerbiriou, C., Redon, L., Carnino, N., Machon, N., et al. (2008). Plant and spider communities benefit differently from the presence of planted hedgerows in highway verges. *Biological Conservation, 141,* 1581–1590.

Limpens, H., & Kapteyn, K. (1991). Bats, their behaviour and linear landscape elements. *Myotis, 29,* 39–48.

Lopez-Garcia, M. C., & Maillet, J. (2005). Biological characteristics of an invasive South African species. *Biological Invasions, 7,* 181–194.

Marini, L., Fontana, P., Scotton, M., & Klimek, S. (2008). Vascular plant and Orthoptera diversity in relation to grassland management and landscape composition in the European Alps. *Journal of Applied Ecology, 45,* 361–370.

McDonnell, M. J., Pickett, S. T. A., Groffman, P., Bohlen, P., Pouyat, R. V., Zipperer, W. C., et al. (1997). Ecosystem processes along an urban-to-rural gradient. *Urban Ecosystems, 1,* 21–36.

McKinney, M. (2008). Effects of urbanization on species richness: A review of plants and animals. *Urban Ecosystems, 11,* 161–176.

MEA. (2005). *Ecosystems and human well-being.* Washington, DC: Island Press.

Merckx, T., & Van Dyck, H. (2002). Interrelations among habitat use, behavior, and flight-related morphology in two cooccurring satyrine butterflies, *Maniola jurtina* and *Pyronia tithonus. Journal of Insect Behavior, 15,* 541–561.

Merriam, G., & Lanoue, A. (1990). Corridor use by small mammals: Field measurement for three experimental types of *Peromyscus leucopus. Landscape Ecology, 4,* 123–131.

Munguira, M. L., & Thomas, J. A. (1992). Use of road verges by butterfly and burnet populations, and the effect of roads on adult dispersal and mortality. *Journal of Applied Ecology, 29,* 316–329.

Nicholls, B., & Racey, P. A. (2006). Habitat selection as a mechanism of resource partitioning in two cryptic bat species *Pipistrellus pipistrellus* and *Pipistrellus pygmaeus. Ecography, 29,* 697–708.

Noordijk, J., Schaffers, A. P., Heijerman, T., Boer, P., Gleichman, M., & Sýkora, K. V. (2010). Effects of vegetation management by mowing on ground-dwelling arthropods. *Ecological Engineering, 36,* 740–750.

O'Farrell, P. J., & Milton, S. J. (2006). Road verge and rangeland plant communities in the southern Karoo: Exploring what influences diversity, dominance and cover. *Biodiversity and Conservation, 15,* 921–938.

Pellissier, V., Roze, F., Aguejdad, R., Quenol, H., & Clergeau, P. (2008). Relationships between soil seed bank, vegetation and soil fertility along an urbanisation gradient. *Applied Vegetation Science, 11,* 325–334.

Penone, C., Kerbiriou, C., Julien, J.-F., Julliard, R., Machon, N., & Le Viol, I. (2013). Urbanisation effect on Orthoptera: Which scale matters? *Insect Conservation and Diversity, 6,* 319–327.

Penone, C., Machon, N., Julliard, R., & Le Viol, I. (2012). Do railway edges provide functional connectivity for plant communities in an urban context? *Biological Conservation, 148,* 126–133.

Ricketts, T. H. (2001). The matrix matters: Effective isolation in fragmented landscapes. *The American Naturalist, 158,* 87–99.

Ridding, L. E., Redhead, J. W., & Pywell, R. F. (2015). Fate of semi-natural grassland in England between 1960 and 2013: A test of national conservation policy. *Global Ecology and Conservation, 4,* 516–525.

Ries, L., Debinski, D. M., & Wieland, M. L. (2001). Conservation value of roadside prairie restoration to butterfly communities. *Conservation Biology, 15,* 401–411.

Roy, V., & de Blois, S. (2006). Using functional traits to assess the role of hedgerow corridors as environmental filters for forest herbs. *Biological Conservation, 130,* 592–603.

Russ, J. M., & Montgomery, W. I. (2002). Habitat associations of bats in Northern Ireland: Implications for conservation. *Biological Conservation, 108,* 49–58.

Russo, D., & Jones, G. (2003). Use of foraging habitats by bats in a Mediterranean area determined by acoustic surveys: Conservation implications. *Ecography, 26,* 197–209.

Saarinen, K., Valtonen, A., Jantunen, J., & Saarnio, S. (2005). Butterflies and diurnal moths along road verges: Does road type affect diversity and abundance? *Biological Conservation, 123,* 403–412.

Stenhouse, R. N. (2004). Fragmentation and internal disturbance of native vegetation reserves in the Perth metropolitan area, Western Australia. *Landscape and Urban Planning, 68,* 389–401.

Taylor, P. D., Fahrig, L., Henein, K., & Merriam, G. (1993). Connectivity is a vital element of landscape structure. *Oikos, 68,* 571–573.

Temple, H. J., & Terry, A. (2009). *The status and distribution of european mammals.* Luxembourg: Office for Official Publications of the European Communities.

Theuerkauf, J., & Rouys, S. (2006). Do Orthoptera need human land use in Central Europe? The role of habitat patch size and linear corridors in the Biaowieza forest, Poland. *Biodiversity and Conservation, 15,* 1497–1508.

Tikka, P. M., Hogmander, H., & Koski, P. S. (2001). Road and railway verges serve as dispersal corridors for grassland plants. *Landscape Ecology, 16,* 659–666.

Vandevelde, J.-C., Bouhours, A., Julien, J.-F., Couvet, D., & Kerbiriou, C. (2014). Activity of European common bats along railway verges. *Ecological Engineering, 64,* 49–56.

Vandevelde, J.-C., Penone, C., & Julliard, R. (2012). High-speed railways are not barriers to *Pyronia tithonus* butterfly movements. *Journal of Insect Conservation, 16,* 801–803.

Verboom, B., & Huitema, H. (1997). The importance of linear landscape elements for the pipistrelle *Pipistrellus pipistrellus* and the serotine bat *Eptesicus serotinus. Landscape Ecology, 12,* 117–125.

Verboom, B., & Spoelstra, K. (1999). Effects of food abundance and wind on the use of tree lines by an insectivorous bat, *Pipistrellus pipistrellus. Canadian Journal of Zoology, 77,* 1393–1401.

Vergnes, A., Le Viol, I., & Clergeau, P. (2012). Green corridors in urban landscapes affect the arthropod communities of domestic gardens. *Biological Conservation, 145,* 171–178.

Voisin, J. F. (2003). *Atlas des Orthoptères et des Mantides de France [Atlas of French Orthoptera and Mantis of France].* Paris: MNHN, Collection Patrimoines Naturels.

Von der Lippe, M., & Kowarik, I. (2007). Long-distance dispersal of plants by vehicles as a driver of plant invasions. *Conservation Biology, 21,* 986–996.

Wehling, S., & Diekmann, M. (2009). Importance of hedgerows as habitat corridors for forest plants in agricultural landscapes. *Biological Conservation, 142,* 2522–2530.

Weiher, E., & Keddy, P. (1999). Assembly rules as general constraints on community composition. In E. Weiher & P. Keddy (Eds.), *Ecological assembly rules* (pp. 251–271). Cambridge: Cambridge University Press.

Williams, N. S. G., Schwartz, M. W., Vesk, P. A., McCarthy, M. A., Hahs, A. K., Clemants, S. E., et al. (2009). A conceptual framework for predicting the effects of urban environments on floras. *Journal of Ecology, 97,* 4–9.

Chapter 17
Wildlife Deterrent Methods for Railways—An Experimental Study

Andreas Seiler and Mattias Olsson

Abstract Since reliable accident statistics and consequent costs have become available, train collisions with wildlife, especially ungulates, have received increasing attention in Sweden. In contrast to collisions on roads, accidents involving wildlife on railways do not entail human injury or death, but can cause substantial train damage and lead to significant delays in railway traffic. Wildlife-train collisions (WTC) are rising in numbers and railways appear as a greater source of ungulate mortality per kilometer than roads. Nevertheless, railways are largely unprotected against wildlife collisions, and mitigation measures that have hitherto been applied to roads are either infeasible or economically unviable for railways. The Swedish Transport Administration is therefore seeking innovative and cost-effective measures for preventing collisions with larger wild animals. In this chapter, we present research on WTC in Sweden that has been used to define the baseline and set up criteria for a new mitigation project. This project aims to develop warning or deterring signals that encourage animals to leave the railway shortly before trains arrive. This will be carried out at several experimental crosswalks for animals along fenced railways where the effect of different signals on animal behaviour can be evaluated. If effective, these deterrent systems could replace fencing and/or crossing structures, and reduce mortality and barrier effects on wildlife. The project was begun in 2015 and will continue for at least 4 years.

Keywords Accident prevention · Animal-deterrent · Crosswalk · Deer-train accidents · Exclusion fences · Hotspots · Wildlife-train collisions · Wildlife-warning

A. Seiler (✉)
Department of Ecology, Swedish University of Agricultural Sciences,
SLU, Riddarhyttan, Sweden
e-mail: andreas.seiler@slu.se

M. Olsson
EnviroPlanning AB, Gothenburg, Sweden
e-mail: mattias.olsson@enviroplanning.se

L. Borda-de-Água et al. (eds.), *Railway Ecology*,
DOI 10.1007/978-3-319-57496-7_17

277

Introduction

Wildlife-train collisions (WTC) on Swedish railways have steadily increased over the past 15 years (Fig. 17.1), totalling about 5,000 reported incidents with moose (*Alces alces*), roe deer (*Capreolus capreolus*), reindeer (*Rangifer tarandus*), and other ungulates (fallow deer, red deer, wild boar) every year (Seiler et al. 2011; Olsson and Seiler 2015). The number of unreported WTCs, as well as the number of collisions with smaller mammals and birds, is unknown. WTCs receive increasing public media attention and their importance is also acknowledged by train operators and the Swedish Transport Administration (Olsson and Seiler 2015). WTC can cause significant disruptions and delays to train traffic; produce considerable repair costs for material damages; and entail further costs related to the retrieval and handling of animal carcasses, the loss of other economic values of

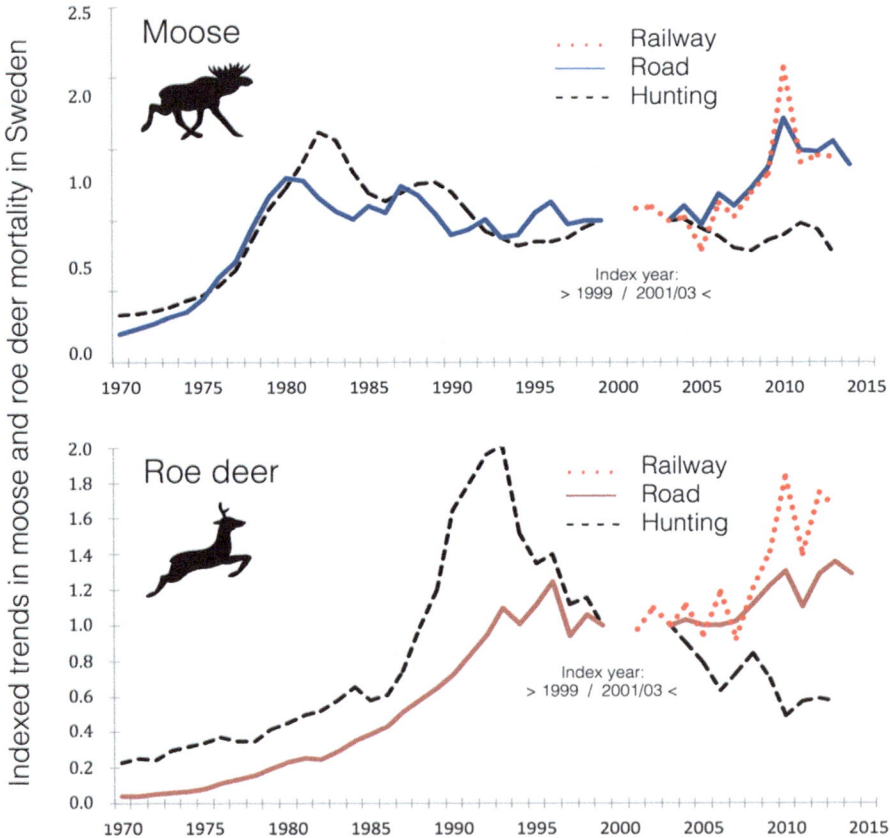

Fig. 17.1 Trends in moose and roe deer reportedly killed on roads, on railways and through hunting. Traffic accounts for about 10–15% of all human-caused mortality, and the proportion is increasing. (updated from Seiler et al. 2011)

wildlife, and the administration of accidents (Child and Stuart 1987; Jaren et al. 1991; Gundersen and Andreassen 1998).

However, since WTCs do not cause human injury or death, and hence are not considered a traffic safety problem, the Swedish railway network is still largely unprotected against collisions with wildlife. Except for a few railway sections in the northern region, where accidents with semi-domestic reindeer lead to expensive reimbursements, fencing has long been regarded as economically unviable for railways, and alternative measures have not been seriously tested. This attitude, however, is changing as train operators upgrade their train systems to modern light-weight multiple-unit trains. These trains are less robust than traditional single train engines and require more expensive repairs after a WTC, sometimes leading to significant delays in railway traffic. The overall socio-economic costs of WTCs in Sweden have recently been estimated at 100,000,000–150,000,000 million Euros per year (Seiler et al. 2014). This is similar to the costs estimated for wildlife-vehicle collisions on roads (250,000,000 Euros per year) (Seiler and Olsson 2015), despite railways comprising less than 2% of the national road network. Thus, the number of collisions with larger wildlife, especially ungulates, per kilometer, is greater on railways than public roads.

Preliminary analyses of the Swedish railway network suggest that there are at least 10 railway sections with very high WTC frequencies where fencing would be economically viable today. On most railway sections, however, fencing still may not be cost-effective. Alternative methods, such as wildlife deterrents or warning systems, need to be developed to keep animals off railways, at least when trains approach. Initial attempts with such wildlife warning systems in other studies have produced promising results (Larsson-Kråik 2005; Werka and Wasilewski 2009; Babińska-Werka et al. 2015; Shimura et al. 2015).

Therefore, the Swedish Transport Administration, in cooperation with Swedish Railways, Enviroplanning AB, and the Swedish University of Agricultural Sciences, initiated a project in 2015 to develop and test methods for deterring wildlife from railways when trains approach. In this chapter, we describe the research that provided the baseline for this unique project and describe its specific settings and objectives.

Background

Wildlife-train collisions (WTC) on Swedish railways were not scientifically studied before 2011 (Seiler et al. 2011) because empirical data was not accessible, and authorities did not perceive WTC as a problem and thus did not finance research. Collisions with semi-domestic reindeer were an exception, probably because rail authorities had to reimburse reindeer owners and therefore maintain detailed statistics (Åhrén and Larsson 1999, 2001). WTCs are reported by train drivers via telephone to a central register. Based on these reports, railway patrols and specially trained hunters visit the accident site to remove the carcasses or take care of the wounded animals. We compiled these reports and analyzed spatial and temporal patterns in the distribution of WTCs. We also surveyed train drivers regarding their experiences with wildlife on railways and wildlife-train collisions. These studies provided the framework for the preparation of the current mitigation project, whose main findings are summarized below. The major objectives were to evaluate the magnitude of the problem, obtain input on potential mitigation measures, analyze spatial and temporal patterns, and identify the hotspots where mitigation might be most urgently required.

Train Driver Survey

In 2010, we conducted a survey with train drivers to map their experiences with WTCs (Seiler et al. 2011). Drivers were asked about how often they observed animals on or near the railway, how often they experienced collisions with ungulates, and how animals typically responded to the oncoming train. Train drivers were also asked for their opinions and ideas on how to reduce WTCs. About 17% of the 1,023 participants took an active part in the survey, and over 65% of the respondents replied that they encountered deer several times a week. Most respondents (91%) had experienced collisions with roe deer (49.3%), moose (23.5%), reindeer (19.4%), and other larger wildlife (7.8%) during the previous year. This matches the overall accident statistics, in which these three species make up over 90% of all reported cases (Fig. 17.2). Collisions with wild boars (*Sus scrofa*), fallow deer (*Dama dama*) or red deer (*Cervus elaphus*) are rare, probably because these species are more restricted in their geographic distributions.

Train drivers perceived poor visibility, thick vegetation, poor light conditions, and deep snow as the main causes for WTCs. These factors may cause the animals to detect approaching trains very late or be unable to escape in time. However, train

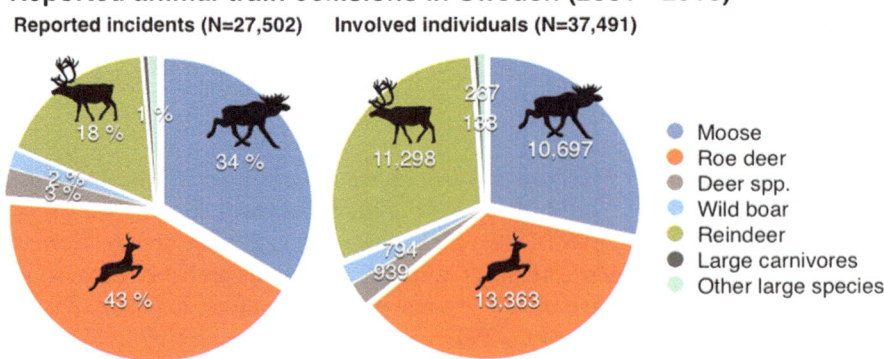

Fig. 17.2 Reported wildlife-train collisions in Sweden during 2001–2010 *Source* Seiler et al. 2011)

drivers also reported that many animals in the vicinity of the railway rarely seemed to react to passing trains. Flight behaviour was not often observed, but if flight was initiated by the train or by warning signals (horn), the animals sometimes fled ahead of the train and along the rail (see also Rea et al. 2010; Child 1983). Animals sometimes even walked up on to the railway track as the train approached, presumably to gain speed during flight (Child 1983).

Most train drivers (90%) were concerned about damage to the train, traffic delays, and unnecessary suffering and non-fatal injuries to the animals. Many drivers considered WTC to be a severe working environment issue because of the psychological stress they cause. Since train drivers are not able to slow down or stop the train to avoid a collision, their only options are to use the horn or flash the headlights in order to alert animals. However, most drivers disagreed about whether these measures are effective.

Collision Statistics

About 2,500–3,000 collision reports were issued every year from 2001–2010, including, on average, 1,070 moose, 1,336 roe deer and 994 other large mammals killed or injured annually (since accidents may involve more than one individual; Fig. 17.1). The number of unreported cases and undetected incidents is not known, but is assumed to be large and more pronounced in the smaller and lighter species. Since 2001, accident reports on roe deer and moose have increased by about 2.5% per year, despite decreasing hunting statistics that suggest smaller populations. The trend in WTC reflects trends in wildlife-accident numbers on roads (Seiler 2011). This follows both a similar diurnal pattern to road accidents with most collisions occurring during dusk and dawn, and a similar seasonal pattern with most accidents reported during autumn and winter (Neumann et al. 2012; Borda-de-Água et al. 2014; Rolandsen 2015). Likewise, as on roads (Seiler and Helldin 2006; Jacobson et al. 2016), WTCs were more frequent on railways with intermediate traffic volumes (50–150 trains per day) and less frequent on both calmer and busier tracks.

Overall, wildlife collisions on railways and roads seem to share many characteristics and patterns. However, collisions with moose appeared to be twice as frequent per kilometer of railway (and with roe deer slightly more frequent) than on public roads, suggesting that railways may be more dangerous than roads on average. We assume that this may be related to the relatively long intervals between trains; animals may be deterred less often by railway traffic and use railway tracks more readily than roads, because trains pass by less often.

In Sweden, WTCs are reported by train drivers not as an actual GPS position, but as incidents occurring between two consecutive nodes in the railway network. Thus, WTCs contain no point statistics, but rather frequencies related to a given section of railway. These sections average 8 km, but range from a few hundred meters to 53 km. A total of 1,377 sections longer than 1 km were included in the study. On average, during 2001–2011, annual reports yielded 0.79 moose and 1.09

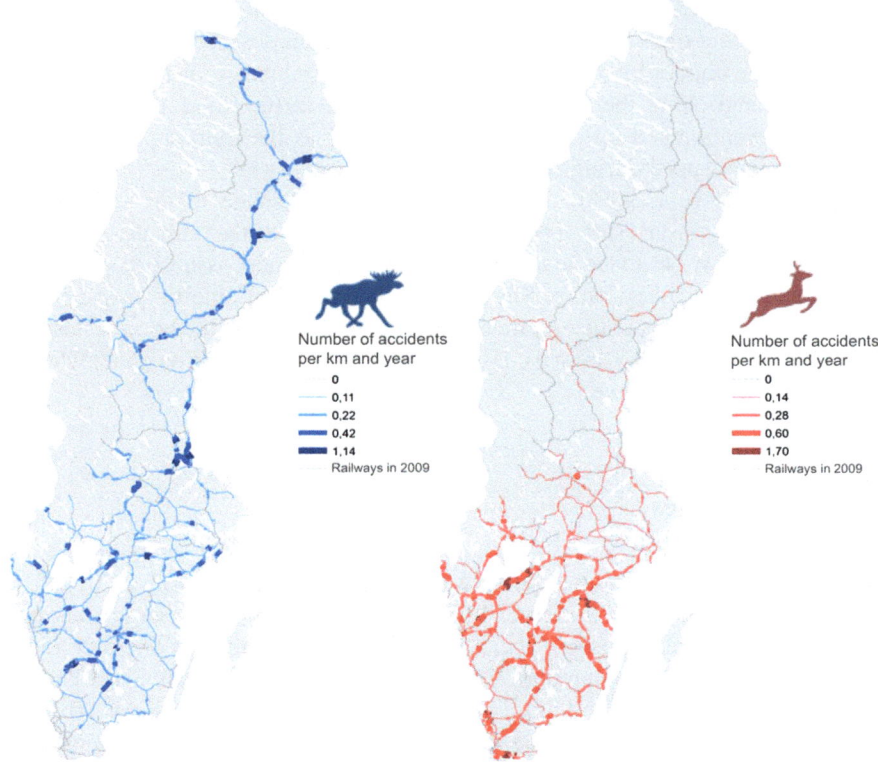

Fig. 17.3 Annual frequencies of reported and positioned train collisions with moose and roe deer per km of railway in Sweden during 2001–2009. (from Seiler et al. 2011)

roe deer incidents per 10 km railway per year. On some sections, however, densities would exceed one accident per km per year (Fig. 17.3).

WTC Hotspots

We identified special aggregations of WTCs or "hotspots" as railway sections with collision frequencies above 2.38 collisions with moose and 3.30 collisions with roe deer per 10 km per year, which corresponded to the top 5% at the national level. We then ranked hotspots according to their stability over time, with high-ranked sections comprising those that were among the top 50 sections in more than four of seven overlapping 5-year intervals during 2001–2011. For comparison, we selected "coldspots" as sections for which no WTC had been reported during the study period. Logistic regression analyses successfully distinguished between "hotspots" and "coldspots" based on the composition of the surrounding landscape, availability of suitable forage, and preferred habitat, and the presence of linear structures, such as minor roads and water courses that would lead animals toward the railway or encourage them to cross (Seiler et al. 2011).

From the top 50 hotspots, we then selected those that contained more than one ungulate species besides moose, had high traffic volumes and were used by SJ passenger trains, and had been subject to vegetation clearance (tree felling) within a 30-m wide corridor alongside the railway. In addition, the railway sections had to run primarily through forest-dominated landscapes and not through built-up areas. With these criteria, we eventually selected three railway sections for the experimental mitigation study (Olsson and Seiler 2015) (see below).

Wildlife-Train Encounters

In cooperation with Swedish Railways, we initiated a new project in 2015 to study the behaviour of animals when they encountered an approaching train. This was carried out using video recordings made by train drivers on a commercial dashcam (DOD-LW730). The dashcam records continuous video in 1–5 min clips, but overrides old recordings when memory capacity is full. If the camera alarm is set off, however, a recording in progress is protected and can be extracted for later analysis (Fig. 17.4). The advantage of this approach is that the alarm does not need to be set immediately, but can be made after the incident in question occurred. In this project, 15 train drivers volunteered to use a dashcam during their daily work routines. They were instructed to set off an alarm whenever they saw an animal on or near the railway. While this project is still ongoing (as of October 2016), preliminary findings from the first 178 recordings confirm the previous driver survey results: many ungulates utilize rail corridors for browsing or transport (see also Jaren et al. 1991). Most individuals were attentive to traffic, but in about 15% of the documented encounters, animals did not respond to the approaching train. The overall kill rate was 5% in both moose and roe deer. Mean flight initiation distances were short in both species (112 and 125 m, respectively), leaving the animals less than 2 s to respond correctly and leave the track when a modern passenger train approached at about 200 km/h. Flight initiation distances varied slightly, depending on whether the animals were on the track or beside the track, but due to the limited number of observations, these differences are not yet significant. They suggest, however, that animals may less easily detect approaching trains when on the railway tracks or be unable to distinguish risks, depending on how close they are to the railway. We could not detect any effect of the horn or the headlights used by the driver to warn animals, but as the study continues, we hope to find out whether acoustic or optic signals help to increase flight distances and reduce the risk for collisions.

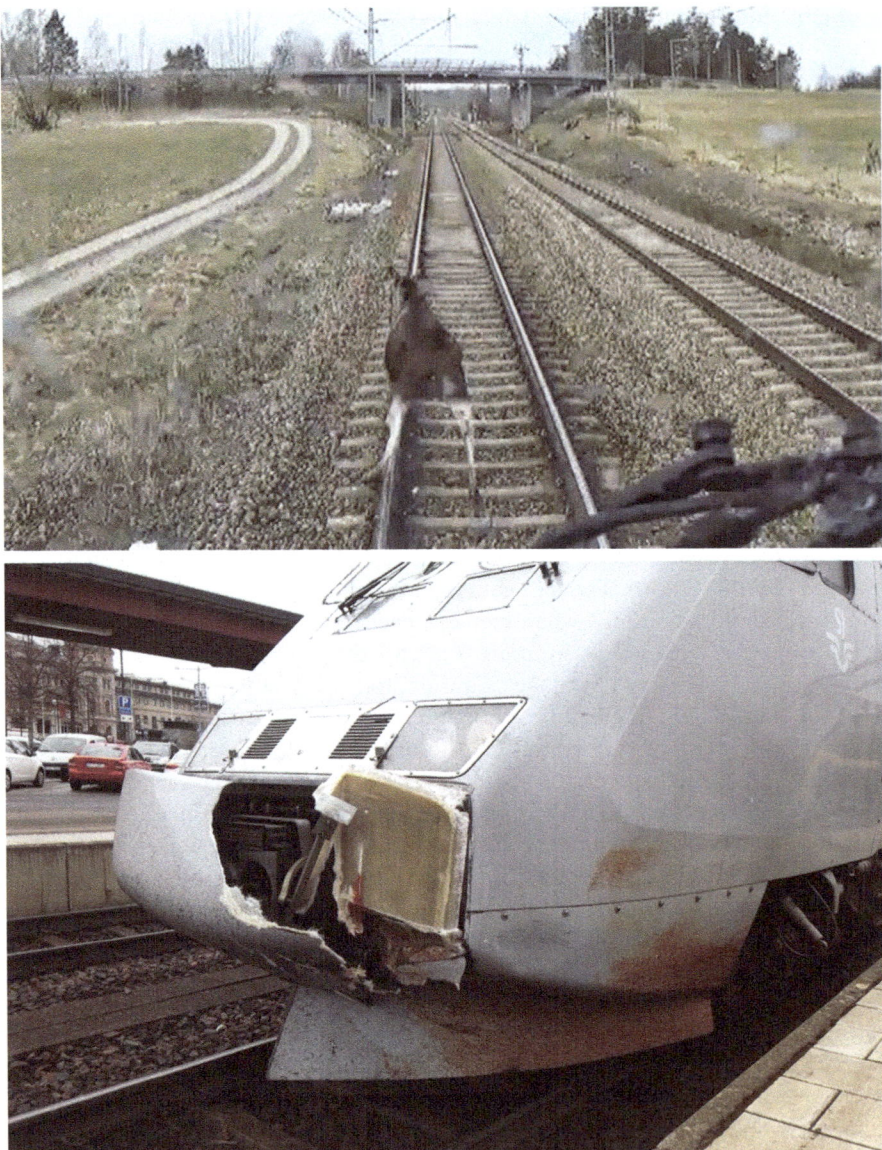

Fig. 17.4 Train driver's view of a fleeing moose shortly before collision (above). The picture was taken with a video dashcam mounted inside the driver's cabin. The damaged front of the X2000 train engine after the moose collision (below) (*photos* Jimmy Nilsson, Swedish Railways)

Ultimate and Proximate Causes of WTCs

Based on these previous studies, we can identify various factors that influence collision risk.

Clearly, WTCs are more frequent in areas with higher wildlife abundances (see also Gundersen and Andreassen 1998; Hedlund et al. 2004; Seiler 2005), or where animals are more likely to cross railways. This relates to landscape composition, food availability, landscape structure, and other regional and local factors typically beyond the responsibility of the train operator or the railway manager (Seiler et al. 2011; Rolandsen 2015). Other factors relate to the attractiveness and accessibility of the railway, including vegetation management alongside railways (Jaren et al. 1991; Rolandsen 2015), the presence of gullies and fences, or by train density (Righetti and Malli 2004; Kusta et al. 2011).

The proximate cause of WTCs, however, must be attributed to the behaviour of the animal itself. In contrast to roads, where motorists carry the main responsibility for avoiding collisions with wildlife (Seiler and Helldin 2006; Litvaitis and Tash 2008), trains are unable to stop or change paths. Therefore, it is only the animal that can avert a collision, but this depends on its chance of detecting an oncoming train in time, its cognitive abilities, and its anti-predator or flight behaviour. As we have seen on video recordings, flight initiation distances were short, leaving little time for the animals to respond appropriately, especially to rapidly moving and silent modern passenger trains (Gundersen and Andreassen 1998; Rea et al. 2010). Therefore, a possible mitigation is to increase the available response time by alerting animals earlier, before the train is too close to be avoided.

Mitigation Project

Objectives

In 2015, the Swedish Transport Administration, in cooperation with Swedish Railways, Enviroplanning AB, and the Swedish University of Agricultural Sciences, initiated a project to develop novel approaches and test available systems for alerting and deterring wildlife from railways shortly before trains arrive (Olsson and Seiler 2015). The goal is to prevent WTCs while still allowing animals to cross the railway when no trains are approaching.

Earlier studies from Poland (Werka and Wasilewski 2009; Babińska-Werka et al. 2015) have been promising, but further technical and methodological development is needed to develop a cost-effective and robust alternative to fences and crossing structures.

Approach

To develop and test wildlife warning systems, we selected three railway sections (8, 11 and 23 km in length) with extreme WTC frequencies ("hotspots") and two other sections for practical reasons. All five sections will be fenced, using 2 m high standard exclusion fences of a type also used along roads. These fences will be perforated by 20–24 experimental crosswalks, i.e., gaps in the fence of about 50 m that are monitored by video and thermal cameras and secured by the alerting or deterring systems (Fig. 17.5). The fences thus serve two purposes: they reduce WTC on most of the railway section by an estimated 80% in moose and 60% in roe deer (Swedish-Transport-Administration 2014), and they lead animals toward the openings at the crosswalks.

These experimental crosswalks serve as test locations where the effect of various stimuli can be studied (Fig. 17.5). They thus provide independent replicates and controls in a BACI setup (BACI = Before, After, Control, Impact). The crosswalks will be placed at feasible locations where animals are known to cross the railway (detected by snow tracking) and are dispersed approximately 2 km apart, providing an estimated sufficient level of permeability to moose and roe deer (Seiler et al. 2015). The crosswalks will be constructed so that terrain conditions, vegetation and ground substrate naturally and effectively lead animals toward the fence gap and

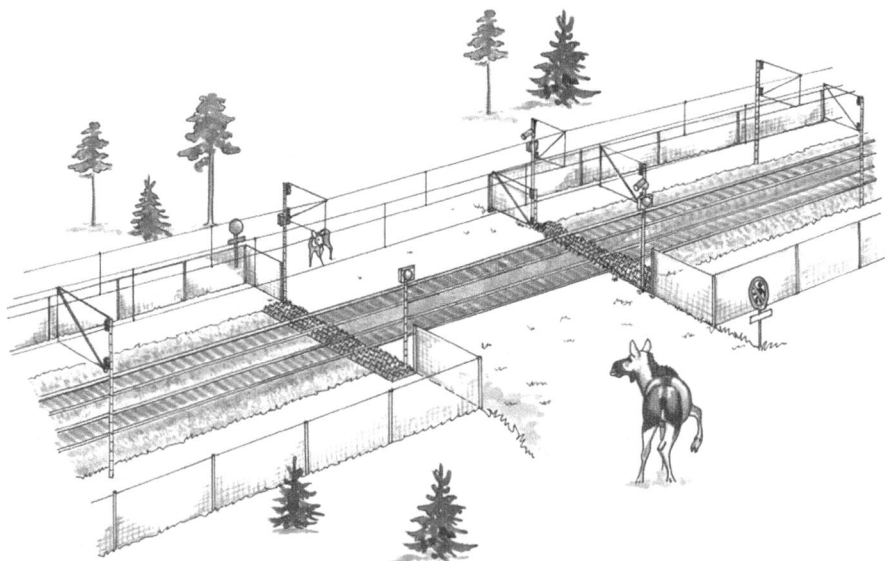

Fig. 17.5 Conceptual sketch of an experimental crosswalk: standard exclusion fences lead animals towards an opening about 50 m wide where movement detectors, thermal cameras, and video cameras monitor the presence and behaviour of animals and trigger the warning system when trains approach. Crushed stone or cattle guards will discourage ungulates from entering the fenced area. Human access to the crosswalk is prohibited (*sketch* Lars Jäderberg)

over the railway. Cattle guards or large crushed stones alongside the crossings will discourage animals from entering the fenced railroad track. Escape ramps or one-way gates will be installed to allow animals to exit if they get trapped inside the fence nevertheless.

Providing the same animal detection and monitoring and surveillance systems, each experimental crosswalk will operate as an independent unit that will be triggered by the same signal from a standard train sensor located several kilometers earlier along the railway. One train sensor will serve several crosswalks, and the delay between the train passing the sensor and the alarm at each crosswalk will be adjusted to the speed of the train and the distance between the crosswalk and the sensor. Warnings will only be displayed when animals are detected near the crosswalk and a train has passed the sensor, i.e., when the risk of a collision is imminent. This basic setup will be identical at all crosswalks and guarantee a standardized basis for the evaluation and comparison of different warning methods.

We budgeted 150,000–270,000 Euros per experimental crosswalk that includes surveillance equipment and warning systems. If such systems were installed on a regular basis, the costs would be substantially lower. The overall budget for the 20–24 crosswalks along 50 km of fenced railways was 5.5 million Euros in total (minimum cost estimate 4,500,000 million Euros and maximum estimate 7,800,000 million Euros) (Olsson and Seiler 2015). Most of the budget is reserved for fences (about 50,000 Euros per km), while research money totalled about 5,500,000 Euros over 4 years. However, if one assumes that over the next 20 years, WTC levels and the average costs for repairs, delays, and reimbursements are similar to those of the last 10 years, and at least 80% of the WTCS could be prevented by fences and/or crosswalks, then the socio-economic costs that could be saved by mitigating the selected hotspots alone would exceed 4,600,000 Euros. However, it is likely that the overall costs for the installation of the experimental study will be outbalanced by the savings, because the benefits are easily underestimated (Olsson and Seiler 2015; Seiler et al. 2016).

The Deterrent System

Traditional attempts to scare animals away from fields, roads, and airports by means of ultrasonic whistles, explosionss or shooting guns have mostly been ineffective (Koehler et al. 1990; Romin and Dalton 1992; Curtis et al. 1997; Belant et al. 1998; Ujvari et al. 2004). The underlying problem is that when the warning signal is not followed by a real threat and is merely a bluff without consequence, animals will soon habituate and learn to ignore the signal (Bomford and O'Brien 1990). In our approach, we rely on the passing train to reinforce the signal. In most cases, as we have seen in our video recordings, the train will be frightening enough when it is nearby. Habituation should therefore not be a problem; instead, learning should instead lead to a conditioning, provided the animals are able to relate the signal to the approaching train. Conditioning will presumably work best in those individuals who have an experimental crosswalk within their home range and thus repeatedly experience the signal and the passing train. However, there will always be a proportion of animals (offspring, dispersers, and migratory animals) that are not yet conditioned or do not know about the danger of traffic. Thus, rather than using abstract stimuli, such as whistles or bangs that would alert humans, it may be more effective to use natural sounds that already communicate a message for animals, such as a human voice indicating the presence of people, or a deer warning or distress call for alarming other animals.

Such approaches have been tested in Poland using animal alarm calls, barking dogs and sounds from hunting scenes (Werka and Wasilewski 2009; Babińska-Werka et al. 2015); in Japan, using sika deer warning calls (Shimura et al. 2015); in Italy, using sounds of dogs and humans to scare wildlife off roads when cars approach (Mertens et al. 2014); and in Northern Sweden, using human voices from a conventional radio to alert semi-domestic reindeer (Larsson-Kråik 2005). Experiences from these studies suggest that such warning systems may be effective in causing the animals to leave the disturbed site.

The basic idea is thus to condition a movement response to an auditory and/or visual stimulus that is strong enough to evoke the desired, subtle response in most individuals, but weak enough to avoid causing a panic reaction and allowing the animals to experience reinforcement through the passing train.

We intend to address questions such as which signals will work best, how quickly animals learn to respond appropriately, and to what extent this reduces the risk of collisions. If the systems prove successful in moving wildlife away from the railway track when trains approach, they could replace more costly crossing structures such as bridges or tunnels and provide the necessary complement to fences in an inclusive mitigation system (Huijser et al. 2009; Seiler et al. 2016). If the system is to replace fencing, however, the deterrent or warning effect must be extendable over several kilometers. This may be evaluated in a later study. If the system does not operate successfully, i.e., if animals show very little response to warning signals, and if collisions in the crosswalks are not reduced, the gaps in the fences may need to be closed.

Complementary Studies

Besides the aforementioned system of crosswalks and fences, the project will involve complementary activities such as in-depth analyses of WTC statistics; field surveys to assess unreported collisions; improvements to the reporting and registration routines of WTCs; studies on the indirect costs of WTCs due to delays in train traffic; exploration of possible animal detection and warning systems that can be mounted on train engines instead of in railway infrastructure; and continued video monitoring of train-animal encounters. Collaborations with similar research projects in Norway and Austria have been initiated. An international reference group will be established together with a group of private companies interested in testing their technical solutions and ideas for a warning system.

Parallel to this project, empirical studies are being conducted to reduce wildlife-vehicle collisions on roads; in combination, these studies will help to develop national objectives and set up a national strategy for reducing wildlife and traffic problems in Sweden.

Acknowledgements This project is being financed by the Swedish Transport Administration and conducted in cooperation with Swedish Railways, Enviroplanning AB, and the Swedish University of Agricultural Sciences. Financial support for the video documentation was provided by the Marie Claire Cronstedt Foundation. We thank Luís Borda-de-Água and Rafael Barrientos for their valuable comments on the manuscript, and Tim Hipkiss at Enviroplanning AB, and Geeta Singh at SLU, for proofreading this paper. The project is being carried out in close collaboration with Anders Sjölund (Swedish Transport Administration), Pär Söderström, and Anders Forsberg (Swedish Railways), Andreas Eklund (WSP), and many others. We are grateful for the help of the train drivers in recording wildlife-train collisions, as well as for the interest of many private companies in developing and testing innovative, technical solutions for warning wildlife on railways.

References

Åhrén, T., & Larsson, P. O. (1999). *Renpåkörningar - En pilotstudie för att hitta förslag till effektiva åtgärder för att minska antalet djurpåkörningar utmed Malmbanan.* Luleå: Banverket Norra Regionen.

Åhrén, T., & Larsson, P. O. (2001). Djur i spår - En nulägesbeskrivning av djurpåkörningar inom Banverket. Swedish National Rail Administration (intern report).

Babińska-Werka, J., Krauze-Gryz, D., Wasilewski, M., & Jasińska, K. (2015). Effectiveness of an acoustic wildlife warning device using natural calls to reduce the risk of train collisions with animals. *Transportation Research Part D: Transport and Environment, 38,* 6–14. doi:10.1016/j.trd.2015.04.021.

Belant, J. L., Seamans, T. W., & Tyson, L. A. (1998). Evaluation of electronic frightening devices as white-tailed deer deterrents. *Proceedings of the Vertebrate Pest Conference, 18,* 107.

Bomford, M., & O'Brien, P. H. (1990). Sonic deterrents in animal damage control: A review of device tests and effectiveness. *Wildlife Society Bulletin (1973–2006), 18,* 411–422.

Borda-de-Água, L., Grilo, C., & Pereira, H. M. (2014). Modeling the impact of road mortality on barn owl (*Tyto alba*) populations using age-structured models. *Ecological Modelling, 276,* 29–37. doi:10.1016/j.ecolmodel.2013.12.022

Child, K. (1983). Railways and moose in the central interior of BC: A recurrent management problem. *Alces, 19,* 118–135.

Child, K. N., & Stuart, K. M. (1987). Vehicle and train collision fatalities of moose: Some management and socio-economic considerations. *Swedish Wildlife Research, (1),* 699–703.

Curtis, P. D., Fitzgerald, C., & Richmond, M. E. (1997). Evaluation of the Yard Gard ultrasonic yard protector for repelling white-tailed deer. *Proceedings of the Eastern Wildlife Damage Control Conference, 7,* 172.

Gundersen, H., & Andreassen, H. P. (1998). The risk of moose *Alces alces* collision: A predictive logistic model for moose-train accidents. *Wildlife Biology, 4,* 103–110.

Hedlund, J. H., Curtis, P. D., Curtis, G., & Williams, A. F. (2004). Methods to reduce traffic crashes involving deer: What works and what does not. *Traffic Injury Prevention, 5,* 122–131.

Huijser, M. P., Duffield, J. W., Clevenger, A. P., Ament, R. J., & McGowen, P. T. (2009). Cost-benefit analyses of mitigation measures aimed at reducing collisions with large ungulates in the United States and Canada: A decision support tool. *Ecology and Society, 14,* 15.

Jacobson, S., Bliss-Ketchum, L., De Rivera, C., & Smith, W. P. (2016). A behavior based framework for assessing barrier effects to wildlife from vehicle traffic volume. *Ecosphere, 7,* article e01345. DOI Ecosphere *7*(4), e01345. 10.1002/ecs2.1345

Jaren, V., Andersen, R., Ulleberg, M., Pedersen, P. H., & Wiseth, B. (1991). Moose-train collisions: The effects of vegetation removal with a cost-benefit analysis. *Alces, 27,* 93–99.

Koehler, A. E., Marsh, R. E., & Salmon, T. P. (1990). *Frightening methods and devices/stimuli to prevent mammal damage—A review.* Paper presented at the proceedings 14th vertebrate pest conference 1990, paper 50. http://digitalcommons.unl.edu/vpc14/50

Kusta, T., Jezek, M., & Keken, Z. (2011). Mortality of large mammals on railway tracks. *Scientia Agriculturae Bohemica, 42,* 12–18.

Larsson-Kråik, P. O. (2005). Utvärdering av djurskrämmer vid plankorsningar. Perioden 2002-02-01 till 2004-12-31. Luleå: Banverket, BRNB Rapport 2005-02-28.

Litvaitis, J. A., & Tash, J. P. (2008). An approach toward understanding wildlife-vehicle collisions. *Environmental Management, 42,* 688–697. doi:10.1007/s00267-008-9108-4

Mertens, A., Ricci, S., Sergiacomi, U., & Mazzei, R. (2014). *LIFESTRADE—A new LIFE Project for the development of an innovative system to prevent road mortality in central Italy.* Paper presented at the IENE 2014 international conference on ecology and transportation, Malmö, Sweden. http://iene2014.iene.info

Neumann, W., Ericsson, G., Dettki, H., Bunnefeld, N., Keuler, N. S., Helmers, D. P., et al. (2012). Difference in spatiotemporal patterns of wildlife road-crossings and wildlife-vehicle collisions. *Biological Conservation, 145,* 70–78. doi:10.1016/j.biocon.2011.10.011

Olsson, M., & Seiler, A. (2015). *Viltsäker järnväg, Utredning av olycksdrabbade sträckor och förslag till åtgärder* (Vol. 2015, pp. 082). Borlänge: Trafikverket Publikation.

Rea, R., Child, K., & Aitken, D. A. (2010). Youtube TM insights in to moose-train interactions. *Alces, 46,* 183–187.

Righetti, A., & Malli, H. (2004). Einfluss von ungezäunten (Hochleistungs-) Zugstrecken auf Wildtierpopulationen. Bern, CH: COST-341 Synthesebericht - PiU GmbH.

Rolandsen, C. M. (2015). Animal-train accidents on the railway in Norway 1991–2014. In Norwegian with English summary: Dyrepåkjørsler på jernbanen i Norge 1991–2014. Retrieved from Trondheim http://www.researchgate.net/publication/277989048

Romin, L. A., & Dalton, L. B. (1992). Lack of response by mule deer to wildlife warning whistles. *Wildlife Society Bulletin, 20,* 382–384.

Seiler, A. (2005). Predicting locations of moose-vehicle collisions in Sweden. *Journal of Applied Ecology, 42,* 371–382. doi:10.1111/j.1365-2664.2005.01013.x

Seiler, A. (2011). *Viltlyckor på väg och järnväg– nya rön om trender, fördelning och kostnader.* Paper presented at the IENE workshop: Viltlyckor, Stockholm. http://sweden.iene.info/2011-viltlyckor/

Seiler, A., & Helldin, J. O. (2006). Mortality in wildlife due to transportation. In J. Davenport & J. L. Davenport (Eds.), *The ecology of transportation: Managing mobility for the environment* (pp. 165–190). Amsterdam: Kluwer.

Seiler, A., & Olsson, M. (2015). Cost-benefit of wildlife mitigation measures on roads. In Swedish: Viltåtgärder på väg - en lönsamhetsbedömning (Vol. 94). Uppsala, Triekol report, CBM Publications. http://www.TRIEKOL.se

Seiler, A., Olsson, M., Helldin, J. O., & Norin, H. (2011). Ungulate-train collisions in Sweden (In Swedish with English summary: Klövviltolyckor på järnväg - kunskapsläge, problemanalys och åtgärdsförslag). (Vol. 2011, pp. 058). Borlänge: Trafikverket Publikation.

Seiler, A., Olsson, M., & Lindqvist, M. (2015). Analysis of the permeability of transport infrastructure for ungulates. In Swedish: Analys av infrastrukturens permeabilitet för klövdjur (Vol. 2015, p. 254). Trafikverket Publikation.

Seiler, A., Olsson, M., Rosell, C., & Van Der Grift, E. A. (2016). *Cost-benefit analyses for wildlife and traffic safety. SAFEROAD Technical report 4.* Brussels: Conference of European Directors of Roads (CEDR).

Seiler, A., Söderström, P., Olsson, M., & Sjölund, A. (2014). *Costs and effects of deer-train collisions in Sweden.* Paper presented at the IENE 2014 International Conference on Ecology and Transportation, Malmö, Sweden. http://iene2014.iene.info

Shimura, M., Ushoki, T., Kyotani, T., Nakai, K., & Hayakawa, T. (2015). Study of behavior of sika deer nearby railroad tracks and effect of alarm call. *RTRI Report, 29,* 45–50.

Swedish-Transport-Administration. (2014). Guideline on wildlife exclusion fences (In Swedish: Viltstängsel). Borlänge: Swedish Transport Administration (Trafikverket) TDOK 2014-0115.

Ujvari, M., Baagoe, H. J., & Madsen, A. B. (2004). Effectiveness of acoustic road markings in reducing deer-vehicle collisions: A behavioural study. *Wildlife Biology, 10,* 155–159.

Werka, J., & Wasilewski, M. (2009). *Animal deterring devices (UOZ-1) monitoring.* Warsaw: Warsaw University of Life Sciences (Sggw), Faculty of Forestry, Department of Forest Zoology and Wildlife Management.

Chapter 18
Commerce and Conservation in the Crown of the Continent

John S. Waller

Abstract While railroads figure prominently in U. S. history and culture, their environmental impacts are often overlooked. Here, I describe situations where operation and maintenance of a section of transcontinental railroad in Montana, USA, resulted in high mortality of grizzly bears (*Ursus arctos*), a threatened species under the U. S. Endangered Species Act, and where proposed avalanche control measures conflicted with the preservation mandates of Glacier National Park. A unique public/private partnership was created to work towards effective solutions to environmental problems. The partnership was successful in decreasing high grizzly bear mortality rates, but less so in finding a consensus on how best to reduce avalanche risk. I suggest that collaborative partnerships, such as that developed here, will be essential to solving future environmental problems associated with railroads.

Keywords Avalanche control · Connectivity · Cooperative conservation · Grizzly bear · Highways · Partnerships · *Ursus arctos*

Introduction

Railroads hold an important place in U. S. history. They served an important function in settling the west and transporting homesteaders and all the goods they needed to build a life on the frontier. Completed in 1893, the Great Northern was one of the last of the frontier railroads. It penetrated the vast wilderness of northwest Montana, opening an area of scenic wonder to a nation hungry to explore itself. The Great Northern railroad was instrumental in the development of what is now Glacier National Park, constructing numerous grand lodges and chalets that now are themselves historic tourist attractions. These lodges and tourist facilities were small islands of civilization within a largely untamed wilderness.

J.S. Waller (✉)
Glacier National Park, West Glacier, MT, USA
e-mail: john_waller@nps.gov

Fast-forward 123 years… Glacier National Park is today an island of wilderness surrounded by various levels of development. The grand lodges and chalets have lots of competition for the tourists that arrive in ever-increasing numbers. The golden age of railroads is long past, as most tourists now arrive by automobile or plane rather than train. Still the trains rumble through, carrying every imaginable type of cargo across this mountain vastness.

Railroads have always operated in a unique public/private capacity. Because they required vast capital to construct, and because they provided great benefits to the nation, they were heavily subsidized by the U. S. Government. During the frontier days, these subsidies often came in the form of land grants that could supply the railroads with the resources they needed for construction and that had other values that could help offset the costs of construction. These subsidies required that railroads also be regulated to counter the natural monopolistic tendencies of such an enterprise.

Nevertheless, the railroads amassed tremendous wealth and political power. During the frontier times, railroads were constructed with little consideration of their environmental impacts, and their power was so great that it's unlikely that anyone could have successfully challenged them on those grounds, had they wanted to. Today, railroads still remain powerful entities, but because they serve an important public function, they are also held to public standards of environmental accountability.

This case study is an example of how a railroad, working with government and public partners, discovered ways to uphold its environmental responsibilities by making meaningful contributions to the recovery of the threatened grizzly bear. However, it also describes how tenuous these partnerships can be, and how not all problems are solved through collaboration.

Geographic Location

The northern-most railroad line in the continental USA is the "hi-line" section of the Burlington Northern–Santa Fe railroad (BNSF, formerly the Great Northern Railroad). It is a primary freight corridor running between Minneapolis, Minnesota, and Seattle, Washington. It also supports the Empire Builder Amtrak passenger line. This railroad line bisects the Rocky Mountains and crosses the Continental Divide at Marias Pass (elevation 1462 m). The portion spanning the Continental Divide and lying between East Glacier and West Glacier, Montana, separates two large protected wilderness areas: Glacier National Park to the north (4079 km^2), and the Great Bear/Bob Marshall Wilderness Complex to the south (6131 km^2). From West Glacier, the railroad follows the Middle Fork of the Flathead River southeasterly to its confluence with Bear Creek, which it then follows northeasterly through the John Stevens Canyon to Marias Pass. East of Marias Pass, the rail line drops down into the South Fork of the Two Medicine River drainage to East Glacier, and crossing into the Blackfeet Indian Reservation. From the Bear Creek/Middle Fork confluence to just west of East Glacier, the railroad is the southern boundary of Glacier National Park (Fig. 18.1).

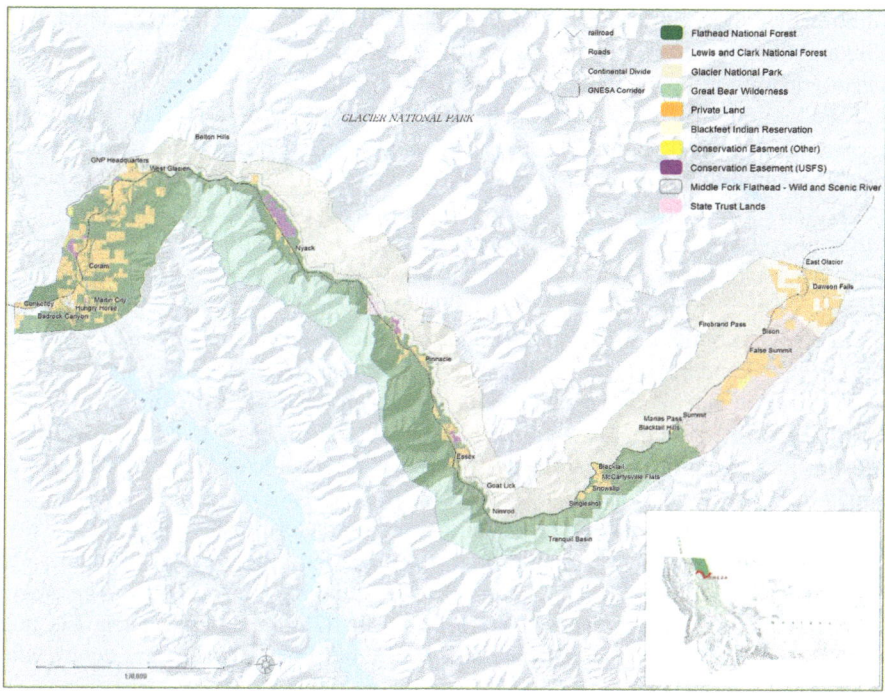

Fig. 18.1 The Great Northern Environmental Stewardship Area between West Glacier and East Glacier, Montana, USA

Land Ownership

Traveling southeast from West Glacier, the railroad line passes through intermixed public and private lands (Fig. 18.1). The private lands are primarily small ranches and recreational home sites, small communities, and campgrounds. The public lands are primarily managed by the U. S. Forest Service for recreation and timber. Also, U. S. Highway 2 parallels the railroad for its entire length between East and West Glacier, as does a natural gas pipeline and electricity transmission line. This narrow ribbon of railroad, highway, utilities, and private land development form what Servheen and Sandstrom (1993) term a "fracture zone" which is a zone where human development impedes the connectivity of wildlife populations within a natural area.

Ecological Characteristics

Glacier National Park and the Great Bear/Bob Marshall Wilderness Complex are highly protected federal lands managed for preservation of their natural qualities and for non-motorized recreation. Glacier National Park has been designated a biosphere preserve and a world heritage site by the United Nations (UNESCO).

Additionally, the Middle Fork of the Flathead River is federally designated as a Wild and Scenic River, which recognizes that it flows on an unmodified flood plain, is ecologically unimpaired, and provides some protection from future development.

The floodplain and its function have been the topic of important ecological research (Harner and Stanford 2003). This area is also home to a number of rare species, including some listed as threatened under the U. S. Endangered Species Act (ESA). These include the grizzly bear (*Ursus arctos*), Canada lynx (*Lynx Canadensis*), wolverine (*Gulo gulo*), and bull trout (*Salvelinus confluentus*). Other species of note are Rocky Mountain elk (*Cervus canadensis*), mountain goat (*Oreamnos americanus*), Rocky Mountain bighorn sheep (*Ovis canadensis*), moose (*Alces alces*), mountain lion (*Puma concolor*), black bear (*Ursus americanus*), wolf (*Canis lupus*), and bald eagle (*Haliaeetus leucocephalus*); in fact, the area retains its entire complement of native fauna with the exception of American bison (*Bison bison*). This diverse fauna is supported by high floral diversity that arises from the areas location within the transition zone between the Pacific coastal and continental climate regimes.

West of the continental divide, the climate is characterized by cold, wet winters, and cool, moist summers, supporting extensive conifer forests, primarily of Engelmann spruce (*Picea engelmannii*) and subalpine fir (*Abies lasiocarpa*). Lodgepole pine (*Pinus contortus*), Douglas fir (*Pseudotsuga menziesii*), and western larch (*Larix occidentalis*) are also well represented. East of the continental

Fig. 18.2 A westbound Burlington-Northern Santa Fe freight train crossing Marias Pass

divide, the climate is characterized by dry, cold winters and hot, dry summers. Here, conifer forests give way to extensive aspen (*Populus tremuloides*) park lands and shortgrass prairie (Fig. 18.2). The entire area has been frequently recognized as an important area for wildlife connectivity, for example, the Yellowstone to Yukon conservation initiative (Levesque 2001), the Western Governors Association (2008), and the National Fish and Wildlife Foundation (Ament et al. 2014). The larger ecosystem, which the BNSF rail line crosses has been variously termed the Northern Continental Divide Ecosystem (NCDE; in a grizzly bear recovery plan context), or the Crown of the Continent Ecosystem (in a trans-boundary management context).

Railroad Attributes

The East Glacier to West Glacier portion of the BNSF railroad line is composed of single and double-track sections, depending on topography. Locomotives are staged along the track to help long freight trains ascend and descend the pass. Freight trains carry a wide variety of goods and materials, especially corn and wheat from the Midwest headed to Pacific export terminals. In recent years, a large volume of crude oil from the Bakken oil fields in North Dakota has been transported as well. Because the railroad traverses extreme mountain topography, it is impacted by snow avalanches, rock slides, floods, heavy snow, and extreme temperatures that can affect rail integrity. Most of the wooden railroad ties over the pass have been replaced with concrete ties to help improve track stability. There are also snow sheds that protect portions of the track from avalanches.

The railroad line in this area generally has a smaller ecological footprint than highways because its width is typically narrower and carries less human traffic. However, as is the case here, railroad lines are typically paired with other developments such as roads and utility corridors. Thus, the cumulative footprint is large. In many areas, wildlife fencing is used to keep animals off rail lines and roads, but in our area, neither the highway nor the railroad line are fenced to exclude wildlife.

A Grizzly Problem Arises

The railroad operated without much environmental controversy through most of the twentieth century. However, a problem arose during the late 1980s and early 1990s. Between 1985 and 1989, at least 134 cars of corn derailed along a 15-mile section of track west of Marias Pass (Fig. 18.3). Each car carried approximately 3500 bushels of corn so the total amount spilled was 469,000 bushels or nearly 12,000 metric tons. In its haste to repair the tracks and to dispose of the spilled grain, BNSF buried it along its right-of-way. As the buried grain began to decompose, it fermented, producing a strong odor that drew black and grizzly bears

Fig. 18.3 Spilled corn covers the slopes below the tracks during cleanup of a freight train derailment near Marias Pass, Montana, USA

into the site from many kilometers away. One observer claimed to have observed at least 20 different grizzly bears feeding at the site. As the bears began to excavate and consume the fermented grain, some observers believed the bears became intoxicated. Regardless, the feeding bears became highly habituated to the close passage of trains. As a result, between 1985 and 1990, at least 9 grizzly bears were hit by trains and killed. Grizzly bears are listed as a threatened species under the U. S. Endangered Species Act; hence, their deaths during operation of the railroad amounted to an illegal taking under that act.

These grain spills and the resulting deaths of grizzly bears brought increased scrutiny to the frequency and causes of grizzly bear mortalities in the corridor. Bears weren't just dying during grain spills–they were dying at relatively high rates in general. Bears were being struck and killed along the tracks at areas other than grain spills and in conflicts with humans at developed sites. Further investigation revealed that many of the hopper-bottom train cars that carry grain had hopper doors that did not seal well; as a result, grain constantly dribbled out along the tracks and created small piles between the rails whenever the train stopped. Also, as train cars were filled at terminals, grain would accumulate along the tops of the cars, which would then fall off along the tracks as the train headed west. This continuous deposition of grain along the tracks acted as a powerful attractant for bears. Most of the bears found killed by trains had grain in their stomachs.

Much of the rail line through the corridor crosses ungulate winter range and areas grazed by livestock. Wild ungulates and domestic livestock are occasionally

Fig. 18.4 A Burlington-Northern Santa Fe freight train crosses the Sheep Cr. Trestle

struck by trains and their carcasses also become attractants for bears. As bears come to scavenge the remains, they themselves become victims.

A third important source of mortality was associated with trestles. The rail line crosses a number of deeply incised water courses with high trestles (Fig. 18.4). As grizzly bears traveled the tracks in search of grain and carcasses, or just as an easy means of travel, they would utilize these trestles. Occasionally, an oncoming train would overtake bears on the trestle where they could not escape and they would be hit and killed.

Working Towards a Solution

In 1990, two environmental activist groups threatened to sue the railroad for continuing to kill grizzly bears. During the summer of 1992, railroad operations caused a small wildfire inside Glacier National Park (GNP) that enabled the park to seek damage reparation payment from BNSF. In order to help prevent future grizzly bear mortalities, then-GNP Superintendent Gil Lusk suggested that rather than pay GNP damages for the fire, BNSF establish a special organization to help facilitate solutions to environmental problems in the rail corridor. BNSF agreed and established the rail line between East and West Glacier as a special operating area called the Burlington-Northern Environmental Stewardship Area (BNESA), along with an organization to administer its operation.

In the agreement, BNSF and the signatory agencies acknowledged the importance of balancing the interests of commerce, agriculture, transportation, natural resources, and wildlife and acknowledged the nationally significant qualities of the corridor. Specifically, the parties agreed to pay particular attention to the operation and maintenance of the railway because of the areas importance to populations of threatened and endangered species. The parties committed to working cooperatively with affected jurisdictions to plan and implement emergency responses to derailments, including providing and placing emergency response equipment to expedite response times and prevent spilled materials from entering rivers and streams. The agreement tasked wildlife biologists from the Montana Department of Fish, Wildlife, and Parks (MFWP) with determining specific sections of track that were hazardous to grizzly bears or sections where bears frequently crossed the tracks or otherwise congregated. BNSF committed to rapid spill response and placement of cleanup material at key points for ready access to these resources; to strict adherence to operating speed limits, and use of continuous welded rail, welded turnouts, and concrete ties to improve rail integrity; the erection of fences at key locations determined by MFWP wildlife biologists; and special training and educational programs for BNSF personnel working in the corridor conducted by the appropriate agency. The BNESA agreement also established a conservation trust fund to be used for the protection of critical grizzly bear habitat and movement corridors, placement of a grizzly bear management specialist to serve as a coordinator in the area, provide education and resolution of management issues and bear conflicts, and research to evaluate the effectiveness of operational changes and develop new techniques to minimize grizzly bear mortalities. Additional actions implemented by BNSF included the regular use of a vacuum truck that picks up spilled grain along the tracks, and every spring after the snow melts, BNSF crews pick up carcasses along the tracks and dispose of them in safe areas. They have also used motion-sensitive electronic noise makers on trestles to discourage use by bears, and experimented with electric fencing and electric "track-mats". It is difficult to assess the success of these efforts in absolute terms, but we do know that even as the number of bears in the ecosystem has increased substantially (Mace et al. 2012), the numbers of bears hit and killed along the railroad tracks has dropped significantly; only four grizzly bears were killed between 2010 and 2015, compared to the six-year running average of 11.

As the BNESA organization grew, its name was changed to the Great Northern Environmental Stewardship Area (GNESA) to reflect its broader mission and purpose beyond just the BNSF railroad. The GNESA organization grew to include representatives of the more than 20 state, tribal, and federal agencies having land management authority within the GNESA corridor (GNP, U. S. Forest Service, MFWP, Montana Department of Transportation, U. S. Fish and Wildlife Service, Flathead Conservation district, and Blackfeet Indian Nation), private land owners, utilities (Flathead and Glacier Electric Cooperatives, Northwestern Energy), and county officials. Over the next several years, GNESA facilitated additional actions to reduce grizzly bear mortalities along the railroad tracks. They helped fund a transition from open garbage containers to bear-resistant models at county refuse sites; worked to establish guidelines for the safe transportation of hazardous materials through the corridor; and facilitated interagency cooperation in planning for responses to accidents involving hazardous materials, including resolution of emergency communications issues.

GNESA has also supported wildlife research in the corridor, including studies of habitat connectivity across the corridor. They worked to preserve local knowledge of wildlife connectivity by hosting workshops that involved citizens who work and live in the corridor, including train engineers, school bus drivers, local law enforcement officials, biologists, and rangers. They helped fund a study of game trail use in the corridor that identified where animals were crossing the highway and railroad (Roesch 2010). This information will be vital, as crossing structures may eventually be required.

BNSF and the U. S. Department of Transportation also supported in-depth research into grizzly bear movement patterns across the corridor. Obtaining fine-scale movement data from grizzly bears was nearly impossible before the advent of satellite global positioning system (GPS) tracking collars in the late 1990s. One of the first studies in the continental U. S. to utilize this technology on brown bears was that of Waller and Servheen (2005), which used a relatively large sample of GPS-collared grizzly bears ($n = 42$), paired with detailed road and traffic counts, to assess the response of grizzly bears to various traffic types and volumes. These researchers found that although the highway and railroad did have measureable effects on the trans-corridor movements of grizzly bears, they still crossed frequently and successfully. Grizzly bears crossing the transportation corridor were faced with traversing both the railroad line and the highway, which closely parallel one another through the GNESA area. The monitored grizzly bears appeared to make behavioral adjustments to daily fluctuations in highway traffic but not for rail traffic. Grizzly bears chose to cross the corridor when highway traffic volumes were lowest, but when rail traffic was the highest. Rail traffic was higher at night because most rail maintenance work is done during daylight hours, thus curtailing daytime freight traffic. Furthermore, freight trains loaded on the coast during the day leave in the evening and arrive in the GNESA area at night, 24–36 h later. It may be that grizzly bears are more easily habituated to rail traffic because it tends to be highly predictable. The consequence, however, is that grizzly bear mortality was much higher on the tracks than on the highway. Between 1980 and 2002, 23 grizzly bears were killed on the tracks in the GNESA corridor, but only two on the highway. The fine-scale movement data did not show that grizzly bears were still feeding on grain accumulations along the railroad tracks, suggesting that grain-control actions were working as intended. Therefore, the authors concluded that the high observed post-grain spill mortality rates were the consequence of high rail traffic volume, low highway traffic volume, and natural grizzly bear movement patterns.

The collaborative environment that fostered the birth of GNESA and innovative problem-solving approaches to environmental issues was recognized as a model for other conservation problems throughout the world. In 2008, GNESA was recognized by the U. S. Department of Interior as an exemplary example of collaborative conservation (USDOI 2008). Such collaboration is often the result of the right people being in the right place at the right time. In 2004, an issue arose that would test the limits of the collaborative efforts embodied in GNESA.

An Avalanche of Problems

On January 28, 2004, heavy snowfall resulted in avalanches in the GNESA corridor that blocked the BNSF railroad for 29 h. These avalanches began in starting zones within the GNP. During the storm, an empty 119-car freight train was struck by an avalanche and derailed. While it was derailed, another avalanche in an adjacent slide path struck the train again, derailing more cars. A third avalanche nearly hit a clean-up crew, and a fourth avalanche hit a truck travelling the highway beneath the railroad. Responding to continued snowpack instability and the need to ensure the safety of its personnel, BNSF requested a special use permit from GNP that would authorize explosive avalanche control in the John Stevens Canyon. After much deliberation, GNP granted a 3-day emergency special use permit, but, as is often the case, moderate weather allowed the snowpack to stabilize, and the avalanche risk abated before any control measures could be implemented.

After the January 2004 incident, BNSF contracted with a company specializing in avalanche control to analyze the avalanche hazard in the canyon. Their report described the avalanche potential in 14 avalanche paths along the railroad in the John Stevens Canyon (Hamre and Overcast 2004). Nine of those paths had some level of protection provided by snow sheds, although seven of them were too short to provide complete protection. Avalanche paths are dynamic, changing over time with changes in vegetation growth. Most of the sheds were constructed around 1900, so for some of them, the paths had widened or altered position, so that the sheds no longer provided complete protection (Fig. 18.5). The report identified avalanche hazard reduction measures ranging from explosive control to snow shed construction. From the viewpoint of BNSF and their consultants, explosive control was the most expedient and cost-effective means to mitigate the avalanche risk. However, the viewpoint of GNP was quite different.

Glacier National Park is congressionally mandated to protect the park from impairment resulting from human-caused activities. Other legal mandates include the protection of wilderness values (the Wilderness Act of 1964), protection of threatened and endangered species, protection of scenic and cultural resources, and maintenance of visitor enjoyment. The use of explosives generally ran counter to these mandates and were inconsistent with the park's status as a biosphere preserve and world heritage site. In order to evaluate alternative courses of action, GNP prepared an Environmental Impact Statement (EIS) as required by the National Environmental Policy Act of 1970. Four alternatives were evaluated in the EIS: a "no action" alternative, by which no action by BNSF to mitigate avalanche hazard

Fig. 18.5 An avalanche partially buries a snow shed on the Burlington-Northern Santa Fe railroad in the John F. Stevens canyon

would be permitted; an alternative where GNP, USFS, and MDOT would recommend constructing and/or modifying snow sheds (the preferred alternative); an alternative similar to the previous alternative but that also included a 10-year special use permit for explosive control during snow shed construction; and, finally, an alternative that allowed explosive avalanche control without concomitant snow shed work (NPS 2008a).

After an extensive analysis and public comment period, GNP selected the preferred alternative and a "record of decision" (ROD) was signed in September 2008. The ROD permitted BNSF to install a weather station and snow depth sensor within GNP, and for BNSF to conduct non-explosive snow stability testing. It allowed the installation of avalanche detection devices, but limited the use of explosives to extenuating emergency conditions where human lives or resources were at risk and after other options, including delays, had been exhausted. The ROD recommended that BNSF construct or modify snow sheds in high-risk slide paths (NPS 2008b).

The search for solutions to the avalanche problem in the GNESA corridor rose quickly to the upper echelons of corporate and government management. As the situation became political, both sides became entrenched in their positions and the collaborative environment of GNESA disappeared. In the end, GNP prevailed in their position but at the cost of some of the close working relationships built within GNESA. Hard feelings remained for some time, and, to date, BNSF has not attempted to construct or modify any snow sheds, instead relying on an avalanche forecasting program to mitigate risk. They have approached the park on at least one occasion for a special permit for avalanche control, but usually high-hazard conditions ameliorate rapidly and so the opportunities for explosion control are very limited (Reardon et al. 2004). The BNSF also purchased a "Daisy Bell" device which, when suspended below a helicopter, can test snow stability by directing a blast downward to the snowpack. This technique eliminates many of the disadvantages associated with explosive control. Fortunately, it appears that, as time passes and as personnel changes take place, working relationships are being restored.

What Next?

The GNESA corridor remains an important locale for commerce and connectivity, and grizzly bears remain an important issue in the corridor. The U. S. Fish and Wildlife Service is working with BNSF to prepare a habitat conservation plan that, under the ESA, will indemnify BNSF against litigation resulting from grizzly bear deaths provided that BNSF continues to implement procedures and practices that work to minimize mortalities. The implementation of the GNESA protocols and continued support of the organization will likely be an important component of that plan.

In general, efforts to control grizzly bear mortality have been successful throughout the ecosystem, and the grizzly bear population rapidly is expanding (Mace et al. 2012). State and federal bear managers are confident about the recovery of the bear population and will likely seek removal of the NCDE grizzly bear from the ESA within the next several years. Collaborative conservation efforts, as embodied by GNESA, have been critical to the attainment of this conservation success.

New issues will continue to surface as development within the corridor continues. Heavy traffic of railroad cars carrying crude oil have been a concern, as has increasing traffic on U. S. Highway 2 (Waller and Miller 2016). As traffic increases, wildlife connectivity across the corridor will diminish to the point where some type of accommodation will have to be made. Crossing structures are the current tool for maintaining connectivity, but other concepts may arise. For example, strategic highway planning might be used to direct higher traffic volumes to better-developed road systems in less ecologically sensitive areas. The avalanche sheds currently used are over 100 years old, in some cases, it might be possible to replace them with new designs that serve to protect the railroad and highway from avalanches and also serve as a crossing structure. Conservation investments of this magnitude will require public/private partnerships and inter-agency collaboration.

New technologies will continue to surface that have the potential to further limit wildlife mortalities and railroads have a responsibility to support the development and application of these technologies. For example, long-range, optical-infrared video alarm systems can warn engineers of wildlife on tracks day or night, far in

Fig. 18.6 A large cut-and-fill railroad bed completely obstructs natural drainage in Glacier National Park

advance, and long-range acoustic devices or other sonic devices may be able to get animals to move out of harm's way (Babińska-Werka et al.2015; Grey 2015).

As connectivity becomes an issue, so will watershed integrity. When the railroad was constructed, massive cuts and fills were used to bridge drainages (Fig. 18.6), these large fills obstruct the natural terrain, alter natural stream courses, and may need to be removed. Modification of this kind will be very expensive and will probably require public investment.

The success of the GNESA partnership was due, in large part, to the participation of individuals who were unafraid to take risks inside corporate and agency cultures. These individuals were committed to using the GNESA group to seek solutions and to resolve issues at the lowest possible levels. They did not use the media to advance their positions; they worked to get to know one another on a personal level; and they respected each other's different mandates. It also required organizational frameworks that were flexible towards this type of participation and tolerant of the risk that comes with trusting others in other organizations. Future success will demand more of the same from individuals and organizations. Building trust takes time and patience; while some conflict may be inevitable, it is important to accept those situations and move beyond them. Building a robust network of trust requires more than just affinitive trust among key individuals; establishing rational and systems-based trust adds resiliency when personnel turnover changes the group dynamic (Stern and Baird 2015). Bridging organizations, such as GNESA, bridge the boundaries between stakeholders that operate at different scales, linking varied constituencies to increase collective capabilities and social capital (Pretty 2003; Pahl-Wostl et al. 2007). With these lessons in mind, the trust and culture will hopefully be in place to allow future collaborative conservation to proceed.

Acknowledgements This study presents the combined contributions of many persons over the course of many years. The author thanks the GNESA partnership for their commitment to environmental stewardship. Major funding for the work described herein was provided by the U. S. Fish and Wildlife Service, U. S. National Park Service, Glacier National Park, and the Federal Highways Administration. Valuable support was provided by the U. S. Forest Service; Montana Department of Transportation; Montana Department of Fish, Wildlife, and Parks; and the Burlington-Northern Santa Fe railroad.

References

Ament, R., McGowen, P., McClure, M., Rutherford, A., Ellis, C., & Grebenc, J. (2014). *Highway mitigation for wildlife in northwest Montana*. Sonoran Institute, Northern Rockies Office, Bozeman, MT, 84 pp.

Babińska-Werka, J., Krauze-Gryz, D., Wasilewski, M., & Jasińska, K. (2015). Effectiveness of an acoustic wildlife warning device using natural calls to reduce the risk of train collisions with animals. *Transportation Research Part D: Transport and Environment, 38*, 6–14.

Grey, E. (2015). *The underdog: Preventing animal casualties on railways*. http://www.railway-technology.com/features/featurethe-underdog-preventing-animal-casualties-on-railways-4532957/. Accessed June 12, 2016.

Hamre, D., & Overcast, M. (2004). *Avalanche risk analysis: John Stevens Canyon, Essex, Montana*. Unpublished Report, Chugach Adventure Guides, LLC., Girdwood, AK, 80 pp.

Harner, M. J., & Stanford, J. A. (2003). Differences in cottonwood growth between a losing and a gaining reach of an alluvial floodplain. *Ecology, 84*, 1453–1458.

Levesque, S. L. (2001). The Yellowstone to Yukon conservation initiative: reconstructing boundaries, biodiversity, and beliefs. In: *Reflections on water: New approaches to transboundary conflicts and cooperation* (pp. 123–162). Cambridge: MIT Press.

Mace, R. D., Carney, D. W., Chilton-Radandt, T., Courville, S. A., Haroldson, M. A., Harris, R. B., et al. (2012). Grizzly bear population vital rates and trend in the Northern Continental Divide Ecosystem, Montana. *The Journal of Wildlife Management, 76*, 119–128.

National Park Service. (2008a). Avalanche hazard reduction by Burlington Northern Santa Fe Railway in Glacier National Park and Flathead National Forest, Montana. Final Environmental Impact Statement.

National Park Service. (2008b). Avalanche hazard reduction by Burlington Northern Santa Fe Railway in Glacier National Park and Flathead National Forest, Montana. Record of Decision.

Pahl-Wostl, C., Craps, M., Dewulf, A., Mostert, E., Tabara, D., & Taillieu, T. (2007). Social learning and water resources management. *Ecology and Society, 12*, 2007.

Pretty, J. (2003). Social capital and the collective management of resources. *Science, 302*, 1912–1914.

Reardon, B. A., Fagre, D. B., & Steiner, R. W. (2004, September). Natural avalanches and transportation: A case study from Glacier National Park, Montana, USA. In *Proceedings of the International Snow Science Workshop* (pp. 19–24).

Roesch, M. J. (2010). *Identifying wildlife crossing zones for the prioritization of highway mitigation measures along US Highway 2: West Glacier, MT to Milepost 193.* Thesis, University of Montana.

Servheen, C., & Sandstrom, P. (1993). Ecosystem management and linkage zones for grizzly bears and other large carnivores in the northern Rocky Mountains in Montana and Idaho. *Endangered Species Technical Bulletin, 18*(3), 10–13.

Stern, M. J., & Baird, T. (2015). Trust ecology and the resilience of natural resource management institutions. *Ecology and Society, 20,* 14.

US Department of Interior. (2008). *Press release.* Available from https://www.doi.gov/sites/doi.gov/files/archive/news/archive/08_News_Releases/080421d.html

Waller, J. S., & Miller, C. S. (2016). Increasing traffic volumes on US Highway 2. *Intermountain Journal of Sciences, 21* (in press).

Waller, J. S., & Servheen, C. (2005). Effects of transportation infrastructure on grizzly bears in Northwestern Montana. *Journal of Wildlife Management, 69,* 985–1000.

Western Governors Association. (2008). *Western Governors' Association wildlife corridors initiative report.* Available from http://www.westgov.org/images/dmdocuments/wildlife08.pdf. Accessed July, 2016.

Part III
Conclusion

Chapter 19
What's Next? Railway Ecology in the 21st Century

Rafael Barrientos, Luís Borda-de-Água, Pedro Brum, Pedro Beja and Henrique M. Pereira

> *Predicting is very difficult, especially about the future,* attributed to Niels Bohr.

Abstract As societies realise the importance of maintaining biodiversity and, accordingly, acknowledge the need for monitoring, minimizing and compensating the impacts of socioeconomic activities, including transportation, more scientists will be called to address these societal challenges. Railway Ecology is emerging in this context as a relatively new field to identify and provide solutions to the specific environmental problems associated with the building and operation of railways, particularly their impacts on wildlife. This is an interdisciplinary field that uses and draws methods from other disciplines, ranging from ecology and genetics, to statistics and computer simulations. Here, we summarize the guidelines to address railway-related biodiversity conservation problems as they were identified in the several chapters of this book, and recommend lines for future research.

Keywords Climate change · Environmental sustainability · Linear infrastructures · Railway ecology

R. Barrientos · L. Borda-de-Água · P. Beja · H.M. Pereira
CIBIO/InBIO, Centro de Investigação em Biodiversidade e Recursos Genéticos, Universidade do Porto, Campus Agrário de Vairão, Rua Padre Armando Quintas, 4485-661 Vairão, Portugal

R. Barrientos (✉) · L. Borda-de-Água · P. Beja · H.M. Pereira
CEABN/InBIO, Centro de Ecologia Aplicada "Professor Baeta Neves",
Instituto Superior de Agronomia, Universidade de Lisboa, Tapada da Ajuda,
1349-017 Lisboa, Portugal
e-mail: barrientos@cibio.up.pt

P. Brum
Faculdade de Ciências Socias e Humanas da Universidade Nova de Lisboa,
Lisboa, Portugal

H.M. Pereira
German Centre for Integrative Biodiversity Research (iDiv)
Halle-Jena-Leipzig Deutscher Platz 5e, 04103 Leipzig, Germany

© The Author(s) 2017
L. Borda-de-Água et al. (eds.), *Railway Ecology*,
DOI 10.1007/978-3-319-57496-7_19

A Bright Look at the Future of Railways

Transportation in the 21st century involves multiple challenges. Transports will have to deal with the increasing needs of a growing and wealthier human population, and the socioeconomic requirements to deliver natural resources, manufactured goods, and transport people. The emergence of new strong economies, and the expectation that economic growth will raise the living standards of vast populations, will be surely accompanied by an increase in transportation infrastructures, given the well-known relationship between transportation activity and Gross Domestic Product (GDP) (Profillidis 2014). How societies will accommodate economic development with the biophysical constrains imposed by the environment is a major challenge, but one where railways will certainly play an important role (Smith 1998, 2003).

One of the reasons to expect a steady development of railways is related to their cost-effectiveness over other means of transportation, particularly for the so-called intermediate range distances (Profillidis 2014). Therefore, it is expected that railways will develop to fill in the niche that lay somewhere in-between the local door to door transportation and, in the other extreme, the long distance transportation of passengers or freight that is better accomplished through aviation or marine transportation. The exact level at which railways will fill in such niche, however, will depend on the characteristics of the region where they operate: its geography, the size of its population and how the population is distributed. Nevertheless, it is likely that in many countries new railway lines will be built and others will be improved, to support, for instance, the commuting of people between the suburbia and city centres, to transport passengers among cities laying a few hundred kilometres apart, and to transport freight among harbours and big cities.

Another reason to expect the expansion of railways is related to its environmental benefits relatively to other means of transportation, which is well illustrated by the role it can play in addressing global warming. For instance, Dulac (2013) analysed infrastructure requirements under different climate change scenarios, concluding that an increase in 335,000 railway track kilometres would be needed to accommodate the projected transportation needs until 2050, while keeping the increase in average global atmospheric temperatures under 4 °C. Under a more strict policy scenario of a 2 °C increase in temperatures, the railway network would have to expand by approximately 535,000 track kilometres, reflecting the higher efficiency of railway over road transportation (Dulac 2013). Once again, this will certainly lead to a steady development of railway networks, thereby calling for the collaboration of railway ecologists to mitigate impacts and to foster the environmental added value of railways.

Emerging Trends and Challenges

While we expect a major development in railway networks in the next decades, technological developments may considerably affect their characteristics, as well as the environmental problems that will be raised. For instance, in regions with high population densities the use of the magnetic levitated ("maglev") trains may become a typical feature (Najafi and Nassar 1996), but some of the problems associated with barrier effects and mortality are still likely to occur. Although maglev trains share several of the characteristics with the present railways, other technologies such as the admittedly farfetched "hyperloop" may differ considerably (Matthews and Brueggemann 2015). In a hyperloop, passengers and freight travel in vehicles enclosed in elevated "tubular" structures. In this case, barrier effects will be greatly reduced, and mortality due to collisions with the trains will disappear, though the possibility of flying animals colliding with the tubular structure will still exist. Another technological advancement that is likely to develop in forthcoming years is electric cars with auto-pilot features. As we write, this technology is being promoted by large companies (e.g., Google) and at least one company already commercializes cars with hardware and software enabling the auto-pilot mode (Tesla Motors), a development that services such as Uber may use automated in the future. These vehicles, or future improved versions, may enable comfortable travels with the flexibility of door to door connections, challenging some of the advantages that trains presently exhibit for some distances. Nevertheless, it is unlikely that railway transportation as we know it today will become obsolete at any time in the near future, requiring a continued engagement of railway ecologists to deal with the current and future challenges associated with this transportation system.

Despite its positive prospects at the global scale, however, it is likely that some railway lines will get out of operation due to technological developments and changes in socioeconomic requirements, as it has happened during the past decades. For instance, railways in the USA after World War II were quickly put aside in favor of other means of transportation of passengers, such as automobiles and airplanes, and by 1990 the 270,000 miles of active rail lines had diminished to 141,000 miles (Ferster 2006), though railways have retained 43% of the freight market (Profillidis 2014). Comparable processes have occurred in other countries such as Portugal, where many lines built during the early twentieth century lost their importance and were progressively abandoned during the second half of the twentieth and early twentyfirst centuries (Sarmento 2002). These changes represent opportunities for conservation and call for the active participation of railway ecologists in collaboration with other specialists, as abandoned railway lines may lead to ecologically-oriented projects as well as to the mitigation of situations of social injustice. The combination of ecological and social minded policies can contribute to a new global pattern of urban transformation, as exemplified by the High Line in New York City or the Sentier Nature in Paris (Foster 2010). The old High Line railroad was disused in 1980 and it then started to develop as a semi-natural habitat that was converted into a public park comprising twenty-two blocks along the lower west side of Manhattan (Foster 2010). Paris' Sentier Nature represents another approach, where an old railway was converted into a nature trail, now enriched with a diversity of types of ecosystems, with 200 species of flora and 70 species of fauna (Foster 2010). Although this field of ecological restoration after railway abandonment was not dealt with in this book, this is an important area where the experience on linear infrastructures of railway ecologists may prove instrumental, namely due to their knowledge on the species and habitats that can persist in railway verges in urban and suburban settings (see Chap. 16).

Railways Travel Towards Environmental Sustainability

As with any other means of transportation, there are environmental issues associated with railways, ranging from wildlife mortality, barrier effects, introduction of exotic species, and several forms of pollution. Some of these are comparable to those found in roads and other linear infrastructures, while others are railway specific. Likewise, some mitigation measures can be adapted from road and power line infrastructures, while others have to be developed specifically for railways. As highlighted throughout this book, information is still in short supply for the evaluation and mitigation of impacts specific to railways, which justifies the need to continue investing in research programs targeting railway ecology on its own right. Based on the collective experience of this book we suggest that the following research lines should be the priority focus of such programs, which will greatly contribute to help keep railways on track towards environmental sustainability:

- First and foremost, studies to evaluate the biodiversity impacts of railways and their mitigation measures need in most cases to be strengthened through the use of adequate designs, sufficiently large sample sizes, and sufficiently long time frames. Ideally, BACI (Before-After-Control-Impact) designs should be used to examine such impacts, avoiding simple designs that have little power to demonstrate causality (Balkenhol and Waits 2009; Corlatti et al. 2009; Soanes et al. 2013; see Chap. 12). Monitoring must be long-term, as some species need time to get used to disturbances associated with railway operation, and to the implementation of mitigation measures (Baofa et al. 2006; Yang and Xia 2008; Corlatti et al. 2009; Soanes et al. 2013).

- Second, we still need better estimates of wildlife mortality in railways, though this is one of the most visible and well-studied impacts. Tackling this problem will require, for instance, field experiments on searcher efficiency and carcass persistence on railways (Barrientos et al. in review). These are expected to be railway-specific because, as commented in Chap. 1, traffic flow on railways is markedly lower than that of roads. This is important because low traffic flows allow time gaps for scavengers to have access to carcasses. On the other hand, trains travel over rails, and corpses falling out of them can persist longer as there is no flattening as it may happen in roads.

- Third, we need a better understanding on the barrier and fragmentation effects of railways, and how these may act together with those of other linear structures such as roads (see, e.g., Chap. 14). Concurrently, we need to know where, when and how can wildlife passes effectively mitigate such barrier and fragmentation effects, by helping to restore connectivity across the landscape. A range of studies may contribute to such understanding, including comparisons of wildlife crossings of railways before and after the building of wildlife passes, in places with and without such passes (i.e., BACI designs; Corlatti et al. 2009; Soanes et al. 2013). Novel molecular techniques may also contribute to tackle these issues, using for instance non-invasive genetics to know which individuals cross the railways and how railway barriers contribute to spatial genetic structuring (Riley et al. 2006; Balkenhol and Waits 2009; Clevenger and Sawaya 2010; Simmons et al. 2010; see also Chap. 4). Modelling approaches may also be useful to explore the demographic and genetic consequences of barrier effects induced by railways, and to disentangle barrier from mortality effects (Borda-de-Água et al. 2011, 2014; Ceia-Hasse et al. 2017).

- Fourth, we need to know if and how mortality and barrier effects translate into population effects, which in turn can affect the persistence of vulnerable species around railway corridors (van der Grift 1999; Dorsey et al. 2015). This might be achieved through population modelling approaches similar to those already used in road ecology (Taylor and Goldingay 2009; van der Ree et al. 2009; Borda-de-Água et al. 2014; Ceia-Hasse et al. in review). The development of these models require careful field studies whereby critical information on fecundity, survival and dispersal are estimated, which can then be used to estimate the conditions under which population viability may be affected by railways. Due consideration should also be given to age- and sex-specific demographic parameters, as huge variation in population responses may occur if mortality affects primarily non-breeding versus breeding individuals, or males versus females. Molecular methods may also contribute to understand population responses to railways, either by helping to estimate population parameters, or to evaluate changes in effective population sizes in relation to mortality and fragmentation effects (Balkenhol and Waits 2009).

- Fifth, we need studies on railway ecology focusing on a wide range of species, which should be representative of life history and behavioural traits potentially affecting vulnerability to railways. To date, most research has focused on species with high socio-economic profile, like large charismatic mammals, or on those with limited mobility, like herptiles (van der Grift 1999; Dorsey et al. 2015). Broadening the scope of research is important, because mitigation measures designed for some species may be inappropriate for others, thus requiring informed adjustment based on scientifically sound information (Clevenger and Waltho 2005; Morelli et al. 2014; Vandevelde et al. 2014; Wiacek et al. 2015; see Chap. 16). To forecast impacts on species with differential traits, pilot field studies and a systematic review of the available evidence should be carried out to investigate what are the key ecological traits, such as dispersal ability or generation time, affecting population responses to railways.

- Finally, there are a number of important issues that remain poorly explored, despite their potential importance to the environmentally-sound management of railways. For instance, little is known about the consequences of vibration and noise on biodiversity living adjacent to the railway bed (see Chap.6). Also, studies are needed on the relative cost-effectiveness of different management tools and mitigation measures to help reconcile environmental protection and socio-economic demands (see, e.g., Andreassen et al. 2005; Ford et al. 2009; Mateus et al. 2011). The potential positive impacts of railways also need to be better explored, including in particular the prospective for the right-of-way to act as a shelter for biodiversity (see Chap. 16), which can probably be improved through well-designed wildlife-friendly management procedures.

Final Remarks

Overall, our book documents the wealth of research that is already available on the ecology of railways, particularly on the evaluation and mitigation of impacts on wildlife, illustrating the considerable progress that has been made over the past decade. However, it also shows that many aspects of railway ecology remain poorly researched, which makes it difficult to understand what are the main impacts of railways on important dimensions of biodiversity such as landscape connectivity and population viability. As a consequence, the evaluation and mitigation of railway impacts continue to be largely driven by the lessons gained from road ecology, which may often be inadequate for application in the railway context. By reviewing the state-of-the-art and presenting a diversity of valuable case studies, we hope that this book can contribute to attract both researchers and practitioners to this interesting field of applied science, thereby promoting the development of new tools and applications for a better integration of the transportation networks with the challenges associated with the protection of biodiversity.

References

Andreassen, H. P., Gundersen, H., & Storaasthe, T. (2005). The effect of scent-marking, forest clearing and supplemental feeding on moose-train collisions. *Journal of Wildlife Management, 69,* 1125–1132.

Balkenhol, N., & Waits, L. P. (2009). Molecular road ecology: Exploring the potential of genetics for investigating transportation impacts on wildlife. *Molecular Ecology, 18,* 4151–4164.

Baofa, Y., Huyin, H., Yili, Z., Le, Z., & Wanhong, W. (2006). Influence of the Qinghai-Tibetan railway and highway on the activities of wild animals. *Acta Ecologica Sinica, 26,* 3917–3923.

Barrientos, R., Martins, R. C., Ascensao, F., D'Amico, M., Moreira, F., & Pereira, H. M., et al. (in review) Searcher efficiency and carcass persistence in field experiments: meta-analysis with management guidelines.

Borda-de-Água, L., Grilo, C., & Pereira, H. M. (2014). Modeling the impact of road mortality on barn owl (*Tyto alba*) populations using age-structured models. *Ecological Modelling, 276,* 29–37.

Borda-de-Água, L., Navarro, L., Gavinhos, C., & Pereira, H. M. (2011). Spatio-temporal impacts of roads on the persistence of populations: Analytic and numerical approaches. *Landscape Ecology, 26,* 253–265.

Ceia-Hasse, A., Borda-de-Água, L., Grilo, C., & Pereira, H. M. (2017). Global exposure of carnivores to roads. *Global Ecology and Biogeography, 26,* 592–600.

Ceia-Hasse, A., Navarro, L., Borda-de-Água, L., & Pereira, H. M. (in review) Population persistence in fragmented landscapes: Disentangling isolation, mortality, and the effect of dispersal.

Clevenger, A. P., & Sawaya, M. A. (2010). Piloting a non-invasive genetic sampling method for evaluating population-level benefits of wildlife crossing structures. *Ecology and Society, 15,* 7.

Clevenger, A. P., & Waltho, N. (2005). Performance indices to identify attributes of highway crossing structures facilitating movement of large mammals. *Biological Conservation, 121,* 453–464.

Corlatti, L., Hackländer, K., & Frey-Roos, F. (2009). Ability of wildlife overpasses to provide connectivity and prevent genetic isolation. *Conservation Biology, 23,* 548–556.

Dorsey, B., Olsson, M., & Rew, L. J. (2015). Ecological effects of railways on wildlife. In R. van der Ree, D. J. Smith, & C. Grilo (Eds.), *Handbook of road ecology* (pp. 219–227). West Sussex: Wiley.

Dulac, J. (2013). *Global land transport infrastructure requirements: Estimating road and railway infrastructure capacity and costs to 2050.* Paris: International Energy Agency. Available at https://www.iea.org/publications/freepublications/publication/TransportInfrastructureInsights_ FINAL_WEB.pdf

Ferster, A. C. (2006). Rails-to-trails conversions: A review of legal issues. *Planning and Environmental Law, 58,* 3–9.

Ford, A. T., Clevenger, A. P., & Bennett, A. (2009). Comparison of methods of monitoring wildlife crossing-structures on highways. *Journal of Wildlife Management, 73,* 1213–1222.

Foster, J. (2010). Off track, in nature: Constructing ecology on old rail lines in Paris and New York. *Nature and Culture, 5,* 316–337.

Mateus, A. R. A., Grilo, C., & Santos-Reis, M. (2011). Surveying drainage culvert use by carnivores: Sampling design and cost–benefit analyzes of track-pads vs. video-surveillance methods. *Environmental Monitoring and Assessment, 181,* 101–109.

Matthews, C. H., & Brueggemann, R. (2015). *Innovation and entrepreneurship: A competency framework.* New York: Routledge.

Morelli, F., Beim, M., Jerzak, L., Jones, D., & Tryjanowski, P. (2014). Can roads, railways and related structures have positive effects on birds? A review. *Transportation Research Part D, 30,* 21–31.

Najafi, F. T., & Nassar, F. E. (1996). Comparison of high-speed rail and maglev systems. *Journal of Transportation Engineering, 122,* 276–281.

Profillidis, V. A. (2014). *Railway management and engineering.* Burlington: Ashgate Publishing Ltd.

Riley, S. P. D., Pollinger, J. P., Sauvajot, R. M., York, E. C., Bromley, C., Fuller, T. K., et al. (2006). A southern California freeway is a physical and social barrier to gene flow in carnivores. *Molecular Ecology, 15,* 1733–1741.

Sarmento, J. (2002). The geography of "disused" railways: What is happening in Portugal? *Finisterra, 37,* 55–71.

Simmons, J. M., Sunnucks, P., Taylor, A. C., & van der Ree, R. (2010). Beyond roadkill, radiotracking, recapture and FST—A review of some genetic methods to improve understanding of the influence of roads on wildlife. *Ecology and Society, 15,* 9.

Smith, R. A. (1998). Global environmental challenges and railway transport. *Japan Railway and Transport Review, 18,* 4–11.

Smith, R. A. (2003). Railways: How they may contribute to a sustainable future. *Proceedings of the Institution of Mechanical Engineers, Part F: Journal of Rail and Rapid Transit, 217,* 243–248.

Soanes, K., Lobo, M. C., Vesk, P. A., McCarthy, M. A., Moore, J. L., & van der Ree, R. (2013). Movement re-established but not restored: Inferring the effectiveness of road-crossing mitigation for a gliding mammal by monitoring use. *Biological Conservation, 159,* 434–441.

Taylor, B. D., & Goldingay, R. L. (2009). Can road-crossing structures improve population viability of an urban gliding mammal? *Ecology and Society, 14,* 13.

van der Grift, E. A. (1999). Mammals and railroads: Impacts and management implications. *Lutra, 42,* 77–98.

van der Ree, R., Heinze, D., McCarthy, M., & Mansergh, I. (2009). Wildlife tunnel enhances population viability. *Ecology and Society, 14,* 7.

Vandevelde, J.-C., Bouhours, A., Julien, J.-F., Couvet, D., & Kerbiriou, C. (2014). Activity of European common bats along railway verges. *Ecological Engineering, 64,* 49–56.

Wiącek, J., Polak, M., Filipiuk, M., Kucharczyk, M., & Bohatkiewicz, J. (2015). Do birds avoid railroads as has been found for roads? *Environmental Management, 56,* 643–652.

Yang, Q., & Xia, L. (2008). Tibetan wildlife is getting used to the railway. *Nature, 452,* 810–811.

Glossary

Barrier effect The combined effects of physical barriers, infrastructure avoidance, traffic mortality and habitat loss which together reduce the railway permeability.

Carcass The corpse of a dead animal (or its remains) that is used in monitoring studies to evaluate mortality rates caused by trains (and other vehicles) collisions.

Carcass persistence time The time each animal carcass remains on the railway (or road) before it disappears, either due to scavengers or to decomposing.

Connectivity The degree to which the landscape facilitates or impedes movement of wildlife among resource patches. From a species' point of view, it is the ability of individuals to disperse through the matrix (e.g., unsuitable habitat).

dB A decibel (dB) is a unit of sound pressure. It is defined as $20 \log_{10}(P/P_{ref})$, where P_{ref} is a value of reference, typically, 20 micropascals in the air.

dB(A) A-weighted decibels. The A-weighting systems attempts to correct to the way the human ear perceives loundness.

Habitat fragmentation The division of contiguous tracts of suitable habitat into progressively smaller patches.

Introduction The deliberate or accidental release of an organism(s) into the wild by human agency.

Mortality hotspot Segments of railways (or roads) with higher concentration or increased probability of wildlife mortality.

Native or indigenous A species that occurs naturally in an area, i.e. whose dispersal has occurred independently of human-mediated transportation. In general, a species thought to have occurred in an area since before the Neolithic is considered to be native.

Naturalised A non-native species that, following introduction, has formed self-sustaining populations in the wild.

'Non-native' or 'alien' A species that occurs outside its native geographical range as a result of human-mediated transportation.

L. Borda-de-Água et al. (eds.), *Railway Ecology*,
DOI 10.1007/978-3-319-57496-7

319

Permeability Describes the degree in which a railway allows wildlife to move freely across it.

Probability of detection Probability of a carcass being present on a railway (or road) being detected by an observer.

Probability of removal Probability of a carcass being removed (or consumed) from the railway, usually by scavengers.

Propagule pressure A measure of the number of individuals introduced and the number of introduction events that took place. Generally, the higher propagule pressure is, the higher the likelihood of establishment will be.

Wildlife pass/passage Designated place for wildlife to safely cross the railway.

Printed by Printforce, the Netherlands